5G Spectrum and Standards

For a complete listing of titles in the
Artech House Mobile Communications Series,
turn to the back of this book.

5G Spectrum and Standards

Geoff Varrall

ARTECH HOUSE
BOSTON | LONDON
artechhouse.com

Library of Congress Cataloging-in-Publication Data
A catalog record for this book is available from the U.S. Library of Congress.

British Library Cataloguing in Publication Data
A catalogue record for this book is available from the British Library.

Cover design by John Gomes

ISBN 13: 978-1-63081-044-3

685 Canton Street
Norwood, MA 02062

10 9 8 7 6 5 4 3 2 1

Contents

Acknowledgments

I was not expecting to be writing this book so soon. Having produced what I thought was a reasonably definitive work on mobile broadband in 2012, I had assumed that it would be at least 5 years before there would be enough new material to justify another venture into the world of technical publishing, but here I am a mere 3 years later writing a book on 5G spectrum and standards.

You might say that this is a symptom of a world in which technology change is happening at an ever faster pace but there is no evidence to support this. In practice for the past 30 years, the cellular industry has reinvented itself on a regular basis with another technology generation becoming visible 5 years after initial network deployment of the prior generation and the first network deployments starting five years later.

The difference this time around is that there is there is an ambition to produce a 5G network for the Winter Olympics in South Korea in February 2018. This implies 5 years of development to deployment compressed into 2 years.

While not impossible, this is ambitious and explains the rush to finalize the detail of a complex spectrum allocation and standardization process. It also leaves little room for wrong decisions. Hence, there is a perceived need to produce a book that translates past implementation experience into this present decision-making process.

Luckily, I am not alone in this task and I would like to acknowledge the generous support and advice that I have received from the vendor community and particularly the help from individuals, many of whom have been working in the industry for 30 years or more. These are people who are able to say "that didn't work well last time" or, more positively, "that's a great idea."

In particular, I would like to thank Sylvia Lu for her contribution to the digital signal processing, digital signal processor, and standards issues addressed throughout the book and in particular in Chapter 11 and for her input and advice on the Internet of Things physical layer standards.

I also thank the team at Artech House, Aileen Storry, Helen Pain, Molly Klemarczyk, and Lindsay Moore for their advice, help, and support through the writing and production process.

1

Introduction

1.1 Fifth Generation (5G) Technology Economics

Before building the Model T, Henry Ford spent a year studying watchmaking and rifle manufacturing. These industries were the best contemporary examples of mass market production, supply chain control, materials innovation, and manufacturing innovation. When the Model T production line opened in 1908, it produced a car that cost less than other cars. The use of lightweight steel and close tolerance engine components meant that the Model T was faster than other cars at the same price and went further on a tank of petrol.

This is the underlying narrative of this book. To be commercially successful, the fifth generation (5G) networks and devices have to cost less than existing mobile wide area systems, has to go further and faster, and has to use less fuel. This is more likely to be achieved if techniques and technologies are borrowed from other industries, including the automotive industry.

The purpose of this book is to analyze the spectrum and standards options presently being discussed for 5G. We study the technology options available and the commercial impact of those options. We also look at what people are doing in other industries, including the satellite industry, point-to-point wireless, and radar industry (automotive radar and military and space radio systems). This chapter serves as an introduction

Chapter 2 reviews the technology costs of the standards process and the relative merits of the proposed 5G physical layer candidates including modulation, coding, and multiplexing options. We review the ongoing Third Generation Partnership Project (3GPP) 5G standards process and the device cost and network cost implications of that process.

Chapter 3 summarizes existing fourth generation (4G) band allocations and potential 5G allocations. In an ideal world, spectrum would be allocated before work started on physical layer standards. For example, present debates about 5G modulation and coding options are largely theoretic given that the choice of modulation and coding should be determined by channel properties that are directly consequent on spectrum and channel bandwidth decisions. In practice, spectrum and standards have to move forward in parallel or, more awkwardly, standards have to be developed before spectrum allocation is agreed upon.

Chapter 4 discusses the technology cost of coexistence. This includes radio systems that cause interference to themselves and those that cause interference to other radio systems.

Chapter 5 studies the theory and practice of the allocation and auction process, in particular the implications of incentive auctions for 5G, and summarizes the outcome of World Radio Congress (WRC) 2015, the proposed agenda items for WRC 2019, and work items now moved to WRC 2023, including the review of the ultrahigh-frequency (UHF) band for Region 1 (Europe and Africa).

Chapter 6 studies the emerging "middle Earth" markets between the 48th parallel north and the 48th parallel south, the technical and commercial requirements of these markets, and the associated cost and performance implications for 5G radio networks and devices.

Chapter 7 argues the case for dividing 5G spectrum by wavelength rather than frequency, dividing the radio spectrum into three wavelength-denominated bands. The reasoning is that wavelength rather than frequency determines the form factor and functional performance of RF devices. More pragmatically, if all the 4G aggregation combinations are added into the mix, there are close to 300-band plan options for 5G which is too complicated; three-band 5G is easier to comprehend.

Chapter 8 covers the meter band but includes a review of longer wavelengths such as long wave, medium wave, short wave, and very high frequency (VHF). VHF and lower frequencies are not a part of the spectrum typically discussed in the context of 5G radio systems, but have relevance for Internet of Things (IoT) applications where range and penetration are key performance objectives. The meter band from 300 MHz to 3 GHz includes the majority of existing cellular networks. The 5G is proposed for deployment into this legacy spectrum, and we discuss the spectral efficiency, coexistence, and economic implications of this option.

Chapter 9 covers the centimeter band from 3 GHz to 30 GHz with a review of existing radio systems in the band including mobile services satellite (MSS) and fixed services satellite (FSS), point-to-point radio, and military and civilian radar, highlighting the increasing bandwidth and power requirements of these systems over time.

Chapter 10 covers the millimeter band from 30 GHz to 300 GHz including point-to-point radio, military wide area radio, and automotive radar and analyzes the potential and proposed 5G allocations in this band. We highlight the importance of beamwidth in 5G physical layer design and the ways in which beamwidth and bandwidth need to be coupled together.

Chapter 11 reviews the impact of digital signal processing and analog-to-digital conversion on wide area 5G systems including the specific challenges of realizing 200-MHz, 500-MHz, 1-GHz, and 2-GHz channel bandwidths over the expected life cycle of the 5G standards process.

Chapter 12 summarizes innovations in the submillimeter band and infrared, optical, and ultraviolet bands. Optical wavelengths are not normally associated with 5G and generally discounted on the basis of noise, linearity, gain, and free-space propagation limitations. We explore the practical limits of these constraints and discuss potentially novel approaches that bring together optical and RF processing techniques. We review optical wireless communication and optical wireless point-to-point systems.

1.2 Technical and Commercial Innovation by Wavelength

Each of these wavelength bands is undergoing a process of continuous innovation.

Digital Radio Mondiale is being introduced into long-wave, medium-wave, short-wave, and VHF radio broadcasting bringing digital audio quality to listeners in remote and inaccessible places. These systems share spectrum with military radio including long-distance and over-the-horizon radar and legacy amplitude modulation (AM) and frequency modulation (FM) broadcasting networks.

In the meter bands (300 MHz to 3 GHz), the cosharing of mobile broadband with TV in the UHF band between 600 and 800 MHz has required innovation both in terms of network and user device radio frequency (RF) hardware and baseband processing to manage and mitigate coexistence issues. At the time of this writing, the industry has been studying cosharing issues in the 600-MHz band prior to incentive auctions in the United States and Puerto Rico designed to achieve a balance between the commercial and technical interests of the mobile broadband and broadcasting community.

The cellular spectrum at 800, 900, and 1,800 MHz is being refarmed to support the introduction of 4G radio systems with a number of networks now deployed. The reallocation of Global System Mobile (GSM) channel bandwidth has to be managed to avoid compromising voice quality and voice coverage. This legacy spectrum is also proposed for 5G IoT applications.

Existing specialist users including public safety agencies are being encouraged to migrate to long-term evolution (LTE). As we shall see in later chapters, this is ambitious and at times contentious.

In L-band, hybrid satellite terrestrial networks have been proposed at 1,500 MHz but to date have proved to be problematically close to Global Positioning System (GPS). Mobile operators in some countries are deploying supplemental downlinks (LTE Band 32) at 1,452 MHz to 1,496 MHz. In the United States, the repurposing of the 1,800-MHz band for 4G has involved the decommissioning of legacy military radio systems. New time division duplex (TDD) cellular bands adjacent to Wi-Fi at 2.4 GHz have needed to be optimized to manage intersystem interference. TDD and frequency division duplex (FDD) cellular systems at 2.6 GHz have had to be planned to mitigate interference with aviation radar.

In the centimeter band (3 to 30 GHz), new iterations of Wi-Fi continue to be introduced into the 5-GHz industrial, scientific, medical (ISM) unlicensed band including support for higher-power outdoor networks [2] and public safety networks [3]. The higher end of the ISM band is proposed for Digital Short Range Communication (DSRC) automotive connectivity based on the Centre European Nationale (CEN, European Committee for Standardization) and European Technical Standards Institute (ETSI) specification TS102-486.

Between 6 and 30 GHz, geostationary orbit (GSO), medium Earth orbit (MEO), and low Earth orbit (LEO) satellite systems are being deployed offering higher data rates based on wider channel bandwidths using higher power and spot beam antennas that provide improved effective isotropic radiated power (EIRP), sensitivity, and selectivity. Additional uplink bandwidth allocated at WRC2015 provides enhanced functionality.

Terrestrial networks are upgrading centimeter-band, point-to-point links to provide improved backhaul capabilities. These existing networks coshare the centimeter band with radar and a range of other radar and military and civilian radar and radio systems including automotive radar at 24 and 26 GHz, immediately adjacent to weather-sensing radar (26 GHz is a water vapor resonance peak).

The satellite industry is currently pursuing a robust advocacy campaign against 5G mobile broadband deployment into the centimeter band. This is understandable and suggests that the mobile broadband community will need to demonstrate that 5G technologies deployed into this band coexist without inflicting technical and commercial harm to these present incumbents or preferably add value to all stakeholders. The alternative is to structure an adequate or possibly generous compensation process.

The same provisos apply though to a lesser extent in the millimeter band from 30 to 300 GHz. For example, there are no large-scale commercial satellite systems deployed in this band. However, there are an increasing number of lightly licensed point-to-point systems being deployed, usually for high-bandwidth backhaul, and we study the potential integration opportunities of these systems with future 5G wide area mobile broadband networks.

There is also substantial market growth in automotive radar at 79 GHz (77–81 GHz) and potentially 122 GHz and 244 GHz. The 79-GHz band has been allocated at WRC 2015 for short-range, high-resolution automotive radar, suggesting that this band in particular is capable of achieving global scale.

We argue that the spectrum adjacency of 79-MHz automotive radar to potential 5G spectrum allocations at 72 to 77 GHz and 82 to 87 GHz could resolve a number of cost and performance issues for millimeter mobile. Automotive radio systems including radar are designed to work across large temperature gradients. They need to be resilient to vibration and moisture and to meet stringent EMC requirements. Automotive radar subsystems have to be manufactured in large volumes with minimal batch to batch or device to device variability, have low failure rates and meet low-cost targets. The radar system front ends must have good linearity, low noise, and adequate power. The signal processing algorithms detect, recognize, and track moving objects. The moving objects are typically up to 200m ahead of the car, a distance equivalent to the proposed intersite distance for urban 5G networks. There is therefore potential commonality between automotive radar three-dimensional (3-D) spatial processing and the adaptive beam-forming proposed for 5G including angle of arrival and angle of departure and angular power calculations. We cover this in Chapter 10.

1.3 RF Performance at Shorter Wavelengths

As radio wavelength reduces, it becomes progressively harder to amplify RF signals and to manage noise and nonlinearity. Increased parasitic effects make it harder to simulate and design RF front ends. RF behavior becomes less predictable. To date, these constraints have precluded serious consideration of 5G deployment in the submillimeter bands between 300 GHz and 3 THz, but the performance issues also apply to the centimeter and millimeter bands. One solution to achieving gain with low phase noise is to use more exotic materials, for example, gallium nitride for power amplification, but this introduces additional cost.

Alternatively, the physical layer can be designed to be more tolerant of phase noise and nonlinearity. This implies the use of lower-order modulation schemes and modulation and multiplexing options that do not produce or depend on large amounts of envelope variation combined with coding schemes that are optimized for channel conditions that may or may not be continuously changing.

In a mobile environment, the dominant distortion mechanism is channel fading, a function of the changing reflections caused by differences in distance between the transmitter and receiver. This includes the distortions introduced by the variable path delays between the transmitter and receiver.

Multipath is used in LTE multiple-input multiple-output (MIMO) systems to increase data rates by treating each path as a separate channel. The time delay spread on the channel translates into phase and amplitude distortion and this has to be accommodated by the channel equalizer and the cyclic prefix overhead in the OFDM symbol stream.

Provided that these equalization processes and guard period overheads are sufficient to prevent intersymbol interference (ISI), the multiple paths can be resolved into multiple channels streams and thus realize a gain in throughput. However, there is an associated cost. The cyclic prefix has a capacity cost and the equalizer has a clock cycle cost and introduces processing delay.

A significant amount of the multipath signal energy arrives as angular reflected energy.

If the beamwidth of the receiver antenna is reduced, the amount of reflected energy received and number of received multipaths will be reduced. This reduces the throughput rate but also reduces the need for channel equalization. The time-domain guard band, in LTE, the length of the cyclic prefix, can also be reduced.

Given that OFDM is specifically chosen for its ability to manage and exploit multipath, an argument can be made that it is not needed or at least produces less efficiency gain in narrow beamwidth radio systems.

Problematically, the channel models used today become progressively less accurate for shorter wavelength propagation modeling, particularly for wide area mobile outdoor propagation in the centimeter and millimeter bands.

Modeling indoor propagation at 60 GHz [4] suggests that a root mean squared delay spread of 18 ns using an omnidirectional antenna will reduce to 4.7 ns using a 10° 3-dB beamwidth antenna.

The matchbox sized adaptive phased array antennas being developed for millimeter-band 5G are capable of producing a 2° beamwidth in elevation and azimuth. This avoids or at least reduces the need to use orthogonal frequency division multiplexing (OFDM) and memory based encoding and decoding. The result should be savings in terms of number of processing clock cycles, which should translate into lower power consumption, reduced cost, and reduced coding and decoding delay and delay variability, although the net gain achieved will depend on the amount of spatial processing overhead. Narrow beamwidth radio systems would also potentially facilitate the repurposing of point-to-point radio hardware and software for mobile wide area systems.

In practice, it may be that OFDM is retained for some but not all 5G channel conditions and, at time of this writing, significant work still needs to be done for modeling mobile wide area propagation, particularly in the millimeter band.

This modeling has to take into account channel bandwidth and modulation bandwidth. The trend to increase channel bandwidth from the 5 or 10 MHz used in present-day systems to 100 MHz of aggregated channel bandwidth in LTE Advanced to 500 MHz and ultimately 1- or 2-GHz channels in passbands of 5 GHz in 5G systems introduce digital signal processing challenges both in terms of throughput rate and analog-to-digital conversion and adds to the argument that a simpler, more noise-tolerant physical layer might deliver an overall gain in system efficiency.

This is an underlying narrative in the next 11 chapters. Radio system design is all about achieving optimized compromise points. Coding and modulation and multiplexing schemes have to be matched in a way similar to the way that we have traditionally used a Smith Chart to calculate optimum reactance, capacitance, and inductance values. Digital signal processing and analog-to-digital conversion have similar compromise points that need to be accommodated.

In second generation (2G), third generation (3G), and 4G radio systems, the focus has been on designing a physical layer that is spectrally efficient. Spectrally efficient networks are generally considered to be economically efficient due to their theoretical ability to support more subscribers per megahertz of auctioned spectrum. From a regulatory perspective, spectral efficiency should translate into high spectral utilization, which translates into a maximization of spectral economic and social gain.

Unfortunately, any increase in spectral efficiency generally results in an equivalent decrease in energy efficiency. This is bad news for base stations (described as e Node Bs in LTE language) and is bad news for user devices with limited battery capacity or IoT devices with extended duty-cycle requirements. An underlying design target for 5G is therefore to reduce the joules per bit energy cost.

1.4 Market and Regionally Specific Requirements

In Chapter 5, we highlight the relatively high cost of energy in developing economies, including the need to provide power to base stations in deep rural areas and the need to support users who may have limited access to main electricity, solar power, or diesel power.

The 5G requirements, including economic and energy requirements, are therefore different from market to market, but there are commonalities. Network capital and operational costs for all mobile broadband networks are based on a set of technical and commercial performance assumptions that have been carefully qualified but are often wrong. Sometimes the technologies work better than expected or improve as technology matures, and sometimes the technologies work worse than expected and fail to improve over time.

The technologies that work worse than expected and fail to improve over time may have been compromised by complex standards, complex band plans, unexpected coexistence costs, intellectual property disputes, or some combination of all of these. Complex standards impose additional test costs and time-to-market delay. An overcomplex band plan results in additional filters and switch path loss in user and network RF hardware. This translates into a need for additional network density with associated capital and operational cost implications.

Coexistence issues, either within networks or between networks, impose additional filtering costs. This can be complex to manage when a new network, for example, a mobile broadband network, is licensed to be spectrally and geographically proximate to a broadcast network or spectrally and geographically proximate to a satellite network or radar system.

Our understanding of these sometimes hard-to-quantify technical and commercial factors has improved over the past 30 years, but we do not really know how these factors scale to shorter wavelengths. In particular, we have limited experience and therefore limited knowledge of how higher-power high data rate 5G systems will behave and perform over and above existing mobile broadband allocations.

This also applies to the various alternative modulation schemes proposed for 5G including filtered OFDM (F-OFDM), filter bank multicarrier (FBMC), universal filtered OFDM (UF-OFDM), and generalized frequency division modulation. FBMC, UF OFDM, and GFDM all have good out-of-band performance and therefore require smaller frequency-domain guard bands. Filtered OFDM uses a subband filter, which is applied to shape the spectrum of the subband OFDM signal allowing the use of different subcarrier spacing and cyclic prefix for each specific subband.

The relative efficiency of these options depends on how and where they are implemented and the relative severity of the problems they solve or mitigate.

Filtering modulation states will generally improve coexistence with other users in the same radio system and/or other systems but will usually introduce throughput performance loss at channel level. Improving out-of-band performance translates into a loss of in-band performance. Adding complexity to modulation and coding schemes does not necessarily result in an overall gain in system or network efficiency.

1.5 Military Millimeter Radio for Wide-Area 5G

There is a general assumption that 5G systems will be predominately local area systems, closer to Wi-Fi than to existing wide area cellular systems. The additional propagation loss at higher frequencies is assumed to preclude wide area coverage. These assumptions are at least open to question and debate.

Military radio systems are being deployed into the millimeter band with a clear weather line-of-sight range of 60 km. The link budget can be addressed by exploiting aperture gain at higher frequencies and adopting noise and distortion tolerant modulation and coding schemes.

There is a parallel assumption that the higher propagation loss of higher-frequency/shorter-wavelength spectrum will minimize interference issues and these bands can therefore be either lightly licensed or unlicensed. This works well for fixed point-to-point systems but our argument is that 5G networks are going to be effectively mobile point-to-point systems, which by definition means that they are likely to be subject to interference and require coexistence management.

A 1° beamwidth antenna will produce between 40 and 50 dBi (isotropic gain) of gain relative to an omnidirectional antenna resulting in a focused but high flux density. This is analogous to shining a laser light into an aircraft cockpit on the approach to an airport. Industrial lasers have a power output of just a few watts but have sufficient focus to burn the retina of a pilot at a distance of hundreds or even thousands of meters [5].

Coexistence issues have exercised the industry ever since the earliest days of broadcasting when U.S. TV broadcasters started raising power output levels to the point at which interference started to become a major problem.

The replacement of the Radio Act of 1927 with the Communications Act of 1934 and the setting up of the Federal Communications Commission (FCC) formalized the process of managing the competing commercial interests of the broadcasting community but with the more broadly stated purpose of "regulating interstate and foreign commerce in communication by wire and radio so as to make available, so far as possible, to all the people of the United States a rapid, efficient, nationwide, and worldwide wire and radio communication service with adequate facilities at reasonable charges, for the purpose of the national defence" (http://transition.fcc.gov/Reports/1934new.pdf).

This relationship between defense communication interests and commercial interests remains a key issue today for 5G both in terms of technology exchange opportunities and commercial common purpose. Common sense suggests that it is sensible for different parts of the industry to work together. This includes the mobile broadband community, terrestrial and satellite broadcasting, fixed point-to-point and fixed and mobile service (FSS and MSS) satellites, the subspace and space communications sector, the automotive radar industry, civilian and military radar, and civilian and military communication systems.

Defense communication systems are deployed across the whole radio spectrum from long wave to light. This includes mobile communication systems at VHF and UHF and L-band and S-band, LEO, MEO, and GSO satellite systems (VHF to E-band), and mobile and fixed radar (VHF to E-band).

Legacy defense systems are being upgraded to provide additional functionality. This requires more rather than less spectrum. Increased radar resolution requires wider channel bandwidths; longer range requires more power and improved sensitivity. Improved sensitivity increases the risk of intersystem interference. Emerging application requirements including unmanned aerial vehicles require a mix of additional terrestrial, satellite, and radar bandwidth. These requirements are geographically and spectrally diverse rather than battlefield and spectrally specific.

The assumption in the 5G community is that the defense industry will be willing and able to surrender spectrum for mobile broadband consumer and civilian use. The advanced wireless service (AWS) 3 auction in the United States is a contemporary example with a $5 billion transition budget to cover legacy military system decommissioning in the Department of Defense coordination zone between 1,755 and 1,780 MHz.

This transition strategy assumes an increased use of LTE 4G network hardware and user hardware in battlefield systems. While this might imply an opportunity for closer coordination and cooperation between the mobile broadband and defense community, it seems likely that an increase in the amount of defense bandwidth needed to support a broadening range of RF-dependent systems could be a problematic component in the global spectral allocation and auction process.

1.6 Coexistence Costs

This is not dissimilar to the issues emerging from the 600-MHz Incentive Auction in the United States, which has been complicated by the recognition by the TV broadcast community that more, rather than less, bandwidth is required to remain competitive with other increasingly high-definition content delivery options. The auction is therefore not a spectrum sale but a compensation process. Given that the bids for AWS 3 spectrum totaled $44 billion, it is likely that the future compensation cost expectations of the defense community may become significantly higher. Satellite operators are likely to have similar expectations.

To date, LTE has (more or less) happily coexisted with existing defense radio and radar and satellite VHF and UHF systems. There have been coexistence issues between air traffic control radar and LTE deployment in Band 7 and

Band 38 (2.6 GHz), which suggest that scheduled auctions at 2.3 and 3.4 GHz and scheduled deployments may have mixed-use challenges that may become more significant over time [6].

Future military mobile communication systems can and probably will make good use of LTE hardware in bands between 700 MHz and 4 GHz, establishing a common interest that should facilitate the resolution of spectrum allocation, sharing and valuation issues. The bigger challenge will be scaling a mutual interest model to the shorter wavelength spectrum needed for 5G deployment.

1.7 5G Definitions and Spectral Implications

Definitions of 5G are many and various with an increasing emphasis on cloud and core technology, but beneath the market fluff there is an assumption that a more effective and efficient physical layer will be required. The design brief and performance expectations for 5G have been summarized as [7]: a 1,000 times increase in mobile data volume, a 10 to 100 times increase in connected devices, a 5 times lower latency, a 10 to 100 times increase in peak data rate, and 10 times battery life extension for low-power devices.

It is hard to see how these capacity and data rate expectations can be met without significant bandwidth allocation above 4 GHz. These are bands that support existing and new generation military high-power radar and radio systems, a combination of terrestrial and subspace systems supported by LEO, MEO, and GSO satellite networks.

The ITU Radio band designations describing these higher bands originated in a Consultative Committee for International Radio (CCIR) meeting in 1937 and were approved at the International Radio Conference in Atlantic City in 1947 (see Table 1.1). Each band was given a number (nine band numbers in total) that is the logarithm of the approximate geometric mean of the upper and lower band limits in hertz, proposed by B. C. Fleming Williams [8].

In 2008, the U.S. military, NATO, and the European Union agreed on a naming protocol for bands into which electronic countermeasure (ECM) RF systems are deployed [9] (see Table 1.2).

However, IEEE descriptions are still generally used for radar and RF-dependent weapons and communication spectrum. This naming system had its origins in World War II when it was classified. It was regularized in a 1984 IEEE standard (see Table 1.3). It is now used for NATO and U.S. electronic countermeasure systems. The terms L-band, S-band, and C-band are often used in present mobile broadband spectrum discussions.

Within the V- and W-bands, there are three bands allocated for fixed (but potentially mobile) services, two 5-GHz bands at 71–76 and 81–86 GHz and a

Table 1.1
ITU Radio Band Designations

Symbol	Description	Band	Frequency	Wavelength
VLF	Very low frequency	4	3–30 kHz	10–100 km
LF	Low frequency	5	30–300 kHz	1–10 km
MF	Medium frequency	6	300–3,000 kHz	100–1,000m
HF	High frequency	7	3–30 MHz	10–100m
VHF	Very high frequency	8	30–300 MHz	1–10m
UHF	Ultrahigh frequency	9	300–3,000 MHz	10–100 cm
SHF	Super high frequency	10	3–30 GHz	1–10 cm
EHF	Extremely high frequency	11	30–300 GHz	1–10 mm
THF	Terahertz (or terrifically) high frequency	12	300–3,000 GHz	0.1–1 mm

Table 1.2
U.S. and NATO Naming Protocols for Electronic Countermeasure (ECM) RF Systems

Band	Frequency	Wavelength
A	<250 MHz	<1.2m
B	250—500 MHz	1.2m—600 cm
C	500 MHz—1 GHz	600 cm—300 cm
D	1—2 GHz	300 cm—150 cm
E	2—3 GHz	150 cm—100 cm
F	3—4 GHz	100 cm—75 cm
G	4—6 GHz	75 cm—5 cm
H	6—8 GHz	5 cm—3.75 cm
I	8—10 GHz	3.75 cm—3 cm
J	10—20 GHz	3 cm—1.5 cm
K	20—40 GHz	1.5 cm—750 mm
L	40—60 GHz	750 mm—500 mm
M	60—100 GHz	500 mm—300 mm

3-GHz band at 92–95 GHz. These are known collectively as the E-band from the waveguide-naming regime for 60 to 90 GHz [10].

E-band was formally established by the ITU at the WARC 1979 World Radio Communication Conference but mostly ignored until 2005 when the FCC issued a light-licensing scheme that permitted E-band radios to operate at up to 3W.

Defense Advanced Research Projects Agency (DARPA) have a mobile hotspot E-band system development project based on gigabit air-to-ground and

Table 1.3
IEEE Standard 521-1984 Radar Frequency Bands

Band	Frequency (GHz)
L-band	1–2
S-band	2–4
C-band	4–8
X-band	8–12
Ku-band	12–18
K-band	18–27
Ka-band	27–40
V-band	40–75
W-band	75–110

ground-to-air links implemented in the E-band between 71 and 76 MHz and 81 to 86 MHz integrated with voice and data support for LTE smart phones. We discuss the functionality of these networks in more detail in Chapter 10 and argue that this is an example of a contemporary military radio system development with direct relevance to the 5G planning process [11].

1.8 5G and Military and Space Communications Research

There has always been a close coupling between military communication technology and civilian radio systems with technology flowing in both directions. Advances in military radio communication in World War I translated into the postwar radio broadcasting revolution, TV receiver technologies in the 1930s translated into World War II radar and combat radios, and the Cold War facilitated solid-state technologies that provided the basis for mass-market transistor radios.

The launch of the Russian Sputnik satellite in 1957 prompted the formation of the National Aeronautics and Space Administration (NASA) and DARPA [12]. Over the next 50 years, the need to develop communication and guidance and imaging systems that could work efficiently at microwave and millimeter wavelengths produced significant material innovation. Some of these innovations became crucially useful for cellular radio, gallium arsenide for microwave power amplification being one example.

Military systems are increasingly looking to leverage the scale advantage of consumer markets in terms of user device functionality. This has motivated military and public protection communication procurement agencies to mandate support for LTE smart phones and where possible to use standard LTE

UHF, L-band, and S-band network hardware. The process has been accelerated by the introduction of relatively rugged smart phones designed for clumsy consumers [13].

However, the mutual interest model extends beyond iPhones for soldiers and iPads for tanks. Defense budgets and telecommunications spending are similar in scale. U.S. military spending peaked in the Cold War at 5.5% of GDP. The present-day budget is $627 billion, 3.4% of the gross domestic product (GDP). The world spends $1.6 trillion per year on defense. The United States accounts for about 40% of this. U.S. and Allied budgets together account for about two-thirds of global spending. An increasing percentage of this budget is being spent on high-data-rate, long-range wide area mobile connectivity. The numbers for the telecommunications industry are similar. Telecommunications spending in the United States is between 3% and 4% of the GDP. That for South Korea is greater than 5% [14].

LTE user device and network hardware development is not inexpensive. Qualcomm has invested $14 billion in LTE baseband development over the past 4 years, Apple's annual spending is on the order of $5 billion, Samsung's annual spending is well over $10 billion, and Intel's annual spending is $10 billion. Huawei is spending more than $5 billion per year. Not all of this is directly related to LTE, but a lot of it is and we have not even started to count other related LTE investment at the component and subsystem levels. It would be conservative to say that LTE global physical layer development investment comfortably exceeds $50 billion per year. Even in military terms, this is significant money focused on a specific outcome with a market volume that provides a cost base that is several orders of magnitude below the cost base of equivalent U.S., Chinese, Russian, or Indian military radio hardware.

In the other direction, there are areas of military research that could potentially reduce 4G and 5G delivery cost and improve existing and future network efficiency. For example, delivery cost to the "middle Earth" markets (markets between the 48th parallel north to the 48th parallel south; see Chapter 6) on either side of the equator would be substantially reduced if and when servicing and in-flight refueling of geostationary satellites become feasible, a current DARPA project [15].

User device costs and network RF hardware costs will be reduced by replacing RF components with silicon. DARPA is presently working to develop digital CMOS amplifiers that can work efficiently at 90 GHz.

Network efficiency gains and latency control are dependent on improved timing accuracy and the capability to distribute highly accurate and stable time references over large distances. This is an emerging problem in high-data-rate wide area networks and addressed by the DARPA Pulse Program [16].

Emerging military/commercial cooperative business models applied in the satellite sector could have a broader terrestrial remit [17] and research into competitive and cooperative spectrum sharing could help resolve potential coexistence issues [18].

Last but not least, there are techniques needed to analyze frequency-agile wide bandwidth radar and electronic warfare systems including time and frequency analysis of complex pulsed signal waveforms, which will become increasingly useful if and when 5G radio systems are deployed into the K-bands and or the E-band spectrum [19].

Every generation of cellular has directly and indirectly benefited from military research. Materials and component innovation have been particularly important and will remain important as the industry moves to realize efficient and effective networks at millimeter wavelengths.

The 4G is proving to be a useful adjunct to existing defense radio systems with military procurement focused on leveraging the scale economics of the consumer mobile broadband industry both in terms of network nodes (terrestrial LTE base stations) and user devices. User devices are being ruggedized to meet consumer expectations of robustness that are not dissimilar to day-to-day military requirements.

Conversely, 4G and 5G systems can benefit from RF innovation in radar and satellite systems including advances in amplifier and antenna design, dynamic beam steering, and interference resilience. These innovations could translate into a much-needed step function improvement in delivery and energy economics.

The LTE of military communication is therefore of specific interest to the 5G community and 5G technology ambitions have considerable relevance to the military procurement community. A closely coupled cooperation would produce clear economic benefits to both parties and their related user communities. The same can be said of the terrestrial broadcast industry and mobile broadband industry, but cooperation between these entities has been frustrated by an adversarial spectral auction process.

Developing successful mixed-use models for military spectrum could prove equally challenging. Coexistence issues between LTE and military and radar and radio systems have to date been managed effectively and efficiently, at least up to the C-band.

Populating 5G into higher bands including the K-bands and E-band will require coordination with next generation RF-dependent defense systems including high-capacity mobile and point-to-point/multipoint connectivity, high-performance wide channel bandwidth radar and satellite systems and unmanned aerial vehicle (UAV) telemetry, telesensing, and telecontrol.

This will be more easily achieved if defense agencies clearly perceive that 5G has a useful role to play in battlefield communication. This implies an ability to support high-data-rate, extended-range, large-cell topologies, not presently a priority within the 5G development community.

1.9 The Real Purpose of 5G: A Reduction in Delivery Cost?

We have highlighted some of the more obvious opportunities to amortize 5G research and development cost across defense and civilian markets and technologies. This is in the context of a present industry in which there is a negative coupling between increased data rates and profitability; as data rates go up, Earnings Before Interest Tax and Depreciation of Assets (EBITDA) and operator margins go down. Partly this is due to the increased maintenance and management costs that are a consequence of increasing network density.

Higher-density networks increase capital spending and reduce return on investment (ROI). In the short term, reduced EBITDA is being addressed by ongoing industry consolidation but in the longer term there is a requirement to achieve step function reductions in cost per delivered user bit. However, this has to be coupled with an ability to support higher data rates.

Since 2007 (the U.S. 700-MHz auction), spectrum discussions have been largely dominated by the battle for broadcasting bandwidth. The U.S. auction was followed in the United Kingdom and parts of Europe by the first digital dividend (closedown of TV in the 800-MHz band and repurposing for mobile broadband) and the ongoing second digital dividend (close down of TV in the 700-MHz band and repurposing for mobile broadband). In the United States there is a third digital dividend process under way to close down TV in the 600-MHz band and repurpose for mobile broadband though this is being met with significant legal challenge by the broadcast community and other stakeholders. Resolving spectrum sharing under 1 GHz (low band) therefore was a priority item for WRC 2015 and included discussion on lower UHF allocations for LTE. In practice, a measure of coordination for coprimary use of the UHF band between 694 MHz to 790 MHz was agreed, including some spectral harmonization, and some countries are closer to considering implementation below 694 MHz, but change does not happen quickly unless all involved parties perceive a clear commercial benefit.

Additional work items after WRC 2015 include the formalization of subset bands (bands within bands) and superset bands (extension of existing bands) for mid-band and high band to support wider channel bandwidth LTE rollout with aggregated bandwidth of 100 MHz and potentially 200 MHz in the meter band and lower end of the centimeter band (3 to 6 GHz). Most of these changes impose a performance cost on the user or IoT device.

1.10 Understanding Radio: Think Wavelength, Not Frequency

We flatter ourselves by talking about the challenge of responding to unprecedented technology change including the difficulties of delivering cost-economic, high-bandwidth, high-frequency radio.

To put this into historical perspective, it is useful to remember that by the 1890s there was a developed practical understanding of radio waves up to and including what we now call the UHF band, L-band, and the centimeter and millimeter bands. In 1888, Hertz used a sheet zinc parabolic cylinder reflector antenna to generate 66-cm radio waves (454 MHz). Marconi used a similar deep parabolic cylinder for 25-cm experiments (1,200 MHz). By the early and mid-1890s, W. C. Bose was demonstrating radio waves at 6 mm (50 GHz) [20] and 5 mm (60 GHz).

The mechanics of high-frequency radio, parabolic reflectors, microwave absorbers, cavity radiators, and round, square, and rectangular waveguides were mastered several years before the theory [21].

Physics has not changed in the last 120 years and our ability to use the centimeter and millimeter bands for efficient, high-capacity local and wide area coverage remains dependent on antenna and waveguide design. For this reason, there are persuasive arguments to going back to using wavelength to describe and define system, product, service, and spectral requirements and investment risk and return for 5G radio systems.

Early wide area/long-range radio systems were described and are still described today as long wave, medium wave, or short wave. The term microwave came into common usage after the World War II to describe wavelengths between 100 cm (300 MHz) and 0.1 cm (300 GHz).

Wavelength was used as a descriptor because wavelength is easier to measure. The Marconi wave meter is an early example of a wavelength measurement device [22]. Accurate measurement of frequency required highly stable quartz crystal oscillators. As these became more readily available through the 1930s, there was a shift to describing radio in terms of frequency: VHF or UHF or other arbitrary naming systems, C-, X-, and K-bands for radar, for example. Forty years on, frequency counters remained expensive, clumsy to use (range switching and cable changing), and only accurate when measuring relatively high power levels. The HP5340 introduced in 1973 solved many of these performance issues and worked from 10 Hz to 18 GHz but cost more than $5,000 [23].

The introduction of cellular radio from 1980 onwards marked a shift to describing systems with a specific frequency description, the 800-MHz AMPS networks (subsequently called Band 5), the 900-MHz TACS/ETACS networks (subsequently called Band 8), the 1,800-MHz networks (subsequently called Band 3), and so on until we arrive 30 years later with 44 LTE bands implemented or proposed, all described by frequency and ranging from Band 31 in Brazil at 450 MHz (0.666m) to Band 43 at 3,800 MHz (0.111m).

LTE band plans are becoming complicated with the complexity compounded by carrier aggregation. Adding in C-band, the K-bands and V- and W-bands (with the E-band subbands) for 5G makes a hard-to-understand landscape increasingly incomprehensible to anyone other than a subject specialist, and we do not want them to be responsible for 5G system design or economic modeling.

The answer is to think in terms of wavelength rather than frequency. We all know that wavelength and frequency are directly related. Radio wavelength is calculated as the speed of light (300 million meters per second) divided by frequency; 300 MHz is therefore conveniently 1m [24]. The theoretically optimum length for an antenna is one-quarter or one-half of the wavelength to be received or transmitted. It is therefore wavelength that defines RF product form factor and RF product functionality rather than frequency.

For example, it is possible to design compact electrically short antennas for smart phones in the 450-MHz band or 700-MHz band, but the wavelengths are 0.666m and 0.428m. These antennas when cramped into an artificially small space with an inefficiently small ground plane introduce a loss of the order of 6 or 7 dB. A 7-dB loss is equivalent to the theoretical propagation gain achieved in a 700-MHz cell when compared to an 1,800-MHz cell.

Put another way, you have just thrown away the additional coverage gain that you thought you might get from a 450-MHz or 700-MHz network and the user experience will be more variable due to unwanted external coupling including hand capacitance effects. Some of this additional variability can be reduced by using adaptive tuning (variable capacitance using RF MEMS), but this adds cost and complexity.

The counter side to this is that as wavelength decreases, antenna size decreases. This means that multiple antenna elements can be fitted in the space formerly occupied by a single longer wavelength antenna. This is the basic physics underpinning 5G smart antenna design and provides the mechanism for delivering high-capacity, high-data-rate, ultradense access or high-data-rate wide area access. At the time of this writing, there were practical, low-cost, 60-GHz Wi-Fi products available that deliver 3 Gbps per square meter of data throughput using beam forming [25].

The 5-mm wavelength is at the peak of oxygen resonance and therefore provides high attenuation. In the Wi-Fi example, this is used to provide user-to-user, channel-to-channel separation and intensive spatial reuse. Beam forming can be used equally effectively to deliver range rather than high capacity per square meter, but radio systems need to be implemented in the transmission windows between the resonance peaks.

The attenuation characteristics of millimeter wave radio links were studied at Rutherford Appleton Laboratory in the 1980s. Figure 1.1 shows the 60-GHz oxygen peak and the three water vapor peaks.

The transmission windows for longer links (up to 60 km) include the 5G E-band options at 72–77 GHz (4.16–3.89 mm), 81–86 GHz (3.7–3.48

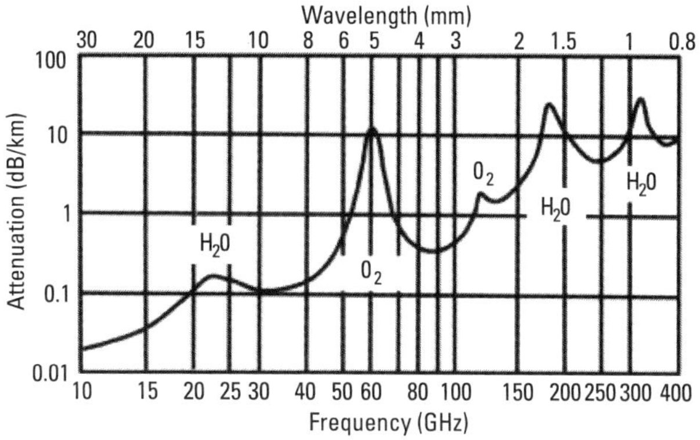

Figure 1.1 Resonant and nonresonant absorption 10 to 400 GHz [26].

mm), and 92–95 GHz (3.25 mm–3.15mm) presently used for lightly licensed fixed point-to-point backhaul and proposed for use for military communication based on rapidly deployable subspace systems to provide high-data-rate, wide area mobile coverage.

1.11 5G Wavelengths of Interest

Our 5G wavelengths of interest can therefore be defined as spanning the meter, centimeter, and millimeter bands, although with potential interest in the wavelength bands below the meter band and submillimeter bands above the millimeter band (see Table 1.4).

Satellites are allocated the following wavelengths for mobile satellite services in the centimeter band (see Table 1.5).

These supplement existing mobile satellite services in the meter band at L-band and proposed services in the S-band adjacent to LTE Band 1 [27] (see Table 1.6).

From these tables it would seem obvious that terrestrial 5G networks for wide area coverage in the centimeter bands will need to integrate with existing and future satellite systems in the centimeter and meter bands. Terrestrial 5G networks for wide area coverage in the millimeter bands will need to integrate with millimeter subspace systems and point-to-point backhaul.

There is minimal recognition of this within the present 5G standards or spectrum discussion process. Gaining consensus from the 3,000 delegates attending WRC2015 proved understandably problematic and it remains hard to get traditionally separate industries to work together.

Table 1.4
Wavelengths and Frequency from 3 kHz to 300 GHz

Frequency	3–30 kHz	30–300 kHz	300–3,000 kHz	3–30 MHz	30–300 MHz	300 MHz–3 GHz	3–30 GHz	30–300 GHz
Wavelength	100–10 km	10–1 km	1,000–100 km	100–10m	10–1m	1–0.1m	10–1 cm	10–1 mm
Name	100-km band	10-km band	Kilometer band	100-m band	10-m Band	Meter Band	Centimeter Band	Millimeter Band
		Long Wave	Medium Wave	Short Wave		Microwave →		
						Terrestrial	Terrestrial?	Terrestrial?
							Satellite	Subspace
Atmospheric noise up to 20 MHz →								
				Galactic noise to 100 MHz				
					Circuit noise →			
					Aperture gain offsets propagation loss →			

Table 1.5
Satellite Mobile Satellite (MSS) Service Spectrum Allocations in the Centimeter Band

	Centimeter Band			
	Ku-Band		**K-Band**	**Ka-Band**
Frequency	12 GHz	14 GHz	18–20 GHz	27–30 GHz
Wavelength	2.49 cm	2.14 cm	1.66–1.49 cm	1.1–0.99 cm

Table 1.6
Meter-Band Mobile Satellite Services

	L-Band			**S-Band**	
Frequency	1,518–1,559	1,616–1,626	1,626–1,675	1,980–2,110	2,170–2,200
Wavelength	19.75–19.22	18.55–18.43	18.43–17.89	15.14–14.20	13.81–13.62

1.12 The Cost of Coexistence and Complexity

The multiplicity of frequency band plan options for LTE is already complicated. The complexity is compounded by regionally specific performance characteristics, out-of-band emissions being one example, and multiple carrier aggregation options. The 5G radio systems proposed for implementation in the K-bands and V- and W-bands add further complexity. The end result will be hundreds of bands and band combinations.

Revisiting potential 5G radio spectrum in terms of wavelength reduces hundreds of band combinations to three wavelength options: the meter band, the centimeter band, and the millimeter band. This includes research and development cross-amortization opportunities with the radio and TV broadcasting industry, fixed point-to-point radio links, satellite and space industry, and even possibly the radio astronomy industry. The automotive radar industry is of particular interest in terms of technology innovation (3-D spatial processing algorithms), production optimization (low-cost, high-performance RF for centimeter and millimeter-band radar), and network innovation (use of cars and trucks and buses and ships as repeaters and relays).

Over the past 30 years, cellular radio has expanded in the meter band from the original allocations at 800 MHz (0.37m) and 900 MHz (0.33m) and now spans 450 MHz (0.666m) to 2,600 MHz (0.111m).

LTE terrestrial networks are proposed for the lower end of the centimeter band between 3,400 MHz (8.81 cm) and 3,800 MHz (7.788 cm). At the time of this writing, Nokia Networks and Deutsche Telekom were demonstrating LTE Advanced three-carrier aggregation combining LTE-FDD in Band 3 (1.8 GHz), with LTE-TDD in Band 42 (3.5 GHz) and (presumably) another band.

The addition of the two TDD bands reflects the allocation of 400 MHz of TDD spectrum in many countries. It is going to take a while to absorb this bandwidth and the supply chain is only just coming to grips with adding 3.5 GHz to user devices (see Chapter 3).

Scaling terrestrial cellular systems to the higher end of the meter band (2,600 MHz) and lower end of the centimeter band (3,400–3,800 MHz) has meant that coexistence with satellite and radar systems has had to be managed. This process has introduced or will introduce additional cost and complexity and performance loss in some markets.

Scaling terrestrial cellular systems to the lower end of the meter band (450–800 MHz) has meant that coexistence with terrestrial broadcasting has had to be managed. This process has introduced or will introduce additional cost and complexity and performance loss in some markets.

Extending this process into the higher end of the centimeter band and millimeter band for 5G terrestrial systems introduces similar coexistence challenges with mobile satellite and subspace service providers and potentially a similar cost and performance impact.

1.13 Profitable Spectrum: The Meter Band

The most profitable parts of the meter band are the bands that have the most users and or highest value users. The scale economy benefits of volume and value produce cost and performance benefits that increase profit and add value to the consumer experience. The technology economics of 5G in the centimeter and millimeter bands are at present far from obvious but would seem to imply a need to provide better capacity economics than present 802.11ad Wi-Fi systems at 60 GHz and better wide area range economics (data reach) than present and future satellite and subspace systems in the centimeter and millimeter transmission windows. The systems also need to support mobility including high-speed users and in some cases highly mobile IoT devices.

This seems unlikely, but then 30 years ago, who would have believed that the meter band would be supporting a terrestrial mobile broadband industry generating several trillion dollars of revenue per year? Thirty years from now, will the centimeter and millimeter bands be generating similar revenues? Which parts of these bands will be profitable and who will own the income?

This book does not provide answers to these questions or at least answers that can be trusted, but analyzing technology economics in terms of three wavelength-denominated bands rather than 300 frequency bands and band combinations helps to clarify the issues.

1.14 Five Processing Domains and Their Relevance to 5G Systems

Over the next 11 chapters, we will be analyzing 5G technology options across five specific processing domains including the following.

1.14.1 The Frequency Domain

Over the past 30 years, the cellular industry has moved from narrowband (25 kHz) to wider-band 200-kHz and 5-MHz channel spacing. The 4G LTE has reintroduced additional frequency-domain complexity with 15-kHz and 7.5-kHz orthogonal subcarriers.

We need to try and answer the question of how efficiently OFDM and OFDMA options scale to higher-bandwidth, higher-frequency spectral allocations in the centimeter and millimeter bands and develop an informed opinion on the optimum compromise points between spectral and energy efficiency. Given that beam-forming minimizes multipath, there may not be a need for OFDM in shorter-wavelength 5G radio systems.

1.14.2 The Phase and Amplitude Domain

Fixed-point radio systems in the millimeter band are now supporting 4,096-QAM modulation. What are the limits and best options for 5G wide area mobile radio systems in terms of implementation complexity and energy and throughput efficiency? Are we better implementing simple low-level modulation with minimal amplitude variation and high tolerance to phase noise and distortion?

1.14.3 The Power Domain

Successive interference cancellation is promoted as a technique for providing user to user selectivity with gains in 5G capacity and coverage including data rate and data reach and is a favored candidate for vendors promoting non-orthogonal modulation and multiplexing schemes, but the additional power domain complexity may be too hard to handle.

Are the alternative filter bank multicarrier options a better option, given that they also provide radio spectrum flexibility or does this just introduce additional unsupportable frequency-domain complexity or clock cycle overhead? Or is additional coding distance the better approach?

1.14.4 The Time Domain

Reduced latency is highlighted as a key objective in 5G radio system design usually on the basis of improved user application performance, but this misses the point. Reduced end-to-end latency directly improves scheduler efficiency and could potentially transform 5G delivery and network economics. Given that beam-forming reduces the delay spread in the end-to-end channel, it becomes possible to avoid the use of memory-based encoding and decoding systems. This reduces coding overhead and coding delay and delay variability.

In parallel, 4G networks are evolving to all IP carried over the carrier Ethernet, which means that the traditional mechanisms for delivering network synchronization are no longer available. Present and future requirements are being addressed in ITU-T (synchronous Ethernet standardization) and within the IEEE through the IEEE 1588-2008 Precision Timing Protocol. The integration of stable distributed time, frequency, and phase synchronization is going to be a significant challenge, particularly for physically and spectrally proximate TDD networks. It has been 100 years since Einstein published his theory of general relativity. Large-scale telecommunications networks over the next 100 years will need to accommodate relativistic effects that are presently imperfectly managed.

1.14.5 The Spatial Domain

One key advantage of millimeter radios is their ability to support highly complex spatial processing. This can either be used to increase throughput (data rate) in local area radio systems or coverage (data reach) in wide area radio networks. There is also a secondary benefit in terms of delay spread.

Last but not least, as we progress through the next 11 chapters, we revisit point-to-point radio systems and their relationship with 5G local area and wide area networks.

1.15 Cost and Performance Economics

The standards, spectrum, and coexistence requirements of local and wide area 5G need to be viewed in the context of their relationship with other spectrally and geographically proximate radio systems. These include Wi-Fi at 2.4, 5, and 60 GHz, satellite and subspace radio systems in the centimeter band, point-to-point radio in the meter band, centimeter band, and millimeter band, civilian, military, and automotive radar systems in the centimeter and millimeter bands and a relatively exotic mix of sensing and telemetry, telesensing, and telecontrol systems, which could be more or less anywhere. We add radio astronomy into the mix as an added extra on the basis of 5G relevant technology innovation. The coexistence issues of 5G and space radio also merit study.

The 5G could potentially be deployed at the meter, centimeter, or millimeter wavelengths but theoretically at least could scale downwards to longer wavelength bands or upwards to submillimeter wavelength spectrum. It can at least benefit from sharing research and development investment across this broader spectral remit.

Given that the computer and communications-added value of a car is expected to move from typically 10% today to more than 60% in 10 years' time, it is reasonable to expect closer cooperation between the automotive industry and the telecommunications and information and communications technology industry.

In a month in which Apple posted an annual income of $234 billion with cash reserves of $200 billion, there was press speculation about the Apple iCar with Apple attempting to recruit or re-recruit engineers from the Tesla car company [28]. A car, after all, is a mobile user device.

Both Apple and Google have plans to introduce autonomous cars over the next 15 years, a program that will hopefully be integrated into the 5G standards process. Sharing research and development investment across this broader industry base would seem to be essential rather than desirable.

We are spoiled for choice. The best technical and commercial options for 5G are the options that offer the best cost economy/performance trade-off, which brings us to Chapter 2.

References

[1] http://www.drm.org/.

[2] https://www.fcc.gov/document/fcc-increases-5ghz-spectrum-wi-fi-other-unlicensed-uses.

[3] https://transition.fcc.gov/pshs/techtopics/techtopics10.html.

[4] Manabe, T., Y. Miura, and T. Ihara, "Effects of Antenna Directivity on Indoor Multipath Propagation Characteristics at 60 GHz," *IEEE Journal on Selected Areas in Communications*, Vol. 13, No. 3, April 1996, pp. 441–448.

[5] http://www.laserpointersafety.com/laser-hazards_aircraft/laser-hazards_aircraft.html.

[6] http://www.microwavejournal.com/articles/22740-co-existence-tests-for-s-band-radar-and-lte-networks.

[7] http://www.ericsson.com/res/docs/2014/5g-what-is-it.pdf.

[8] Fleming Williams, B. C., Letter to the Wireless Engineer, 1942.

[9] http://en.citizendium.org/wiki/EU-NATO-US_frequency_bands.

[10] http://www.microwaves101.com/encyclopedias/rectangular-waveguide-dimensions.

[11] http://www.microwavejournal.com/articles/23121-darpas-mobile-hotspot-program-drives-e-band-performance-benchmarks.

[12] http://www.darpa.mil/default.aspx.

[13] http://www.extremetech.com/computing/190698-apples-iphone-6-is-more-durable-less-breakable-than-galaxy-s5-and-one-m8.

[14] http://www.statista.com/statistics/270565/ict-share-of-telecommunications-expenditures-in-the-gdp/.

[15] http://www.darpa.mil/Our_Work/TTO/Programs/Phoenix.aspx.

[16] http://www.darpa.mil/program/program-in-ultrafast-laser-science-and-engineering.

[17] http://www.defense.gov/home/features/2011/0111_nsss/docs/NationalSecuritySpaceStrategyUnclassifiedSummary_Jan2011.pdf.

[18] http://www.darpa.mil/Our_Work/I2O/Programs/Spectrum_Challenge.aspx.

[19] https://www.keysight.com/main/editorial.jspx?cc=GB&lc=eng&ckey=2510838&id=2510838&cmpid=47318.

[20] https://www.cv.nrao.edu/~demerson/bose/bose.html.

[21] Rayleigh, John William, "On the Passage of Electric Waves Through Tubes or the Vibrations of Dielectric Cylinders," *Phil. Mag.*, Vol. 43, 1897, pp. 125–132.

[22] https://www.mhs.ox.ac.uk/marconi/collection/glossary.php.

[23] http://historycenter.agilent.com/exhibit1.

[24] www.unitconversion.org/unit_converter/frequency-wavelength.html.

[25] http://www.wi-fi.org/discover-wi-fi/wigig-certified.

[26] Gibbins, C. J., "Statistical Studies of Millimetre-Wave, Infra-Red and Optical Propagation over a 500 Metre Path," Rutherford Appleton Laboratory, 1989.

[27] http://www.inmarsat.com/news/inmarsat-to-deliver-in-flight-connectivity-over-new-european-aviation-network/.

[28] http://appleinsider.com/articles/15/02/05/apple-trying-very-hard-to-nab-tesla-engineers-as-van-sightings-proliferate.

2

The Technology Cost of Standards

2.1 The Purpose of Standards

"Any color as long as it's black," the Henry Ford approach to producing the Model T at low cost, was based on the principle that customer choice introduces complexity and complexity introduces cost.

This chapter analyzes the contemporary cost of telecoms standards using the LTE standards process as a case study (Releases 11 through to 16) to provide a technology and economic context for the 4G to 5G transition. We explain how the Third Generation Partnership Project (3GPP) Release process has introduced additional cost multipliers into the industry supply chain (including test cost) and forced and enabled consolidation (supply chain efficiency for the few not the many). We cross-reference the IEEE Wi-Fi standards process and its relative merits and demerits.

We do not cover Internet Protocol standards, which you might argue is a gaping hole in a book on 5G and certainly key to end-to-end performance in any Internet-based connection. All the protocols above the physical layer potentially introduce delay and delay variability. The Internet was designed for resilience rather than throughput with resilience achieved by routing flexibility, individual packet addressing, and send-again protocols. However, the main objective in this book is to review 5G radio physical layer standards and the impact that these standards have on spectral requirements and to review spectral allocation policy and its impact on 5G radio requirements. If the radio part of an end-to-end link does not work adequately well or costs too much, then you have not even left the starting block. We would also argue that physical layer standards in the telecoms industry dictate the technologies used for end-to-end communication. There is no point in having closely controlled latency in a wireless link if the core network introduces variable delay.

In terms of commercial viability, physical layer radio standards are an important mechanism for achieving network scale economy. They are an essential mechanism for achieving interoperability and intersystem coexistence and have been fundamental to the evolution of the cellular industry for the past 30 years. Physical layer standards are also used as a mechanism for achieving competitive advantage usually based on a claim that a particular implementation is sufficiently different to merit patent protection. This makes this aspect of standards making an imperfect process but nevertheless essential. Physical layer standards add cost and save cost. The trick is to ensure that the savings come out on top.

2.2 5G Internet of Things

"If I had asked people what they wanted, they would have said faster horses," another Henry Ford piece of advice, warned about overreliance on market research and market forecasts.

Standards need to respond to changing market needs, but it can be difficult to forecast future market requirements and technology adoption rates. Subjective speculation about future user cases can be a frustratingly imperfect process. Mobile TV was promoted as a market opportunity 10 years ago but failed to match forecast expectations. As with mobile TV, market projections for the growth of Internet of things (IoT) connectivity should be treated with caution. Irrespective of potential market demand, market and technology uptake can be frustrated by multiple competing standards, particularly if the competing standards introduce additional cost or compromise performance.

Two proposals for a cellular IoT physical layer (see Table 2.1) provide a contemporary example, an evolved narrowband LTE physical layer (NB-LTE), supported by Ericsson, and a narrowband cellular IoT (NB-C IoT) supported by Huawei/Neul. The desired outcome is a new LTE category, Release 13 Category M (M for Machine), also known as Category 00.

The narrowband LTE proposal includes ternary phase shift keying (TPSK), a combined modulation and coding scheme where three symbols are mapped together in the code domain and transmitted during a single symbol period.

A single proposal on a converged solution was cosigned by more than 30 companies and was agreed in RAN1#83 (November 2015) to be approved by 3GPP RAN plenary#70 in December 2015.

The LTE-based machine type communication (MTC) standards have evolved from Rel-8 LTE Cat-1, Rel-12 Cat-0, and most recently Rel-13 Cat-M (also known as Cat-00), optimized for machine-to-machine (M2M) devices

Table 2.1
C IoT Physical Layer Candidates

	NB-C IoT Proposal	NB-LTE Proposal	NB-IoT (Release-13 Work Item)	
			Single Tone	Multitone
Bandwidth	200 kHz	200 kHz	200 kHz	200 kHz
DL multiplexing scheme	OFDMA	OFDMA	OFDMA	
DL subcarrier spacing	3.75 kHz	15 kHz	15 kHz	
DL modulation	BPSK, QPSK	BPSK, QPSK	BPSK, QPSK	
UL multiplexing scheme	FDMA	SC-FDMA	FDMA and/or single tone SC-FDMA	Multitone SC-FDMA
UL modulation	GMSK	TPSK	For example, $\pi/2$ BPSK	For example, BPSK, QPSK, TPSK
UL subcarrier spacing	5 kHz	2.5 kHz	3.75/15 kHz	15 kHz
Operation mode	Standalone, guard-band, in-band	Standalone, guard-band, in-band	Standalone, guard-band, in-band	

that require low cost, longer battery life, and extended coverage. The Release-13 NB-IoT was designed to meet a similar goal but with better coverage due to a lower noise floor, a product of the narrower channel bandwidth (200 kHz instead of 1.4 MHz for Cat-M). As can be seen from Table 2.1, the compromise is based on adding both standards together. While this is understandably pragmatic, it makes cost and performance targets harder to achieve.

The requirements of the new physical layer for IoT and local and wide-area mobile broadband will be addressed by a new 5G requirement study item (RAN70) with specification work targeted to be completed in two phases, phase 1 by the second quarter of 2016 (end of 3GPP Release 15) and Phase 2 by December 2019 to coincide with the end of 3GPP Release 16 with outputs fed into the IMT2020 submission process. Telecommunication standards involve hundreds of engineers spending thousands of hours in standards meetings. As with spectrum allocation, achieving a standards consensus can be a challenging and time-consuming process.

2.3 The Cost of Complexity

We cannot capture every nuance of this process in a book. However, we can draw on past experience to highlight areas where standards and market requirements are less than perfectly aligned. The nonalignment is often simply that the cost metrics needed to develop a new unproven market are never in practice realized.

Additional unnecessary costs are introduced as and when standards become overcomplex. The standardization of 29 classes of GPRS modem in the 1990s provides a past example of overcomplexity. The combination of uplink and downlink time slot configurations had a direct impact on RF front-end design including filter implementation, RF power amplification, and the switch path. Uncertainty as to which options should have priority absorbed the engineering effort. Only two of the 29 classes achieved sustainable market scale.

Today there are 17 categories of LTE user devices differentiated by peak bit rate throughput [1]. Each category requires a different RF front end to accommodate different channel combinations. RF economies of scale are not the same as digital economies of scale. Radio frequency behavior is defined by Ohm's law, not Moore's law.

2.4 Global Scale as a Precondition for Commercial Success

To be efficient, standards need to scale globally. In the late 1980s, Japan decided to standardize a country-specific digital cellular standard implemented on country-specific frequency bands. This was partly influenced by the national cellular operator, NTT, and reflected an ambition to develop a local market ecosystem that would provide Japanese vendors with a degree of market protection.

In practice, it meant that Japanese vendors had to give development priority to their local market requirements. This weakened their competitive position in other markets including at that time the emerging global system mobile (GSM) market at 900 MHz and later 1,800 MHz. China has followed a similar strategy with the development of time division single carrier multiple access (TD SCDMA) standard with some deployments also in India. The China market is large enough to support a nationally focused standard (and India adds another 500 million users), but time division duplex (TDD) does not scale well to large radius cells unless competing operator base stations are cosited and synchronized. This may change as the 3.5-GHz TDD networks are built and loaded, but TDD to date has not been a financial success. It is an understandable ambition to build patent value from physical layer standards, but generally there will be additional cost projected on other parts of the industry supply chain.

2.5 3GPP and Wi-Fi Standards

Additional unnecessary costs are also introduced when standards overlap. There is an ever-present risk of this happening with the 3GPP and IEEE standards processes.

Parallel standards processes work well when they are focused on achieving different outcomes. The IEEE Wi-Fi standards process has traditionally focused on local area high data rate radio systems deployed into unlicensed spectrum. The 3GPP process has traditionally focused on lower-data-rate, high-mobility, wide-area coverage deployed into licensed spectrum. Problems arise when either camp tries to do a land grab across these two naturally separate application domains. The commercial failure of the IEEE WiMAX standard for wide-area coverage is an example. In terms of contemporary overlaps, it is hard to see how the 3GPP Licensed Assisted Access for Unlicensed Spectrum (LAA) or LTE-U initiatives to deploy LTE into Wi-Fi spectrum can be technically or commercially successful. Attempts by the IEEE to standardize wide-area Wi-Fi for low-cost rural access will similarly overlap parallel low-cost LTE initiatives. These overlaps are industrially inefficient.

Tension points can also arise as and when the commercial ambitions of user device vendors do not align with operator commercial policy. Apple iPhone support for making voice calls over Wi-Fi is an example where some but not all operators are keen to adopt Wi-Fi voice. Operators have different voice delivery strategies that can include legacy GSM or voice over LTE and it is challenging for standards groups to accommodate these operator-specific requirements.

2.6 Standards and Supply Chain Efficiency

In an ideal world, standards facilitate industrial efficiency. In the real world, standards are used to achieve competitive advantage. There is neither surprising nor necessarily a bad thing, although it can result in parts of the industry supply chain achieving a near monopoly for periods of time, which can result in cost and price distortion. The LTE component supply chain has been serviced for at least 5 years by one dominant baseband vendor. This is efficient in terms of the return on investment for that one vendor, but could hardly be described as competitive.

Eventually, market forces prevail and new vendors move in to fragment and diversify over consolidated parts of the supply chain. This often coincides with the emergence of either new country-based or regionally based supply chains, with China being a contemporary example.

Standards therefore facilitate change but also have to respond to shifts in market dominance. First generation cellular in the 1980s was built on U.S. market growth and to a large extent enabled by U.S.-sourced component and network technology with Motorola as a dominant vendor.

Second generation cellular represented a fundamental technical shift from analog to digital processing and was built on European and Asian growth with

Nokia and Ericsson as dominant vendors servicing European and rest of the world markets.

Third and fourth generation cellular has marked a shift back towards U.S. dominance in the component supply chain despite the U.S. market becoming relatively smaller in global terms, today accounting for less than 10% of overall market value. This relative decline has been more than offset by the dominance of Apple in the smart phone market. Apple manufactures offshore but many of the components are U.S.-sourced and much of the intellectual added value is U.S.-based though Korea and Japan have similarly sized patent portfolios.

2.7 Regional Harmonization of Global Standards

Most industry commentators point towards China as the emerging leader of the mobile broadband market. While this is self-evidently true, the market reality is more nuanced and likely to be increasingly dominated by developing economy markets, a topic addressed in Chapter 6.

China is well placed to service these markets but not uniquely so. There is substantial playing space for U.S., European, Japanese, Indian, and Brazilian vendors and new developing economy players that are presently below the industry radar.

The 3GPP standards release process is relentless. LTE Advanced is divided into three phases. Phase A coincided with Release 10 and 11 and overlapped the 2012 World Radio Congress, WRC-12. Phase B started at the beginning of 2013 and coincided with Releases 12 and 13 and WRC 15. Phase C starts in early 2016 and coincides with Releases 14, 15, and 16 to coincide with WRC 19.

At least in theory, this should encourage coordination between the spectrum allocation process (covered in more detail in the next chapter) and the standards process.

The 3GPP has the unenviable but critically important task of bringing seven telecommunications development organizations together including the Association of Radio and Businesses in Japan (ARIB) [2], the Alliance for Telecommunication Industry Solutions in the United States (ATIS) [3], the China Communication Standards Association (CCSA) [4], the European Telecommunication Standards Institute (ETSI) [5], the Telecommunications Standards Development Society of India (TSDSI) [6], the Telecommunications Technology Association of Korea (TTA) [7], and the Telecommunications Technology Committee of Japan (TTC) [8].

These seven organizations between them have a reasonable claim to be representative of regional interests across Europe, the United States, Latin America, Japan, and Asia. Other entities such as the operator led Next Generation

Mobile Network Alliance (NGMN) [9] influence standards making but do not have a formal drafting role. Other industry consortia are focusing on network architecture concepts [10].

The ITU process is coupled specifically to the outputs from the 2015 World Radio Congress agenda and the agreed work items for WRC2019 with a proposed completion deadline of 2020 hence the title (IMT-2020 in ITU-R).

2.8 5G Technology Standards

ETSI has a new industry standardization group working on millimeter-wave transmission focusing initially on V-band (57–66 GHz), E-band (71–76 and 81–86 GHz), and future higher-band options up to 300 GHz [11].

There is also a group of vendors and operators promoting 5GPP as a standards body [12] and a cross section of research institutions with a 5G research remit, the 5G Innovation Centre at Surrey University in the United Kingdom is one example [13].

We could spend this whole chapter reviewing the views and ambitions of these various standardization and research entities, but it is going to be more useful to summarize what is actually happening in LTE Advanced standardization and use that as a basis for assessing the relative technical and commercial viability of the 5G options being presently discussed. This means that we are focusing on the 3GPP standards process and the impact of that process on 5G technology options.

The 3GPP standards are developed by four technical specification groups: Radio Access Networks (RAN), Service and System Aspects (SA), Core Networks and Terminals, and GSM EDGE Radio Access Networks (GERAN); the GERAN group will be merged into RAN 6 in June 2016 [14].

Terminals are a generic term used to describe anything connected to the network, including smart phones and IoT modems. We would argue that terminal cost and performance are fundamental to the overall economics of mobile broadband network technologies.

Our general interest is in LTE Advanced, and our specific interest is in LTE Advanced Phase B and C and the likely cost and performance of LTE Phase B and Phase C user devices.

LTE Advanced has been part of the Release process from Release 10 onwards and includes carrier aggregation, advanced multiple input multiple output (MIMO), coordinated multipoint transmission and reception, relay nodes and repeaters, heterogeneous networks, and self-organizing networks. The initial focus was a response to a marketing assumption that peak data rates were a dominant design brief. This resulted in the specification of eight-antenna

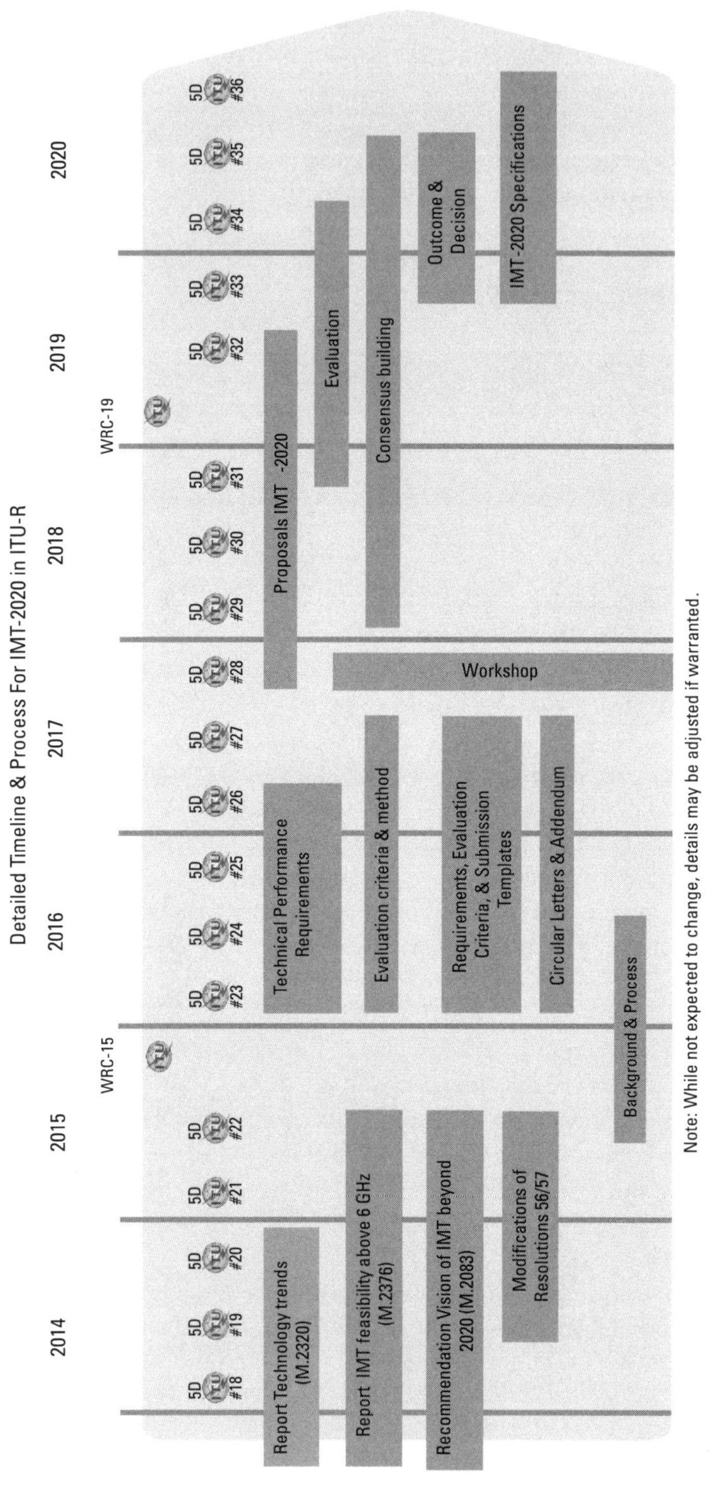

Figure 2.1 ITU IMT 2020 timeline. (Reproduced with permission from the International Telecommunications Union.)

port MIMO and 100-MHz carrier aggregation, which was fine in theory but undeliverable in a low-cost, low-energy budget terminal.

LTE B shifts the focus to lower cost more power efficient terminals, a 30 times capacity boost relative to LTE Release 8, a 10 times increase in throughput at the cell edge and support for machine-type communication (MTC), which confusingly is more or less the same as IoT connectivity. The increase in cell edge throughput requires a parallel increased emphasis on interference management including coordinated multipoint transmission, network-assisted interference cancellation, and interference mitigation in the terminal.

The three-dimensional (3-D) beam forming at the base station allows for control of the downlink coverage both in azimuth and elevation (hence, the 3-D description). This can be regarded as a precursor to the beam-forming adaptive phased array antenna systems that are emerging as a core part of the 5G centimeter- and millimeter-band product offer.

2.9 Centimeter-Band and Millimeter-Band 5G: Energy Efficiency as a Requirement

The priorities within LTE C are subject to final agreement, but we would argue that they should focus on the scalability or lack of scalability of the existing orthogonal frequency division multiplexing (OFDM) physical layer into the centimeter and millimeter bands, including coexistence with Wi-Fi in the centimeter band (Wi-Fi at 5 GHz) and the millimeter band (Wi-Fi at 60 GHz). There is also an emerging consensus that formal targets should be agreed for energy efficiency described in terms of power per unit area (watts per square meter), throughput per unit area (bits per square meter), and energy (joules) per bit [15]. An 800% improvement in energy efficiency is included in the LTE C performance target wish list.

It may be easier to meet these energy-efficiency targets by adopting a simpler physical layer in 5G. OFDM is effective at countering multipath and supporting wider channel bandwidths but introduces large amounts of envelope variation, which makes it hard to get good power-added efficiency out of RF amplifiers. The delay spread from multipath requires clock intensive power-hungry channel encoding/decoding, with the decoder and memory overheads being a particular issue. Mitigating multipath by beam forming potentially means that OFDM does not need to be used and channel coding can be relatively lightweight, although spatial processing overhead will increase.

Closely integrated backhaul is also an essential mechanism for improving network and user device energy efficiency. In particular, there is a close relationship between backhaul delay and delay variability on scheduler efficiency, which we explore in Chapter 9.

Improved energy efficiency at the network level reduces operational costs and makes solar power more practical for deep rural base station sites. At user device level, improved scheduler efficiency reduces power drain which improves the user experience (increases user value) but also potentially increases uplink and downlink offered traffic. Users or IoT devices cannot use a network if the device battery is flat.

Improved scheduler efficiency also increases network capacity. The impact of the scheduler on mobile broadband network economics is therefore significant and is one of the key differentiators in the 4G and 5G vendor offer. This is relevant when we come to consider 5G modulation and coding schemes, specifically modulation and coding schemes, which are spectrally efficient but introduce end-to-end delay and delay variability. Modulation and coding also need to be closely coupled with multiplexing. The three functions are separate but closely interrelated and together have a fundamental impact on system performance.

No one set of options is perfect. The objective is to arrive at an efficient compromise between spectral efficiency and power efficiency, and the optimum choice is likely to be different for the meter band (300 MHz to 3 GHz), centimeter band (3 GHz to 30 GHz), and millimeter band (30 GHz to 300 GHz) and will be influenced by the channel spacing used within each passband and the channel delay spread.

For example, a train traveling at 300 km per hour supported by a radio system at 2.6 GHz will have a constant channel for about 1 ms, equivalent to a subframe in LTE. To have an equivalent radio channel at 60 GHz, the train would have a maximum speed of 15 km per hour, slower than an elite marathon runner.

It is therefore clear that adaptive modulation and coding schemes and frame structure for 5G, particularly 5G mobile wide-area systems implemented in the centimeter and millimeter bands, will need to be different from existing options. If existing options prove capable of being scaled to these higher bands, then optimization will be needed.

2.10 2G to 5G Modulation and Channel Coding

The 5G modulation and coding and multiplexing options are the subject of significant debate within the vendor community and particular options and combinations are claimed to be superior. The veracity of these claims is dependent on the overall performance objectives.

The use of simple Gaussian minimum shift keying (GMSK) modulation in GSM 2G radio systems, for example, is not particularly spectrally efficient but performs well in the power domain (efficient power amplifiers). Spectral efficiency is achieved through a combination of bandwidth-efficient voice codecs and digital coding gain based on block and convolutional coding.

The 3G systems introduced higher-order QPSK and 16 QAM modulation. This improves spectral efficiency but performs less well in the power domain. Coding gain is similar to 2G but with more resources (clock cycles of processing) dedicated to equalizing the wider (5-MHz) channel bandwidth.

The 4G systems add in OFDM as a mechanism for managing the increased time dispersion of wider bandwidth channels (>5 MHz). Time dispersion occurs when a transmitted signal propagates to the receiver via multiple paths with different delays due to varying path lengths. In the frequency domain, a time-dispersive channel translates into a nonconstant frequency response. The channel becomes frequency-selective. As bandwidth increases, time dispersion increases.

OFDM is useful because it can be implemented using a computationally efficient FFT. Time dispersion is managed by the use of a cyclic prefix in which the last part of the OFDM symbol is copied and inserted at the beginning of the next symbol; it also helps with circular convolution reconstruction of the channel at the detection part of the receiver. This is a time-domain process that results in improved frequency diversity.

It also allows for interference coordination for broadcast channels in which identical time aligned signals are transmitted from multiple cells in a single frequency network. At the terminal, intercell interference appears as time dispersed signal corruption. Provided that the cyclic prefix is long enough, broadcast rates are only limited by noise.

Time dispersion can be exploited in multiantenna systems. A simple example is two-antenna space-time coding in which modulation symbols are transmitted on the first antenna and then modulated on to the second antenna with the order of the modulation symbols reversed. This is supported within 3G wideband code division multiple access (W-CDMA) systems.

In 4G LTE, the use of OFDM allows for more complex space-frequency time coding where a block of frequency-domain modulation symbols are mapped to specific frequency subcarriers and then are applied to an antenna element.

The performance gain achieved is a function of the spatial separation between the antennas, which is a function of the wavelength/frequency of the carrier. If the antennas are relatively close together, for example, in meter-band antenna arrays, there will be high mutual antenna correlation (minimal wavelength separation). The transmission beam can be steered by applying different phase shifts to the different antennas, but the beam will move slowly and be relatively wide. The result should be an increase in the received signal strength but no additional diversity against radio channel fading.

If the antennas are relatively wide apart in terms of wavelength separation, for example, in centimeter-band and millimeter-band antenna arrays, there will be low mutual correlation. Combined with polarization diversity, this allows the antenna weights to have complex phase and amplitude values. This enables antenna precoding in which the phase of the transmitted signal is rotated to compensate for the instantaneous channel phase. The result should be fast beam forming and/or highly directive beam forming, which together should provide more protection against radio channel fading and provide the ability to support high data throughput to multiple mobile users and devices.

It is also possible to spatially multiplex the antenna array to utilize high signal-to-noise ratios to support high data rates (MIMO antennas) and or to implement successive interference cancellation.

In successive interference cancellation, the receiver demodulates and decodes one of the spatially multiplexed signals then subtracts that time and frequency signal energy from the next spatially multiplexed signal. The first signal decoded will need to be more robust in order to counter the higher interference floor.

This is achieved by using a lower-order modulation and lower coding rate. It could also be achieved by applying different power levels to each successive symbol stream. This is described in the technical literature as nonorthogonal multiple access (NOMA). The power efficiency and linearity requirements of NOMA require additional study.

The existing LTE physical layer is in many ways elegant. The disadvantage is that the cyclic prefix absorbs bandwidth and power. If the frequency selectivity of the channel exceeds the span of the DFT then the inverse DFT will be unable to reconstruct the original block of transmitted signals. The subcarriers are also sensitive to narrowband interferers. The scheduler will work its way around this, but usable bandwidth will decrease.

More problematically, the OFDM subcarriers have a habit of ganging up together so large resource block allocations can cause issues with out-of-band emissions. The highest-order modulation option at the moment is 256 QAM. The composite waveform, including its envelope variation, is a product of the modulation and multiplexing and coding and scheduling. Out-of-band emissions can be managed proactively by not using resource blocks at the edge of the channel but as with narrowband interference, there will be a loss of usable bandwidth. This can be particularly problematic if wide band channels, for example, 10- or 20-MHz LTE channels, are adjacent to narrowband LTE (5 or 3 or 1.4 MHz) or other narrowband radio systems.

There are also doubts as to whether a conventional Fourier transform will be able to scale to the higher data rates expected from 5G wide-area radio systems or (as stated earlier) whether it will be needed.

These potential limitations provide the basis for alternative modulation and multiplexing schemes including filter bank multicarrier where each subcarrier is individually shaped in the frequency domain by an individual subband filter. The claimed benefits are higher out-of-band attenuation from the subband filtering with a low post sampling filter rate, better spectral efficiency due to the eradication of the cyclic prefix, and asynchronous allocation of empty subchannels to new users as they become available. This would be particularly useful in dynamic spectrum access schemes. The performance of the subband filtering will be crucial to the overall efficiency of this option.

The various coding options are next.

2.11 Channel Coding: The Simple Explanation

The technical literature on coding tends to leap into post graduate mathematics at the first opportunity, but essentially the job of all coding schemes is to answer two questions. Is this a 0 or a 1? And is this my 0 or 1? The correct answer to the first question delivers sensitivity gain. The correct answer to the second question delivers selectivity gain.

Seventy years of coding theory and practice can be summarized as follows. In the beginning, parity checks were used to detect bit errors. If the parity check sum failed, the coded bit stream would be retransmitted. Parity checks are still used today.

Block codes developed from parity checks. They work on a similar basis but the parity check number points to where an error has occurred in a code block, which can then be error corrected.

Convolutional codes are different because they use memory. A 0 or a 1 travels through the encoder influencing the output code word at every multiplication point. This is analogous to asking multiple times if it is a 1 or a 0. There is no bandwidth expansion. The cost is coding and memory delay and delay variability and clock cycle overhead as the code stream passes through the convolutional encoder and decoder.

Turbo coders are two convolutional coders working in parallel and use soft decision information, for example, instantaneous channel conditions, to weight the decision matrix in the decoder. Each codec encodes the entire input block of data bits.

Low density parity check codes (LDPC) are an alternative to turbo codes and are proposed in several 5G coding schemes. They use similar decision trellis methods, but the data is encoded in short blocks across multiple encoders.

Signal-to-noise ratio and or signal-to-interference-and-noise ratios are measured and described by the channel state indicator. As these ratios deteriorate, the coding overhead will increase and lower-level modulation options will be used. A 10% error threshold is usually used to determine the choice of modulation and coding. The objective is to avoid triggering higher-level send-again protocols. An out-of-sequence transport block caused by errors or coding error extension will incur an automatic repeat request round trip delay of 8 ms, which is tolerable when admission control is being managed at frame level (every 10 ms) but destructive if admission control is being managed at subframe level (every 1 ms). It is not just the absolute delay but the delay variability which can be difficult to manage, including at higher layers of the protocol stack where deterministic processes such as timeout-based challenge and response authentication have to be supported.

2.12 Modulation and Coding for Wide-Area Mobility

Modulation and coding requirements for wide-area mobility are therefore different from fixed point-to-point radio links. By definition, in fixed point-to-point radio links, there is no need to code for Doppler shift (frequency shift) and there will be minimal time dispersion, assuming that the link is line of sight. Combined with the >40 dBi of directional gain available from a dish antenna in the centimeter or millimeter bands, this allows lightly coded higher-order modulation to be implemented.

However, this may change in 5G networks. Mobile platforms, for example, cars, boats, trains, planes, and ships (the Internet of slow-moving and fast-moving objects), could act as repeaters and relays and would need to be channel coded for Doppler shift. If both ends of the radio link are moving, Doppler spread will increase.

Active multiple antenna arrays in the millimeter band could be producing similar gains to existing dish antennas with existing 1-km to 2-km hop lengths being significantly extended. Wide-area coverage in the millimeter band will therefore be based on narrow beamwidth links with moderate line-of-sight time dispersion and with near-line-of-sight time dispersion exploited to provide additional delivery bandwidth. The physical layer requirements for wide-area mobile coverage and backhaul will therefore become more similar over time.

It is thus not implausible to consider using similar modulation, coding, and multiplexing schemes for backhaul and wide-area mobile connectivity. If large amounts of contiguous bandwidth are available, for example, 1-GHz or 2-GHz channel spacing within 5-GHz passbands, and if beam forming is used to increase throughput, then simpler constant envelope modulation schemes could be used. This would improve RF amplifier efficiency.

The proviso is that even with the link gain available from narrow beamwidths, there will still be higher noise floors and more nonlinearity in the front end of centimeter- and millimeter-band transceivers. The addition of an extra

3 dB for every doubling of frequency serves as a rough rule of thumb excluding other factors, for example, the use of more exotic materials in higher-frequency devices.

This adds additional weight to the argument that simpler modulation options might be better suited to the centimeter and millimeter bands due to their resilience to phase noise and distortion.

2.13 Implementation Loss: Simulation Versus Real-Life Performance

Caution needs to be applied at this point. The MATLAB simulations for 4G LTE suggested that significant throughput gains could be achieved with an OFDM downlink and single-carrier frequency division multiplexing access (SC-FDMA) uplink coupled with adaptive modulation and coding.

While these have been at least partially realized, the practical gains have not been as big as the theory predicted. This should not be a surprise as simulations do not generally account for implementation loss: the difference between a simulation, a specification datasheet, and a real-life device. The performance difference has included higher-than-expected out-of-band emissions. These are easy enough to control through additional filtering, but this introduces additional insertion loss (reduced coverage). The alternative is to reduce power, which reduces coverage and throughput or reduce resource block allocation (15-kHz subcarriers mapped on to the millisecond frame structure), which reduces capacity and per-user throughput. There is also the overhead of the cyclic prefix, which critics argue incurs an unacceptably high spectral utilization cost.

The question is whether other schemes are significantly better. Bear in mind that alternative options are often predicated on assumptions of a shift in application requirement. For example, filter bank multicarrier is promoted for its ability to handle fragmented chunks of spectrum including narrow slices of spectrum that might become available at any time at any place, usually described as white space or more recently as dynamic spectrum [16].

The counterargument to this is that the filter is only one part of the transceiver chain. At present, the only way to process multiple slices of noncontiguous spectrum of differing bandwidths is either to have multiple transmit/receive chains with multiple frequency-specific matching networks or to have a wideband front end that introduces significant dynamic range limitations. You could argue that it would be better to allocate 5G spectrum in discrete bands. Most if not all RF designers if given the choice of 5 GHz of contiguous spectrum in the millimeter band or 5 GHz of noncontiguous spectrum spread here and there across the meter, centimeter, and millimeter bands would opt for the contiguous bandwidth option. This is unlikely to change unless someone invents a well-matched, compact, low-cost, broadband antenna. In the process they will have discovered some previously unknown physics.

We would therefore argue that filter bank multicarrier is solving a problem, that of fragmented spectrum, which need not and should not exist. It is technically and commercially more efficient to solve the problem though the regulatory spectral allocation and auction process.

A similar argument is advanced to support a shift from a synchronous to nonsynchronous physical layer. The 5G NOW project (nonorthogonal waveforms for asynchronous signaling) is an example [17].

The assumption here is that 5G networks will be supporting a large amount of machine initiated communication. The closed loop time synchronization used in LTE random access imposes a signaling overhead, which is considered to be insufficiently energy-efficient for these devices. The alternative is to either implement an open loop scheme where the devices calculate their own timing advance or have a completely asynchronous scheme, sometimes described as Pure Aloha (Hawaiian for "hello"). This will have to coexist with legacy synchronous physical layers, which is potentially problematic.

2.14 The Economic Impact of Implementation Loss

Any physical layer is going to suffer some implementation loss. In practice any loss of RF performance in a user device or IoT device will be partly compensated by the network. Networks are also generally planned on worst-case interference assumptions and at least when partially loaded will operate at lower noise floors. Put another way, implementation loss only becomes a problem when networks become fully loaded.

2.15 Thirty Years of Cellular Standards

At this point, it is worthwhile to revisit the technical rationale behind each of the physical layer options so far for cellular radio.

First generation analog cellular was in many ways an obvious choice. The concept of cellular radio had been around since 1947 [18], but this was 30 years before low-cost microcontrollers became available capable of handling the handover protocols. There was also a need for low cost high performance FR4 circuit board material to support devices working at 800 and 900 MHz. The physical layer was more or less identical to existing two-way radio with FM modulated 25-kHz or 30-kHz channels.

These networks were hard to manage in the frequency domain. The 25 by 25 MHz Advanced Mobile Phone System (AMPS) band allocation at 800 MHz in the United States meant that 833 channel pairs had to be frequency planned across the network and user devices had to have frequency synthesizers and phase locked loops that could support dynamic access to any duplex-spaced channel pair across the band.

GSM simplified that RF planning and processing overhead by replacing 25-kHz and 30-kHz channel spacing with 200-kHz channel spacing but could only be implemented once sufficiently low-cost digital signal processor (DSPs) came available to support the voice codec, channel coding, and equalization.

Some of the complexity therefore reduced in the frequency domain but was replaced by at least equivalent additional complexity in the time domain. In the standards process, time-domain complexity was responsible for many thousands of pages of specification.

The 3G theoretically reduced RF domain complexity by relaxing channel spacing from 200 kHz to 5 MHz but replaced RF complexity with code-domain complexity. The original orthogonal variable spreading factor (OVSF) code structure, designed partly to work around intellectual property issues, was a masterpiece of elaborate physical layer design that performed well in simulation and poorly in practice and only began to work adequately well when simplified down to the 16-code structure used in high-speed downlink packet access (HSDPA). The power budget of these devices also proved to be a challenge. The code division multiple access (CDMA) system in the United States also had initial problems calibrating user device power outputs. Achieving user to user selectivity in the code domain depends on the symbols being delivered to the base station at similar power levels. User device power output calibration was an underestimated and relatively expensive requirement.

With 4G LTE, the decision on the physical layer was largely determined by the decision to support wider bandwidth channels (5, 10, 15, and 20 MHz and aggregated bandwidth to 100 MHz) and narrower 3-MHz and 1.4-MHz channels. This involved the reintroduction of RF domain complexity with the use of 15-kHz and 7.5-kHz subcarriers, which in turn relaxed the channel timing issues implicit in the wider channel spacing options. Note that the wider channel bandwidths were adopted partly due to their increased multiplexing gain but also to support the wider passbands being introduced globally (see Chapter 3).

The fast Fourier transform (FFT) for the orthogonal transform is computationally efficient. The downside of OFDM (the division of the channel into subcarriers) and orthogonal frequency division multiplexing access (OFDMA) (the mapping of multiple users across frequency-domain and time-domain denominated subcarriers) is that it requires linear amplification and creates relatively high levels of out-of-band emission, which needs to be managed at network level with a consequent loss of spectral efficiency, power efficiency, or range.

Note that the linearity requirements are determined by the modulation used and the OFDM and OFDMA process. The modulation can be anything from four-level QAM to 256-level QAM so at least the linearity requirements (and resilience to noise and channel distortion) can be adaptive. The higher order modulation options in particular require close attention to noise floors and amplitude modulation (AM) to phase modulation (PM) distortion.

First and second generation cellular devices both use constant envelope modulation, FM in the case of first generation and GMSK in the case of second generation. Third generation introduced 14-level to 16-level QAM and fourth generation increased the modulation complexity up to 256 QAM with additional unwanted amplitude modulation created by the aggregation of OFDM waveforms within a channel and or multiple aggregated channels.

In first and second generation systems, amplifiers could run as Class C devices and could deliver power-added efficiencies of greater than 50%. For third and fourth generation systems, it is relatively straightforward to back off amplifiers to avoid clipping and distortion in third and fourth generation user devices but the intrinsic efficiency will reduce and can be as low as 10%. This can be managed by a number of correction methods including predistortion and envelope tracking but these techniques are bandwidth limited, introduce noise, and absorb DSP clock cycles. In other words, the correction techniques improve RF efficiency but introduce a baseband processing overhead.

In practice, vendors have used a mix of techniques including envelope tracking, fixed backoff and adaptive backoff in addition to network-level mitigation (managing out of band by reducing resource block allocation). Whatever methods are used, there is an associated performance cost somewhere in the processing chain.

Given that gain and linearity are harder to achieve as frequency increases, then it is reasonable to question whether OFDM is an optimum choice for the centimeter and millimeter bands, particularly if the multipath problem is managed by other mechanisms such as beam forming. The problem of RF power efficiency in LTE user devices is partially solved by an additional inverse transform, which helps to randomize the envelope variation, the underlying principle of SC FDMA, but this only partially solves the RF efficiency problem.

To reiterate, there is a body of opinion that suggests that OFDM will not scale to higher frequencies and that better candidates are available. However, some of the other candidates, for example, filter bank multicarrier, are solving problems that are better solved in other ways, such as by avoiding spectrum fragmentation.

The same argument applies to nonorthogonal and nonsynchronous physical layer options. For certain, the cyclic prefix is an overhead but in practice it acts as a guard band in the time domain and is therefore no different to a frequency-domain guard band. It has a purpose and it realizes that purpose adequately well.

A nonsynchronous physical layer would be undeniably more bandwidth and power-efficient for low bit rate and sporadic data but present assumptions that machine type communication (MTC) will be a dominant part of the 5G traffic mix are not based on any present market reality. Meanwhile, General Packet Radio Service (GPRS) modems get cheaper by the day.

OFDM and OFDMA work well for 5-GHz Wi-Fi, but the initial implementations of 802.11 ad at 60 GHz have not implemented OFDM and the power efficiency of the devices remains a challenge that to date has only being partially solved with low output devices (10 dBm). The RF system efficiency of a 2-W user device at 70 GHz with an OFDM physical layer would be particularly challenging.

The combination of lower output power and less dynamic range has generally allowed the IEEE physical layers to evolve faster than 3GPP physical layers, which have needed to accommodate wider area mobility, which means a bigger dynamic range. Supporting wide-area mobility introduces additional channel signaling overhead, which needs to be accommodated alongside the traffic channels. This increases the amount of envelope variation on the composite channel.

2.16 3GPP and IEEE Standards: Wide-Area Wi-Fi?

The 3GPP is sometimes criticized for moving more slowly than the IEEE process, but there are practical reasons for a more conservative approach to physical layer innovation. However, there is some truth in the assertion that the IEEE standards process can respond faster to changing market conditions. This is partly because the process has traditionally been driven by the computer industry rather than the telecommunications industry.

To an extent this remains true today, although companies such as Intel are active in both. There has also been a traditional focus within the IEEE on unlicensed spectrum, whereas 3GPP has been focused on licensed allocations. This separation is becoming less obvious with 3GPP producing proposals for LTE in the ISM bands and IEEE producing proposals for wide-area Wi-Fi for rural areas and wide-area IoT connectivity. Being faster is not a guarantee of market success, and the failure of WiMAX to scale globally is an example of that.

However, it is worthwhile to briefly compare the IEEE and 3GPP standards process. The allocation of the 2.4-GHz band for unlicensed low-power communication by the FCC in 1985 prompted a major standardization effort. The allocation was linked to the proposed use of spread spectrum techniques to provide interference resilience to and from other devices and machines used in the band.

In 1997, the first version of the 802.11 standard was finalized with first generation products delivering data rates of 1 or 2 Mbps. Although modest by today's standards, such data rates were an order of magnitude higher than contemporary cellular radio networks.

Cellular networks could do a lot of other clever things including power control, 35-km radius cells, and seamless handover, a range rather than rate proposition. Fifteen years on, the extensions of the original 802.11 standard are delivering data rates of more than 100 Mbps. Cellular networks struggle to deliver 10 Mbps. This 10:1 ratio for headline data rates remains constant over time.

Putting Wi-Fi and cellular together therefore seems like a good idea. Most smart phones now include Wi-Fi. A seamless wide-area "best connect" user experience remains elusive. This is partly because defining best connect can be difficult. Best connect for an operator may not be the same as that for a user. The user connectivity experience is inconsistent. Automatic search algorithms flatten batteries. Manual network selection is not as easy as it could or should be.

Wi-Fi works remarkably well most of the time, with path loss mitigating many if not all of the potential coexistence issues. If operators want a denser network, then that is what they get; in Figure 2.2, 400 Wi-Fi sites (on the right) provide equivalent coverage to 20 cellular sites (on the left) with a significant theoretical gain in peak data rates.

The issue is whether this is economic once site acquisition and site management costs and backhaul costs are factored in to the equation.

Public Wi-Fi networks, for example, in train stations, airports, and shopping malls, can also become bandwidth-limited if multiple uncoordinated networks are deployed. This is because the contention protocol uses direct frequency sensing. If contention is detected on a channel, then that channel cannot be used for 30 minutes.

Down at the physical layer, many of the features being specified for LTE in Releases 10, 11, and 12 including higher-order modulation, channel bonding, high-order MIMO, and multiuser MIMO are already supported in Wi-Fi.

In the 5-GHz band, 802.11ac supports RF channel bandwidths of 160 MHz using 256 QAM. In the 60-GHz band, 802.11ad uses simpler modulation but supports four 2.16-GHz channels.

The 802.11ad is now being replaced or at least supplemented with 802.11ay [19] with a timeline for ratification by 2017 and a target peak data rate of 20 Gbps intended for short-range, high-bandwidth applications such as TV and monitor displays. There is also an IEEE work group developing

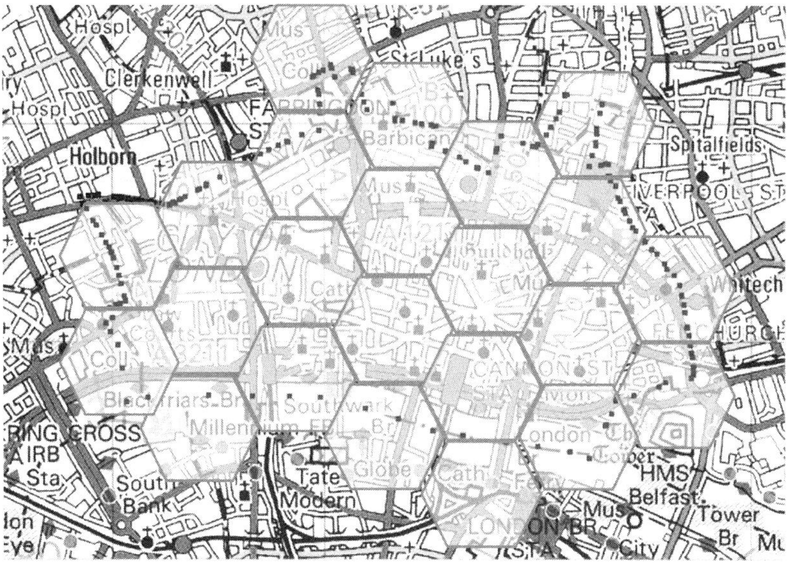

(a)

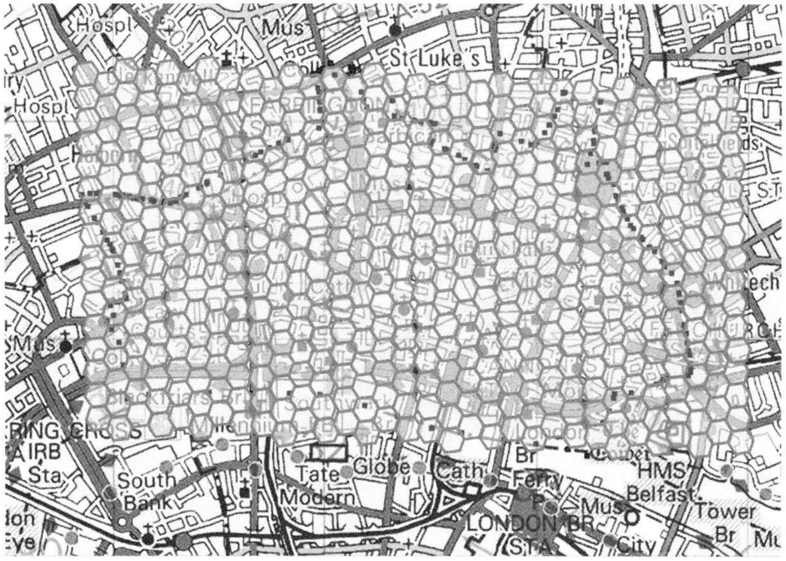

(b)

Figure 2.2 Wi-Fi/cellular site density comparison. (Courtesy of Plum Consulting.)

802.11ah, which repositions Wi-Fi for larger cell applications including rural coverage and low-energy IoT connectivity.

Figure 2.3 shows the parallel development of LTE, Wi-Fi, and 5G standards with associated peak data rates and expected timelines.

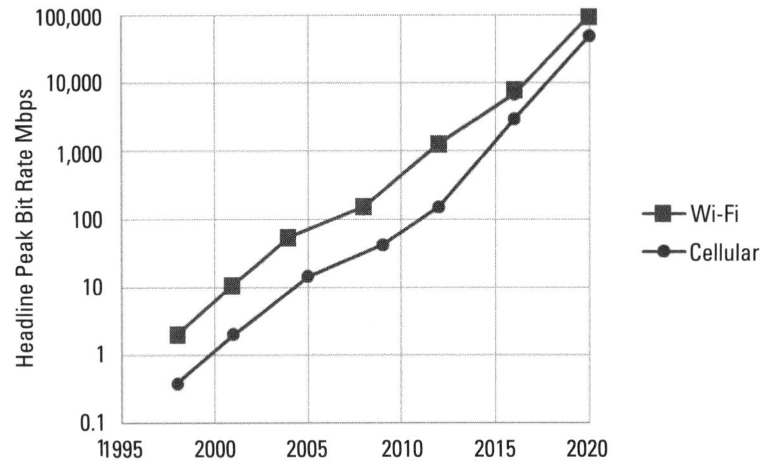

Figure 2.3 LTE and Wi-Fi evolution. (Courtesy of Aegis and Plumn Consulting.)

There is an additional work item within 3GPP to develop an LTE U standard to support LTE in the 5-GHz unlicensed band, adding duty cycling to existing Release 12 functionality to facilitate LTE Wi-Fi coexistence.

LTE-U does not meet ETSI coexistence requirements and cannot be used in applications where dynamic sensing is presently used. LTE-LAA (Licensed Assisted Licence) would work its way around this restriction by incorporating Wi-Fi "politeness protocols" to become compliant with ETSI requirements and is expected to be supported in Release 13. This is covered in more detail in Chapter 4.

In the context of 5G, our main interest in the 802.11 standards process is that it provides early visibiity to the the challenges of implementing power efficient low-cost, high-order modulation coupled to high-order MIMO combined with an OFDM uplink and downlink, implemented in each of the three bands of particular interest for 5G (2.4 GHz in the meter band, 5 GHz in the centimeter band, and 60 GHz in the millimeter band).

In particular, it illustrates that 5G represents a significant increase in spatial domain processing complexity over and above existing 4G systems with the gain available from that complexity significantly dependent on a move to the higher-frequency/shorter-wavelength centimeter and millimeter bands.

2.17 Adaptive Antenna Arrays

The assumption here is that 5G will follow the example of Wi-Fi and will implement active phase arrays where each element is coupled to its own dedicated power ampifier on the TX path and LNA on the receive path both on the downlink (base station to user device) and uplink (user device to base station).

Within user devices, it can be seen that even at millimetric wavelengths a 32 by 32 element array (proposed for WiGig at 60 GHz to support 100-Gbps data rates) will be relatively large with a half-wavelength at 30 GHz being 5 mm and 2.5 mm at 60 GHz. Typical antenna configurations for user devices are therefore likely to support lower array counts with array sizes varying from 10 by 10 mm for a 2 × 2 array at 30 GHz or 5 by 5 mm at 60 GHz through to 10 mm by 10 mm for a 4 × 4 16-element array at 60 GHz. The 16-element array would provide an antenna gain of about 12–14 dBi with a half-power beamwidth of 30° in elevation and azimuth. Higher array counts could be supportable at the base station to provide an overall link budget gain.

The mechanical spacing is arguably the simplest part of the puzzle. The 802.11ay specifications are largely exercised initially by issues of peak to average ratio and cumulative spectral density, which in turn dictates RF power efficiency and out-of-band emission.

The specification data sheets from 60-GHz Wi Gig vendors are claiming that single-chip phased array transceivers supporting 16 TX and RX chains are feasible working across 2-GHz channel bandwidths at a power level of +10 dBm per power amplifier (PA)with RX noise figures of the order of +6 dB receiver supporting 16-level QAM modulation [20, 21].

These devices are based on deep-submicron Complementary metal-oxide semiconductor (CMOS) and silicon germanium (SiGe) BICMOS semiconductors.

The challenge for 5G will be to scale this to higher power and up to 100 dB of dynamic range to support wide-area mobility. The assumption is that the air interface for the centimeter and millimeter bands will be TDD so that the RX and TX channels are reciprocal. This will simplify channel sounding but may have a disproportionate cost in terms of system-level efficiency and may introduce major challenges in terms of time coordination between networks that are geographically and spectrally proximate. We revisit this FDD versus TDD debate in subsequent chapters, but it is worth noting that most fixed point-to-point systems are deployed in traditional duplexed spaced bands. The sensitivity gain achievable from duplex spacing and a duplex gap is likely to outweigh any additional processing complexity in an adaptive antenna system, at least for wide-area deployment.

Note that there is also an important difference between the job that adaptive array antennas have to do in a local area Wi-Fi network and the job that adaptive array antennas have to do in a wide-area mobile network. Users in a Wi-Fi network will be stationary or slow-moving over short distances. The antenna design and beam-forming algorithms are therefore optimised to create multiple paths per user.

In a wide-area mobile network, for example, in the millimeter band, the antenna system and beam-forming algorithms are tracking users who will generally be, although not always, moving further and faster. The algorithmic requirement is therefore closer to the algorithms used in millimeter-band automotive radar in which moving objects are continuously tracked in terms of their distance and direction of travel and require real-time calculation of angle of arrival, angle of departure, and angular power. It is likely that a user will be supported on a single narrow beamwidth with the higher data rate achieved though increased effective isotropic radiated power (EIRP) and lower visibility to interference and isotropic noise.

2.18 Big Radios Need Small Radios

2.18.1 Evolving Bluetooth Standards

Most of this book is looking at 5G as a wide-area mobile technology, but any new physical layer has to coexist technically and commercially with other physical layers including Wi-Fi for local area connectivity and Bluetooth for personal area connectivity.

The Bluetooth standard had a major update in 2010 with the introduction of Bluetooth Smart, although most of the standard's work was focused on Internet Protocol optimization for IoT connectivity. The most recent specification update, Bluetooth 4.2, takes this work forward with support for energy-efficient access protocols. Underneath all this upper-layer optimization is a nice simple physical layer using frequency modulation (FM)/frequency shift keying (FSK) constant envelope modulation to maximize RF efficiency.

There has been considerable debate about Wi-Fi/Bluetooth coexistence in the 2.4-GHz band. In practice, the radios work together adequately well most of the time. The 4.1 and 4.2 specifications address Bluetooth/LTE interoperability and it should be assumed that this work stream should at some stage include 5G compatibility, preferably reasonably early in the 5G standardization process.

2.18.2 ZigBee

Bluetooth was designed originally as a cable replacement and like many other standards has undergone a broadening of its original application remit. A similar evolution has occurred with the ZigBee standard developed initially to replace infrared-based remote controls but extended to support lighting, curtains, and domestic appliances and meters within the home and or industrial buildings.

ZigBee is claimed to provide optimized low-latency mesh deployment including multihop mesh topologies with latencies of around 30 ms compared to 100 ms or more from Bluetooth or Wi-Fi. Although the ZigBee standards forum [22] claims an installed base of around 500 million devices, the standard needs to scale at a significant rate to remain competitive. Vendor-specific implementaions of the standard are making that process harder. If 5G vendors are serious about IoT markets, there will need to be be a closer coupling with existing systems including ZigBee radios.

2.18.3 Proprietary Narrowband Physical Layers for IoT Connectivity

There are also proprietary narrowband connectivity solutions for IoT connectivity including a range of products from SigFox [23]. The LoRA Alliance also has an increasing visibility, particularly in U.S. markets [24]. It is possible although unusual for proprietary standards to become adopted globally.

2.19 Summary

Each successive generation of mobile cellular network has been standardized on the basis of assumed future improvements in device performance including packaging and substrate performance, a process sometimes described as technology interception.

The standards process for first generation analog cellular in the 1970s assumed that low-cost microcontrollers and low-cost FR4 printed circuit board material would be available to support seamless handover and an RF front end that would be adequately efficient in terms of noise and gain at 800 and 900 MHz.

The standards process for second generation cellular in the 1980s had to assume that low-cost, low-power budget digital signal processing would become available to support digital voice encoding and time-domain channel equalization.

The standards process for 3G cellular in the 1990s had to assume that sufficient baseband processing would be available to support code-domain processing and more advanced channel coding including turbo encoders/decoders.

Higher-level (QAM/16 QAM) modulation with symbols mapped in phase and amplitude required closer control of phase noise and or additional coding distance and improved linearity to minimize PM to AM errors.

The addition of OFDM and OFDMA in 4G cellular standards increased the unwanted amplitude in composite (multiuser) wider-band channels and the use of frequency subcarriers required careful system design to ensure orthogonality across a wide range of operational and offered traffic conditions. This has introduced additional system complexity in user devices in order to minimize a potentially significant loss in RF throughput efficiency with an associated cost in terms of additional baseband processing.

The 5G standards process has included a number of candidate physical layers positioned as being more efficient than OFDM and more scaleable to higher frequencies. The efficiency gain is predicated on a number of assumptions.

Filter bank multicarrier filter implementation efficiency gains are predicated on the assumption that there will be an increasing need for spectrum flexibility to support increasingly fragmented spectrum.

Nonorthogonal asynchronous physical layer efficiency gains are predicated on the as-yet-unproven assumption that an increasing amount of offered traffic will be machine initiated.

The counterargument is that spectrum fragmentation is better addressed at a regulatory level. In particular, there is no intrinsic reason why bandwidth in the millimeter band should not be contiguous. Coexistence issues may make contiguous allocation more problematic in the centimeter band, but that in itself is not an argument in favor of band fragmentation.

Scaling an OFDM and OFDMA wide-area physical layer to the centimeter and millimeter band is challenging due to the additional dynamic range needed over above existing Wi-Fi systems at 5 GHz (lower end of the centimeter band) and 60 GHz (lower end of the millimeter band).

The solution may be to make the physical layer more tolerant to noise and linearity, which would suggest that lower-order modulation might be a more RF-efficient option. OFDM has been introduced in to 4G to accommodate wider bandwidth channels. These deliver improved multiplexing gain but are not inherently power efficient. The 60-GHz Wi-Fi system design is initially struggling to make OFDM sufficiently power-efficient at 60 GHz and higher-power implementation of OFDM above 60 GHz must be considered as a significant challenge and may not be necessary in a narrow beamwidth wide bandwidth 5G millimeter band radio.

Last but not least, it is entirely possible that operators will wish to deploy 5G into the meter band. This means that 5G systems will need to coexist with legacy 4G systems, which brings us back to the title of this chapter: the technology cost of standards.

Standards are generally a good thing. We cannot live without them. They facilitate scale economy and are essential to interoperability and coexistence management.

Standards have a habit of becoming overcomplex. This can generally be ascribed to an understandable motivation to respond to assumed changes in the offered traffic mix and application mix, which introduces additional options. This happens equally in the 3GPP and IEEE standards process. As complexity increases, test overhead increases. More fundamentally, signaling overheads increase as well. This absorbs bandwidth and power. This test and performance cost overhead is compounded when new systems have to coexist with legacy systems.

Only the bravest of RF network engineers would suggest it would be a good idea to mix an orthogonal synchronous physical layer (4G LTE) with a nonorthogonal 5G asynchonous radio system at the network level. Only the bravest of RF design engineers would confidently assert that a nonsynchronous nonorthogonal radio layer could coexist with a synchronous orthogonal radio layer within the cramped spatial confines of a smart phone.

On balance, the evidence to date would therefore suggest that simpler modulation and multiplexing might be more appropriate for shorter-wavelength spectrum with issues of channel delay spread addressed in the spatial domain, which suggests it is time to look at the technology cost of spectrum.

References

[1] http://www.3gpp.org/keywords-acronyms/1612-ue-category.

[2] http://www.arib.or.jp/english/.

[3] http://www.atis.org/.

[4] http://www.ccsa.org.cn/english/.

[5] http://www.etsi.org/.

[6] http://www.tsdsi.org/.

[7] http://www.tta.or.kr/English/new/external_relations/IMT-Advanced.jsp.

[8] http://www.ttc.or.jp/j/std/committee/wg/3gpp/.

[9] https://www.ngmn.org/home.html.

[10] https://5g-ppp.eu/5g-norma/.

[11] http://www.etsi.org/news-events/news/861-2014-new-industry-specification-group-on-millimetre-wave-transmission-at-etsi.

[12] https://ytd2525.wordpress.com/category/5gpp-5g-public-private-partnership-association/.

[13] http://www.surrey.ac.uk/5gic.

[14] http://www.3gpp.org/about-3gpp.

[15] http://www.comsoc.org/best-readings/topics/green-communications.

[16] http://www.dynamicspectrumalliance.org/.

[17] http://www.5gnow.eu/.

[18] http://www.theatlantic.com/technology/archive/2011/09/the-1947-paper-that-first-described-a-cell-phone-network/245222/.

[19] http://www.ieee802.org/11/email/stds-802-11/msg01472.html.

[20] http://www.sibeam.com/en/Products.aspx.

[21] https://www.qualcomm.com/news/releases/2016/01/05/qualcomm-80211ad-products-lead-way-multi-band-wi-fi-ecosystem.

[22] http://www.zigbee.org/.

[23] http://www.sigfox.com/en/#!/technology.

[24] https://www.lora-alliance.org/S.

3

Technology Cost Spectrum: The Cost of Band Complexity

3.1 44 LTE Bands: Cost Multiplication

In this chapter, we use the 44 LTE bands specified in 3GPP Release 12 as a case study of band complexity. We highlight the performance loss that this imposes on user devices and cost and inefficiency that this imposes on the supply chain and the implications for 5G in terms of spectrum options and bandwidth configuration.

In Chapter 2, we pointed out that RF components obey Ohm's law rather than Moore's law. Higher peak rate user device categories require more filtering, more highly specified switch paths, and more efficient power amplifiers. Noise and intermodulation need to be more aggressively managed. More bands, wider bands, wider channels within bands, and higher-frequency operation make these processes harder.

The present 5G discussions include the idea of 5G being deployed into existing meter-band spectrum including existing mobile broadband allocations below 3 GHz and planned deployments between 3 and 4 GHz. We need to qualify whether this is likely to be a realistic option.

If 5G evolves as a new physical layer implemented into the centimeter and or millimeter bands, then we need to assess the related impact on 4G user device form factor, cost, and performance.

We need to take a view on whether 5G is better deployed as a frequency division duplex (FDD) physical layer, time division duplex (TDD) physical layer or as a full duplex system in which transmit and receive are accommodated on the same radio channel.

Fragmented nonharmonized spectrum can be shown to be adding cost to existing 4G user devices and can be shown to be compromising 4G radio frequency (RF) performance. High device cost has slowed market adoption into low-income developing countries.

3.2 Component Costs and Performance Costs

How much cost is added, how much performance is lost, and how much market delay is introduced depend on how many bands are included in a user device, what those bands are, and the mix of band and technology combinations.

Adding a band might seem trivial: just add another switch path and it is done. The real enemy is the lack of space or, rather, volume. The addition of a switch path means that the pin count increases in an already-complex package. If the device stays the same size, then the pins will be closer together as will the switch paths through the device. Switch technologies including silicon on insulator/silicon on sapphire have improved significantly with an insertion loss on paper of fractions of a decibel and impressive linearity. However, implementing, for example, a 16-throw switch, into a complex RF front is a nontrivial design challenge.

As with switches, acoustic filter vendors have done great work miniaturizing the packaging of surface acoustic wave (SAW) and bulk acoustic resonator (BAR) filters, but these are frequency-dependent devices that have to handle hundreds of milliwatts of power across large temperature ranges so they have to meet close tolerance limits on frequency drift. Some space can be saved by packing filters together in a filter bank, but filters still take up valuable real estate, add cost, and introduce insertion loss. Wider passbands also make acoustic filters more badly behaved. The distance between specified and real-life performance becomes larger.

Power amplifier manufacturers can produce wideband amplifiers that cover multiple bands, but the matching networks are frequency-specific. The devices have to be efficient and linear and this requires signal processing for predistortion and envelope tracking. These adaptive mechanisms are bandwidth-limited, introduce delay, and create noise so an improvement in RF transmit efficiency translates into a loss of receive sensitivity.

Antenna designers can produce electrically short antennas or tunable antennas, but this means constantly changing the capacitance of the devices. One answer is to use RF microelectrical mechanical systems (MEMS) devices to switch across multiple capacitance values but achieving an overall system efficiency gain can be elusive and other dynamic changes such as hand capacitance effects have to be accommodated. RF MEMS require a high voltage that creates problems with internal noise that have to be addressed. The digital control lines needed to support these adaptive mechanisms add noise and occupy additional space.

As bands are added in, the front end of the phone becomes progressively wider and therefore exposed to more noise. A requirement to handle wider passbands and wider channels within those passbands needs more dynamic range from the active devices in the front end. This requires more power. Wider channels become problematic for the digital signal processor (DSP).

Legacy bands have to support legacy technologies including Global System Mobile (GSM), General Packet Radio Services (GPRS), wideband code division multiple access (W-CDMA), and code division multiple access (CDMA). This produces awkward trade-offs. The use of direct conversion receivers for W-CDMA and CDMA is one example. Direct conversion receivers mix the incoming RF with the same frequency but with a phase offset. This reduces the component count by avoiding intermediate frequency (IF) filtering but the offset process is problematic for GSM.

3.3 FDD and TDD Complexity Cost Including Test Cost

Combining FDD and TDD similarly adds complexity and cost. In FDD bands, mobile transmit channels in a handset are separated (duplex spaced) from mobile receive channels by typically several tens of megahertz, roughly 5% of the center frequency. The frequency spacing provides the separation needed to minimize desensitization in the receive path.

The duplex gap protects user device receive channels at the bottom end of the upper duplex from other user device transmit channels at the top end of the lower duplex. The function of the duplex gap is therefore to provide protection between users who are physically close to each other.

Producing a design for an FDD transceiver and or FDD becomes progressively more difficult as more bands are added. The design overhead also has to include conformance testing. The more band technology combinations, the greater the risk of conformance test failure and the greater the risk of time to market delay.

The rule of thumb in the test industry is 100 hours of testing per band/technology combination. The cost per hour in 2010 was £450 to £500 [personal communication with Anite, Anritsu and Rohde and Schwartz]. By 2015, this had reduced to £150 per hour for 2G GSM, £200 to £250 for 3G, and £450 per hour for LTE. This might seem trivial, but there are usually at least 12

Table 3.1
Duplex Spacing and the Duplex Gap

Guard band	Operational bandwidth lower duplex	Duplex gap	Operational bandwidth upper duplex	Guard band
		Duplex spacing		
◄──────────────────────────────────────►				
Allocated spectrum				

band/technology combinations even in a low-end smart phone. However, this hides a more awkward reality.

3.4 Conducted Domain Tests Versus Anechoic Chamber Testing

Conformance tests are carried out in what is called the conducted domain with test equipment directly connected to a test port. Conformance tests therefore do not capture the additional losses that are incurred by poorly matched high return loss antennas.

To create a closer approximation to real life, devices need to be performance tested in an anechoic chamber, an expensive and time-consuming process. On the receive path, the tests measure total isotropic sensitivity (TIS) also described rather misleadingly as total radiated sensitivity (TRS). On the transmit path the tests measure total radiated power (TRP). The gap between conformance tests made directly from an antenna port and TIS and TRP tests carried out in an anechoic chamber is presently at least 5 to 7 dB.

This performance gap is widening over time due to the increased antenna count in small-form factor devices. Antennas are now needed for Global Positioning System (GPS), Wi-Fi, cellular low band (below 1 GHz), cellular mid-band (1–2 GHz), and cellular high band (>2 GHz, including 3.5 GHz) with additional antennas for diversity and multiple input multiple output (MIMO).

Considering that every 3 dB represents a halving of available power, it can be seen that this amount of RF performance loss has a profound impact on the user experience (range and throughput) and operator economics (loss of coverage and capacity).

3.5 Frequency Bands and Scale Economy: The Cost of Band Aggregation

There are other commercial issues. Intuitively, you might assume that scale economy increases with market size and that with a market volume of over 1 billion user devices per year, almost anything should be possible. In practice, spectrum fragmentation imposes additional cost on the supply chain, including the RF supply chain. A user device component manufacturer has to prioritize the biggest markets by volume and value and the most important operators within those markets. Individual operators often have unique requirements.

This is currently an emerging problem for devices required to support carrier aggregation where a sub-1-GHz band is required to be paired with a mid-band, for example, 1,800 or 1,900 MHz, and a high band, for example, 2,100 MHz or 2.6 GHz. Each unique mix of band combinations produces a unique set of intermodulation products where the radio channels mix together to unwanted third-order frequency components. These have to be accommodated and managed within the RF front-end design.

Figure 3.1 shows the components in the RF front end of an FDD user device that are band-specific and or frequency-dependent. The diplexer provides separation between low band and high band, the duplexer provides separation between transmit and receive channels and with some additional filtering defines the passband of the radio channel. The local oscillator, in this direct conversion block diagram, mixes the incoming RF with the same frequency but with a phase offset to resolve the RF signal down to baseband.

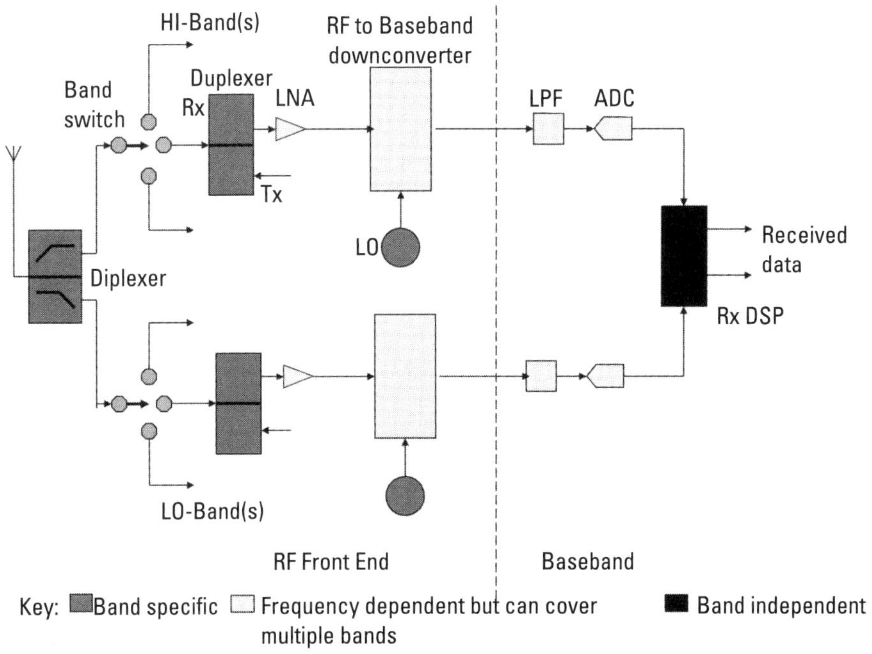

Figure 3.1 Band-specific and frequency-specific components in the front end of a smart phone.

3.6 Band and Technology Combinations as a Cost Multiplier: The Global Picture

Some LTE bands sit within wider passbands. These are known as subset bands and can usually share a filter configured for the wider band. Even with this re-use of filter paths, a modern LTE phone will typically have between 12 and 16 multiple transmit receive chains, which between them will be supporting less than half of the presently allocated LTE bands.

Figure 3.2 shows the complexity introduced into the front end of a contemporary smart phone by multiple band and technology combinations including GSM, W-CDMA, and LTE. In this example, the 14 switch paths are divided across low band, mid-band, and high band coupled to a low-band and high-band antenna with the high-band antenna covering Japanese Band 21 at 1.5-GHz (Band 21) and 3.5-GHz LTE.

The design and product planning and prioritization process associated with these front end designs is becoming more complex over time. Table 3.2 shows the core bands that have been added in Region 1 (including Europe) and Region 2 (including the United States) with each successive generation of cellular radio. Note that generally it has proved more technically and commercially efficient to deploy a new technology into new spectrum. It is technically more efficient because the new technology does not need to coshare spectrum with legacy technologies. It is commercially more efficient because it avoids writing down previous and existing technology investment.

The global requirement has always been more complex with additional regional and country specific band plans, with Japan and China being particular examples. The scale of this complexity as at Release 12 in March 2015 becomes evident when we look at the full list of LTE band options.

Table 3.3 shows the 44 LTE bands in numerical order [1], their uplink and downlink frequencies, bandwidth, duplex spacing, guard band, and region into which they are deployed or planned to be deployed.

Bands 1 to 28, Band 30, and Band 31 are duplex-spaced FDD bands. Bands 29 and 32 are downlink-only bands. Bands 33 to 44 are TDD (time division duplexed) bands. TDD-only devices do not need to have duplex filters, although the need to support FDD as well means that this is only a marginal benefit.

User device LTE 4G RF design teams generally divide the meter band into low band up to 1 GHz, mid-band from 1 to 2 GHz but including Band 1 and the TDD bands around 2 GHz, and high band above 2 GHz to include TDD Band 40 at 2.3–2.4 GHz, Wi-Fi at 2.4 GHz, Band 7 FDD, and Band 41 TDD at 2.6 GHz.

Sorting the LTE band list into low band <1 GHz, mid-band 1–2 GHz, and high band produces values are shown in Table 3.4.

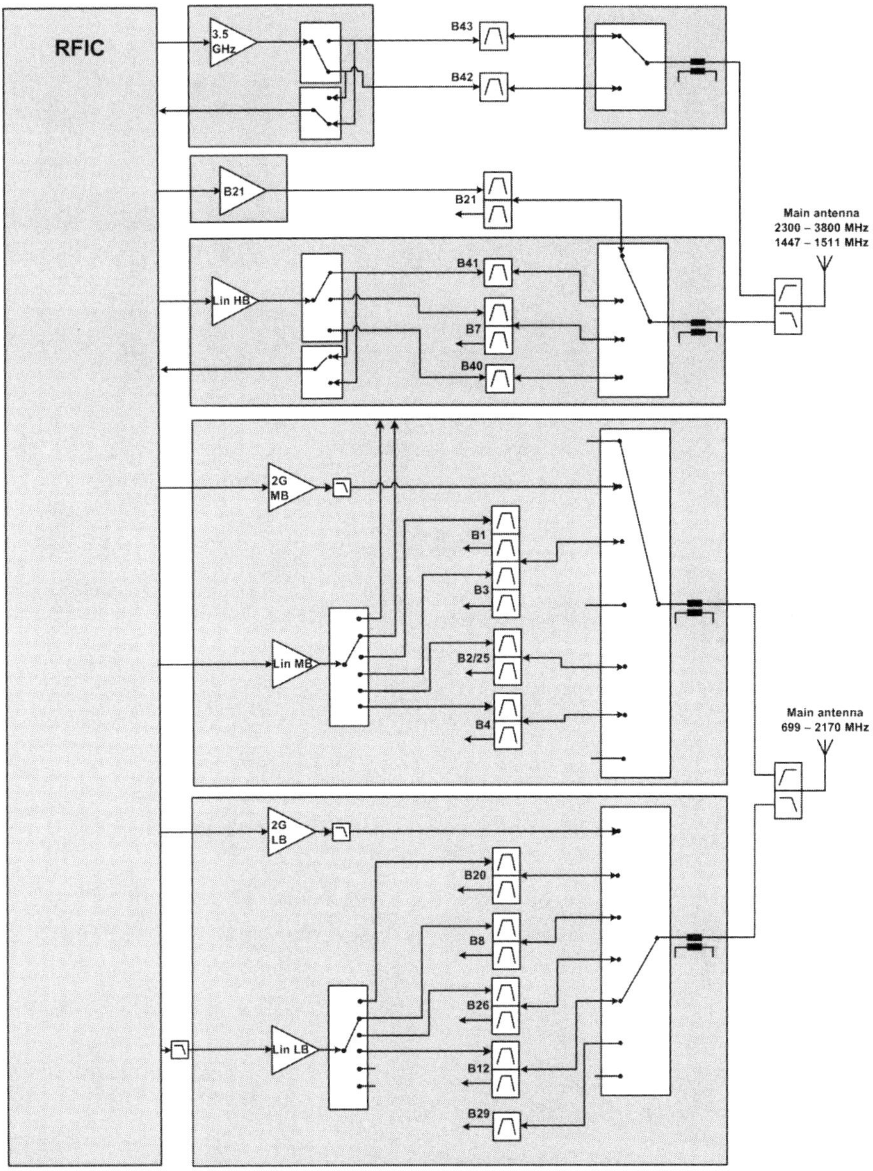

Figure 3.2 Typical front-end configuration showing switch paths needed to support multiple bands and multiple technologies. (Courtesy of TDK Nordic OY.)

Band 5 is the original U.S. AMPS band from 30 years ago and supports a mix of GSM and CDMA. It now forms a subset of Band 26, which increases the Band 5 25+25 MHz passband to 35+35 MHz.

Band 6 is a subset UMTS-only band within the passband of Band 19 used in Japan.

Table 3.2
Additional Bands over the Past 30 Years

Generation	Region 1 (Including Europe)	Region 2 (Including the United States)
First	900 MHz (880—915/925—960)	850 MHz (824—849/869—894)
Second	1,800 MHz (1,710—1,785/1,805—1,880)	1,900 MHz (1,850—1,910/1,930—1,990)
Third	2 GHz (1,920—1,980/2,110—2,170)	Advanced Wireless Service (AWS) (1,710—1,755/2,110—2,155)
Fourth	800 MHz (832—862/791—821), 2.6 GHz (2,500—2,570/2,621—2,690), 3.5 GHz (3.4—3.6 GHz)	700 MHz (698—716/728—746), (777—788/746—757), 2.6 GHz (2,495—2.690) TDD (unpaired), 3.5 GHz (3.4—3.6 GHz and 3.6—3.7 GHz FDD and TDD)

Band 8 is an evolution of the original ETACS cellular allocation in the United Kingdom with an original pass band of 25+25 MHz subsequently extended to 35+35 MHz.

Band 12 is one of the bands auctioned in the U.S. 700-MHz TV band in 2007. It is awkwardly close to terrestrial TV at 699 MHz. Band 17 sits within the Band 12 passband but effectively has a guard band providing protection to and from terrestrial TV. Band 17 is used by AT&T and supports a commercially successful and technically robust 10 MHz +10 MHz LTE FDD network.

Band 13 is the band used predominantly by Verizon to compete with AT&T 700-MHz LTE. It is implemented as a reverse duplex with mobile transmit in the upper duplex. Traditionally, mobile transmit is in the lower duplex. This is because the mobile uplink is more power limited than the downlink from the base station and propagation conditions are theoretically more favorable in the lower duplex. Reverse duplex is used in Band 13 to facilitate coexistence with Band 14 also implemented as reverse duplex. Band 14 is intended to be used for an LTE public safety network. The bands are also described by their Block designation from the original auction process; Block A is Band 12, Block B is Band 17, and Block C is Band 13.

Band 18 and 19 are specific to Japan. Band 20 is the digital dividend band released by moving TV channels to the lower end of the UHF band. It is implemented as a reverse duplex with mobile transmit moved to the upper duplex to provide additional protection to TV below the lower duplex. This might seem odd to put high mast-mounted, higher-power base stations spectrally next to TV receivers and lower power mobiles further away, but the good thing about base stations is that you know where they are and they do not usually move, which makes coexistence planning easier. Mobiles can be anywhere and theoretically at least could be directly pointed at a TV receive antenna.

Band 26 we have already covered as the extension of Band 5. There is potentially some performance loss as a result of the wider passband, includ-

Table 3.3
LTE 3GPP 36.101 Release 12, March 2015

Band	Name MHz	Uplink		Downlink		Bandwidth MHz	Duplex spacing	Guard band	Area
FDD									
1	2,100	1,920	1,980	2,110	2,170	60+60	190	130	All
2	1,900 PCS	1,850	1,910	1,930	1,990	60+60	80	20	NAR
3	1800+	1,710	1,785	1,805	1,880	75+75	95	20	All
4	AWS 1	1,710	1,755	2,110	2,155	45+45	400	355	NAR
5	850	824	849	869	894	25+25	45	20	NAR
6	UMTS	830	840	875	885	10+10	45	35	APAC
7	2,600	2,500	2,570	2,620	2,690	70+70	120	50	EMEA
8	900	880	915	925	960	35+35	45	10	All
9	1,800	1,749.9	1,784.9	1,844.9	1,879.9	35+35	95	55	APAC
10	AWS1+	1,710	1,770	2,110	2,170	60+60	400	340	NAR
11	1,500 lower	1,427.9	1,447.9	1,475.9	1,495.9	20+20	48	28	Japan
12	700 a	699	716	729	746	17+17	30	13	NAR
13 (RD)	700 c	777	787	746	756	10+10	−31	21	NAR
14 (RD)	700 ps	788	798	758	768	10+10	−30	20	NAR
17	700 b	704	716	734	746	12+12	30	18	NAR
18	800 lower	815	830	860	875	15+15	45	30	Japan
19	800 upper	830	845	875	890	15+15	45	30	Japan
20 (RD)	800 dd	832	862	791	821	30+30	−41	20	EMEA
21	1,500 upper	1,447.9	1,462.9	1,495.9	1,510.9	15+15	48	33	Japan
22	3,500	3,410	3,490	3,510	3,590	80+80	100	20	EMEA
23	2,000 S-band	2,000	2,020	2,180	2,200	20+20	180	160	NAR
24 (RD)	1,600 L-band	1,626.5	1,660.5	1,525	1,559	34+34	−101.5	67.5	NAR
25	1,900+	1,850	1,915	1,930	1,995	65+65	80	15	NAR
26	850+	814	849	859	894	35+35	45	10	NAR
27	800 SMR	807	824	852	869	17+17	45	28	NAR
28	700 APT	703	748	758	803	45+45	55	10	APAC
29	700 d			717	728	11	Downlink only		NAR
30	2,300 WCS	2,305	2,315	2,350	2,360	10+10	45	35	NAR
31	450	452.5	457.5	462.5	467.5	5+5	10	5	CALA
32	1,500 L-band			1,452	1,496	44	Downlink only		EMEA

ing softer filter rolloff. The wider passband makes LTE implementation less problematic in terms of refarming the legacy spectrum.

Table 3.3 (continued)

Band	Name MHz	Uplink		Downlink		Bandwidth MHz	Duplex spacing	Guard band	Area
TDD									
33	TD 1,900	1,900	1,920	1,900	1,920	20			EMEA
34	TD 2,000	2,010	2,025	2,010	2,025	15			EMEA
35	TD PCS Lower	1,850	1,910	1,850	1,910	60			NAR
36	TD PCS Upper	1930	1990	1930	1990	60			NAR
37	TD PCS Center Gap	1910	1930	1910	1930	20			NAR
38	TD 2600	2,570	2,620	2,570	2,620	50			EMEA
39	TD 1,900+	1,880	1,920	1,880	1,920	40			China
40	TD 2,300	2,300	2,400	2,300	2,400	100			China
41	TD 2,500	2,496	2,690	2,496	2,690	194			All
42	TD 3,500	3,400	3,600	3,400	3,600	200			
43	TD 3,700	3,600	3,800	3,600	3,800	200			
44	TD 700	703	803	703	783	100			APAC

APAC = Asia and Pacific, EMEA= Europe. Middle East and Africa, NAR= North America Region, CALA =Central Latin America, APT= Asia Pacific Telecommunity, DD = Digital Dividend, PS = public safety, and RD = reverse duplex (mobile transmit in upper duplex).

Table 3.4

FDD <1 GHz

Band	Name (MHz)	Uplink		Downlink		Bandwidth (MHz)	Duplex spacing	Area
5	850	824	849	869	894	25+25	45	NAR
6	UMTS only	830	840	875	885	10+10	45	APAC
8	900	880	915	925	960	35+35	45	All
12	700 a	699	716	729	746	17+17	30	NAR
13 RD	700 c	777	787	746	756	10+10	–31	NAR
14 RD	700 ps	788	798	758	768	10+10	–30	NAR
17	700 b	704	716	734	746	12+12	30	NAR
18	800 lower	815	830	860	875	15+15	45	Japan
19	800 upper	830	845	875	890	15+15	45	Japan
20 RD	800 DD	832	862	791	821	30+30	–41	EMEA
26	850+	814	849	859	894	35+35	45	NAR
27	800 SMR	807	824	852	869	17+17	45	NAR
28	700 APT	703	748	758	803	45+45	55	APAC
29	700 d	Downlink only		717	728	11		NAR
31	450	452.5	457.5	462.5	467.5	5+5	10	CALA

Band 27 overlaps Band 26 and is a candidate band for LTE for specialized mobile radio LTE.

Band 28 was a surprise outcome for most delegates at WRC 2012 and was the result of an Asia Pacific Telecommunity [2] submission of a 45 by 45-MHz band plan with potentially a global footprint through Asia, Latin America, and Africa representing an addressable market of over 2 billion subscribers.

As acoustic filters are bandwidth limited to 30 MHz at 700 MHz (4% of the center frequency), the passband is covered by two 30-MHz duplex filters with a 15-MHz overlap. The lower filter covers the subband described as APT (a); the upper filter covers the upper subband described as APT (b). The duplex gap of the combined (a) and (b) band plan is a relatively ambitious 10 MHz, particularly if wider channel bandwidths (>10 MHz) are deployed.

APT (b) overlaps LTE Band 20 in Europe. European operators have therefore (more or less) decided to implement only the APT (a) passband. This means that there will be a 25-MHz duplex gap. However, CEPT decided that the APT out-of-band limits provided insufficient protection to the DTT multiplex and should be increased.

Band 29 is the legacy of Qualcomm's $2 billion investment in Media Flo (Forward Link Only) including the spectrum formerly occupied by U.S. TV channels 54 and 55 and is intended for use by AT&T as a supplementary downlink for eMBMS (enhanced Multimedia Broadcast Multicast Service). It could potentially be coupled to downlink only Band 32 in Europe at L-band designated as a supplemental downlink at 1,452–1,496 MHz particularly as these L-band allocations have now been globally harmonized (at WRC2015). Last but not least, there is Band 31 allocated for an LTE 5 MHz + 5 MHz network in Brazil at 450 MHz.

3.7 Sub-1-GHz Low-Band Summary

To summarize the present position with spectrum in the sub-1-GHz part of the meter band, there have been technically and commercially successful LTE deployments at 700, 800, and 900 MHz, although regional variations in band allocation and technical requirement have frustrated potential economies of scale. This has not been a problem for operators servicing high-value markets such as AT&T and Verizon in the United States. It has been more problematic for European operators addressing a smaller market by value with specific regional technical requirements. Band 20 is an example.

Transition to LTE in the legacy bands (Band 5 850 MHz in the United States, Band 8 900 MHz in Europe and Asia) has been slow due to underamortized high-speed packet access (HSPA) 3G investment and some doubts about the efficiency and effectiveness of voice over LTE (VoLTE) and or simultaneous voice and data (SV LTE).

Any deployment of 5G into legacy 4G sub-1-GHz bands would rely on 4G networks being fully amortized to avoid painful write-downs. It is also unclear what additional functionality a 5G network could deliver over and above evolved <1 GHz networks if deployed in the same band with equal channel spacing. The UHF band in Region 1 (Europe and Africa) will be reviewed at WRC 2023 at which point spectrum might become available for 4G and potentially 5G below 694 MHz.

3.8 vMid-Band L-Band Allocations Between 1 and 2 GHz

Table 3.5 shows the LTE band allocations between 1 and 2 GHz.

3.8.1 Europe 1,800-MHz Band 3

Band 3 has been the rather unexpected big success for LTE. In retrospect, it should have been obvious that a 75 by 75-MHz passband with a 20-MHz duplex gap was relatively ideal for 10-MHz LTE deployment. The band is also less heavily loaded than Band 8 at 900 MHz, which has made refarming less technically and commercially problematic. The W-CDMA networks in Band 1 are not fully amortized, so LTE transition is fiscally trickier due to the need to write down existing investments.

Historically (late 1980s), 1,800 MHz was chosen for the first dual-band networks because of the harmonic relationship with 900 MHz. This relationship remains useful when developing optimized dual-band antennas, but the

Table 3.5
FDD 1–2 GHz

Band	Name MHz	Uplink		Downlink		Bandwidth MHz	Duplex spacing	Area
1	2,100	1,920	1,980	2,110	2,140	60+60	190	All
2	1,900 PCS	1,850	1,910	1,930	1,990	60+60	80	NAR
3	1,800+	1,710	1,785	1,844.9	1,879.9	75+75	95	All
9	1,800	1,749.9	1,784.9	1,844.9	1,879.9	35+35	95	APAC
10	AWS 1+	1,710	1,770	2,110	2,170	60+60	400	NAR
11	1,500 lower	1,427.9	1,447.9	1,475.9	1,495.9	20+20	48	Japan
21	1,500 upper	1,447.9	1,462.9	1,495.9	1,510.9	15+15	48	Japan
24 RD	1,600 L-band	1,626.5	1,660.5	1,525	1,559	34+34	–101.5	NAR
25	1,900+	1,850	1,915	1,930	1,995	65+65	80	NAR
32	1,500 L-band	Downlink only		1,452	1,496	44		EMEA

main attraction is available bandwidth and a comfortable passband and duplex gap.

The Band 3 mobile TX part of the duplex filter can be reused as an AWS 1 handset TX lower duplex filter for U.S. AWS 1 devices. The downlink upper duplex part of the duplex filter for Band 1 can be reused as the upper duplex filter for AWS.

3.8.2 U.S. Band 4 AWS (Advanced Wireless Service)

The AWS band has always been awkward in terms of scale (U.S.-centric) and band plan (large duplex gap means uplink and downlink are nonreciprocal). These factors have made it harder to build an economic and competitive Band 4 mobile broadband offer in the United States, although the opportunity to reuse filters from Band 3 and Band 1 in smart phone RF front ends has been helpful. The 45+45 MHz passband does not offer particular challenges but is less flexible than Band 2 (PCS 1,900 65+65 MHz) for LTE.

An auction in the United States in 2015 gathered bids of $44 billion for additional bandwidth or blocks of a new super set band known as Band 10. This increases the 45 by 45 MHz passband to 60+60 MHz. AWS 3 and EWS are potential additional supersets, although they do not presently have 3GPP numbers but would extend the band to 70+70 MHz making it more equivalent to the European and Asian LTE band (Table 3.6). The 70-MHz lower duplex sits within the 75-MHz LTE 1,800 lower duplex. The upper duplex has the same lower-band edge as Band 1 but extends 10 MHz higher at the top end (a 70-MHz passband overlaid on a 60-MHz passband).

It is possible to extend the upper passband of Band 1 by 10 MHz, which would mean that it could be used as a common filter with AWS 3. The 10-MHz extension could be used as an LTE channel but this would require agreement

Table 3.6
AWS 1, AWS 3, and EWS

Band 4 Mob TX	**Band 10 Mob TX**	**Band 3 Mob TX**	**Band 4 Mob RX**	**Band 10 Mob RX**	**Band 1 Mob RX**
1,710–1,755	1,710–1,770	1,710–1,785	2,110–2,155	2,110–2,170	2,110–2,170
Passband					
45 MHz	60 MHz	75 MHz	45 MHz	60 MHz	60 MHz
	AWS 3		AWS 3		
	1,755–1,780		2,155–2,180		
	EWS		EWS		
	1,710–1,780		2,110–2,180		
Passband					
	70 MHz		70 MHz		

with European satellite operators who have priority access to an uplink at 1,980 to 2,010 MHz and a downlink at 2,170 to 2,200 MHz. The band can be used for mobile satellite or terrestrial but a terrestrial network can only be deployed on a noninterference basis or as a complementary ground component for a Mobile Services Satellite (MSS) system. Such an arrangement would need to be made commercially appealing to the satellite community including Inmarsat, which has plans to use the band for in flight Wi-Fi [3].

3.8.3 U.S. Band 2 PCS 1,900 and the Band 25 Superset

U.S. Band 2 is configured as a 65 by 65-MHz passband with a 20-MHz duplex gap. Band 25 extends the passband by 5 MHz (see Table 3.7). An additional 5 MHz is potentially available known as Block H AWS2, but this would mean the duplex gap is reduced to 10 MHz. The original duplex gap of 20 MHz is also designated as a TDD band, Band 37. As with Band 1 in Europe, there is also an issue with spectrally adjacent mobile satellite spectrum.

Within L-band there is also a Band 24 from 1,525 to 1,559 MHz and 1,626 to 1,660 MHz. This is immediately adjacent to the Iridium MSS satellite uplink at 1,616 to 1,626.5 MHz [4] and Inmarsat MSS allocation between 1,525 and 1,660 MHz. Attempts to implement mobile terrestrial services into this band, for example, by Light Squared in the United States, have been frustrated by the spectral proximity to GPS at 1,575 MHz. There are two existing LTE bands in L-band, Band 11 and 21, used in Japan (see Table 3.8).

New supported candidate bands at WRC 15 in L-band for LTE are at 1,427–1,452 MHz and 1,452–1,492 MHz with further consideration being given to 1,350–1,375 MHz, 1,375–1,400 MHz, and 1,492–1,518 MHz. These new band allocations could be LTE or 5G. As with the sub-1-GHz allocations, it is not entirely clear what differentiation 5G would offer if implemented into similar passbands and channel bandwidths.

Band 7 is implemented in Europe for FDD LTE (see Table 3.9). There have been some issues with aviation radar coexistence that have needed to be

Table 3.7
Band 2, Band 25, and AWS 2 Block H

Band	Mob TX	Duplex gap	Mob RX
Band 2	1,850–1,910	20	1,930–1,990
	60 MHz	Band 37 TDD, 1,910–1,930	60 MHz
Band 25 Sprint	1,850–1,915	15	1,930–1,995
	65 MHz		65 MHz
AWS 2, Block H	1,850–1,920	10	1,995–2,000
	70		70

Table 3.8
Bands 11, 21, and 24

Band	Mob TX	Mob RX
Band 11, Japan Lower PDC	1,427–1,447	1,475–1,495
Band 21, Japan Upper PDC	1,447–1,462	1,495–1,510
Band 24	1,626.5–1,660.5	1,525–1,559

Table 3.9
FDD 2–4 GHz

Band	Name (MHz)	Uplink		Downlink		Bandwidth (MHz)	Duplex Spacing	Area
7	2,600	2,500	2,570	2,620	2,690	70+70	120	EMEA
22	3,500	3,410	3,490	3,510	3,590	80+80	100	EMEA
23	2,000, S-band	2,000	2,020	2,180	2,200	20+20	180	NAR
30	2,300 WCS	2,305	2,315	2,350	2,360	10+10	45	NAR

resolved and coexistence with TDD networks implemented in the duplex gap needs to be carefully coordinated .

Band 23 and Band 30 are specific to the United States. Band 23 is proposed for wider deployment in ROW markets but the proximity to Wi-Fi has to be managed and the spectrum is more commonly deployed as Band 40 TDD.

3.9 TDD Bands

TDD bands theoretically span allocations from 700 MHz (Band 44 100 MHz contiguous TDD) through to Band 42 and Band 43 between 3.4 GHz and 3.8 GHz. This includes bands specified for implementation into the duplex gap of existing FDD bands, Band 37 being an example (TDD in the duplex gap of Band 2).

In practice, once FDD networks are deployed it is difficult to implement spectrally and geographically located TDD networks unless they are synchronized and preferably cosited. TDD to FDD interference could be managed at subcarrier level, but to date there has been no wide-scale implementation of this.

The most widely deployed TDD networks are in China and are listed in Table 3.10. As of the time of this writing, there have been no large-scale deployments of TDD Band 42 or 43 at 3.4 to 3.8 GHz, although there are WiMAX fixed access networks still serving subscribers in some countries [5].

Table 3.10
TDD < 1 GHz, 1–2 GHz, 2–3 GHz

Band	Name MHz	Uplink		Downlink		Bandwidth MHz	Duplex spacing	Area
TDD <1 GHz								
44	TD 700	703	803	703	803	100 MHz		
TDD 1-2 GHz								
34	TD 2000	2,010	2,025	2,010	2,025	15		EMEA
33	TD 1900	1,900	1,920	1,900	1,920	20		EMEA
37	TD PCS Center Gap	1,910	1,930	1,920	1,930	20		NAR
39	TD 1,900+	1,880	1,920	1,880	1,920	40		China
35	TD PCS Lower	1,850	1,910	1,850	1,910	60		NAR
36	TD PCS upper	1,930	1,990	1,930	1,990	60		NAR
TDD 2–3 GHz								
38	TD 2,600	2,570	2,620	2,570	2,620	50		EMEA
40	TD 2,300	2,300	2,400	2,300	2,400	100		China
41	TD 2,500	2,496	2,690	2,496	2,690	194		All
TDD 3–4 GHz								
42	TD 3500	3,400	3,600	3,400	3,600	200		
43	TD 3,700	3,600	3,800	3,600	3,800	200		

3.10 Present Band Support in LTE Smart Phones and Related Supply Chain Economics

In practice it is not technically feasible or commercially viable to support all 44 LTE bands in an LTE smart phone. The component count would be unrealistically high, and the smart phone would be large, fat, heavy, and expensive and have poor RF performance and unacceptable battery life.

User device vendors quite understandably give research and development priority to the most important bands servicing their most important markets and most important customers within those markets.

As a contemporary example, Table 3.11 lists the LTE bands and technologies supported by four model variants of the Apple iPhone 6/iPhone6 Plus and the addressed markets by band.

The first two variants are for Europe, the United States, and Asia and support either GSM or CDMA (for the U.S. market). The second two variants are for China and include support for the TDD bands 38, 39, 40, and 41 for China Mobile, China Telecom, and China Unicom. Table 3.12 reorders the list into low band, mid-band, and high band.

Table 3.11
Apple iPhone 6-Supported LTE Bands in Numerical Order

Model numbers A1549/A1522 (GSM) A1549/A1522 (CDMA)		
LTE Band Support	**Name**	**Area**
1	2,100 MHz	All
2	1,900 MHz	NAR
3	1,800 MHz	All
4	AWS	NAR
5	850 MHz	NAR
7	2,600 MHz	EMEA
8	900 MHz	All
13	700 c MHz	NAR
17	700 b MHz	NAR
18	800 MHz	Japan
19	800 MHz	Japan
20	800 DD	EMEA
25	1,900+ MHz	NAR
26	850+ MHz	NAR
28	700 APT MHz	APAC
29	700 MHz Downlink only	NAR
A1586/A1524 (GSM) A1586/A1524 (CDMA)		
All of the above+		
38	TD 2,600	EMEA
39	TD 1,900	China
40	TD 2,300	China
41	TD 2,500	All

There are some commonalties that help reduce component count. In low band, Band 5 sits within the passband of Band 26 and the AWS band in mid-band can share the filter path of the Band 3 lower duplex and Band 1 upper duplex, but most other bands are supported on unique switch paths. There is some sharing of power amplifiers, but this is still a complex process to optimize. The phone has to work efficiently at all bands and across all blocks within those bands and all channels within those blocks.

The fact that iPhone supports these band technology combinations does not preclude other vendors supporting other band technology combinations, but to do that, they need the support of the RF component supply chain.

This supply chain dynamic explains why operators can find themselves owning orphan band allocations. An orphan band is a band that is not scale economic. If the band is technically challenging, the scale economy threshold will be significantly higher. Scale economic in this context means not being

Table 3.12
Apple iPhone 6-Supported LTE Bands: Low, Middle, and High

Model numbers A1549/A1522 (GSM) A1549/A1522 (CDMA)						
LTE Band Support	**Name**					**Area**
Low band		**MHz**	**MHz**	**MHz**	**MHz**	
5	850 MHz	850	824	849	869	894
8	900 MHz	880	915	925	960	All
13	700 c MHz	777	787	746	756	NAR
17	700 b MHz	704	716	734	746	NAR
18	800 MHz	815	830	860	875	Japan
26	850+ MHz	814	849	859	894	NAR
19	800 MHz	830	845	875	890	Japan
20	800 DD	832	862	791	821	EMEA
28	700 APT MHz	703	748	758	803	APAC
29	700 MHz Downlink only			717	728	NAR
Mid-band						
3	1,800 MHz	1,710	1,785	1,844.9	1,879.9	All
4	AWS 1,700	1,710	1,755	2,110	2,155	NAR
25	1,900+ MHz	1,850	1,915	1,930	1,995	NAR
High band						
1	2,100 MHz	1,920	1,980	2,110	2,170	EMEA
7	2,600 MHz	2,500	2,570	2,620	2,690	EMEA
A1586/A1524 (GSM) A1586/A1524 (CDMA)						
All of the above+						
38	TD 2,600	2,570	2,620	2,570	2,620	EMEA
39	TD 1,900	1,880	1,920	1,880	1,920	China
40	TD 2,300	2,300	2,400	2,300	2,400	China
41	TD 2,500	2,496	2,690	2,496	2,690	All

able to range list premium smart phones due to lack of band support in the user device.

This is a spectrum technology cost that is not necessarily initially apparent but has a major impact on network viability. Owners of the highest end most popular smart phones tend to be the highest revenue subscribers delivering the highest per user margin.

The decision process is driven by which countries the phone is going to be sold into and then which operator within the country is going to be supported. Adding Band 12 into the iPhone, for example, would seem to be sensible but it is an awkward band to accommodate with tricky filter requirements, which add cost and increase insertion loss. This project's cost and performance loss on to Band 13 operators, for instance, Verizon, and Band 17, for instance, AT&T. Component vendors have to consider these local market dynamics. This is technology-determined market tension.

3.11 The Supply Chain Economics of Carrier Aggregation

The tension becomes more evident as and when user devices start supporting carrier aggregation.

Carrier aggregation requirements are country-specific and operator-specific. The technical rationale is seductive. An operator owns blocks of bandwidth within low band, mid-band, and high band. In an ideal world, the low-band channels can be combined with mid-band and high-band channels to provide multiplexing gain with the low-band channels used for signaling and large cell rural and deep rural coverage, the mid-band blocks used for semidense rural and urban and high band used for dense urban integrated with Wi-Fi. This is called interband aggregation.

If an operator owns or has access to multiple channels within a block of spectrum, these can be aggregated together to increase channel bandwidth. The channels can either be contiguous (next to each other) or noncontiguous (not next to each other). This is called intraband aggregation.

Base station RF hardware has to be capable of supporting all channels within a passband, so intraband aggregation is reasonably easy to deliver. Similarly interband aggregation involves separating traffic streams and delivering them across multiple RF channels. This happens already on a multiuser basis so the requirement is to scale this to support multiple users to single channels.

The challenge for user device designers is harder. User devices have not been traditionally designed to receive or transmit multiple simultaneous RF channels. Interband aggregation inevitably produces intermodulation products that have to be handled within a small confined area. The increase in dynamic range increases power drain. Intraband aggregation increases the load on the DSP and raises the noise floor of the receiver front end.

Operators have unique requirements defined by which blocks they own within a band and which bands they own at low band, mid-band, and high band. Only the largest operators have sufficient market scale to justify the design, development, and test cost of an operator-specific multiband aggregation combination. Table 3.13 lists the Release 12 aggregation options, the operator who has requested the aggregation, and the market into which the device would be sold.

The options include two downlink/one uplink interband combinations, interband three downlink aggregation including intraband noncontiguous ag-

Table 3.13
Carrier Aggregation Bands in 3GPP Release 12

Band	Name	Uplink		Downlink		Operator	Region
Interband two downlinks, one uplink							
1+3						China Unicom, China Telecom	China
3	1,800+	1,710	1,785	1,805	1,880		
1+7						LGU+	Korea
1	2,100	1,920	1,980	2,110	2,170		
7	2,600	2,500	2,570	2,620	2,690		
1+8						Softbank	Japan
1	2,100	1,920	1,980	2,110	2,170		
8	900	880	915	925	960		
1+11						Softbank	Japan
1	2,100	1,920	1,980	2,110	2,170		
11	1,500 lower	1,427.9	1,447.9	1,475.9	1,495.9		
1+18						KDDI	Japan
1	2,100	1,920	1,980	2,110	2,170		
18	800 Lower	815	830	860	875		
1+26						KDDI	Japan
1	2,100	1,920	1,980	2,110	2,170		
26	850+	814	849	859	894		
2+4						T Mobile	US
2	1,900 PCS	1,850	1,910	1,930	1,990		
4	AWS 1	1,710	1,755	2,110	2,155		
2+5						AT&T	US
2	1,900 PCS	1,850	1,910	1,930	1,990		
5	850	824	849	869	894		
2+12						US Cellular	US
2	1,900 PCS	1,850	1,910	1,930	1,990		
12	700 a	699	716	729	746		
2+13						Verizon	US
2	1,900 PCS	1,850	1,910	1,930	1,990		
13 (RD)	700 c	777	787	746	756		
3+19						NTT DoCoMo	Japan
3	1,800+	1,710	1,785	1,805	1,880		
19	800 upper	830	845	875	890		
3+20						Telekom Austria	Austria
3	1,800+	1,710	1,785	1,805	1,880		
20 (RD)	800 dd	832	862	791	821		
3+26						Korea Telecom	Korea
3	1,800+	1,710	1,785	1,805	1,880		
26	850+	814	849	859	894		
3+27						Korea Telecom	Korea

Table 3.13 (continued)

Band	Name	Uplink		Downlink		Operator	Region
Interband two downlinks, one uplink							
3	1,800+	1,710	1,785	1,805	1,880		
27	800 SMR	807	824	852	869		
3+28						eAccess	Japan
3	1,800+	1,710	1,785	1,805	1,880		
28	700 APT	703	748	758	803		
4+12						T Mobile	US
4	AWS 1	1,710	1,755	2,110	2,155		
12	700 a	699	716	729	746		
4+27						NII Holdings, Mexico, Argentina, Brazil	CALA
4	AWS 1	1,710	1,755	2,110	2,155		
27	800 SMR	807	824	852	869		
5+7						LG U plus	Korea
5	850	824	849	869	894		
7	2,600	2,500	2,570	2,620	2,690		
5+25						US Cellular	US
5	850	824	849	869	894		
25	1,900+	1,850	1,915	1,930	1,995		
7+20						Telekom Austria	
7	2,600	2,500	2,570	2,620	2,690		
20 (RD)	800 dd	832	862	791	821		
7+28						Telefonica	CALA
7	2,600	2,500	2,570	2,620	2,690		
28	700 APT	703	748	758	803		
8+11						Softbank	Japan
8	900	880	915	925	960		
11	1,500 lower	1,427.9	1447.9	1,475.9	1,495.9		
8+20						Vodafone	Europe
8	900	880	915	925	960		
20 (RD)	800 dd	832	862	791	821		
12+25						US Cellular	US
12	700 a	699	716	729	746		
25	1,900+	1,850	1,915	1,930	1,995		
19+21						NTT DoCoMo	Japan
19	800, upper	830	845	875	890		
21	1,500, upper	1,447.9	1,462.9	1,495.9	1,510.9		
20+32						Orange	Europe
20 (RD)	800 dd	832	862	791	821		

Table 3.13 (continued)

Band	Name	Uplink		Downlink		Operator	Region
Interband two downlinks, one uplink							
32	1,500, L-band			1,452	1,496		
23+29						Dish Holdings	US
23	2,000, S-band	2,000	2,020	2,180	2,200		
29	700 d			717	728		
39+41						China Mobile	China
39	TD 1900+	1,880	1,920	1,880	1,920		
41	TD 2,500	2,496	2,690	2,496	2,690		
41+42						China Unicom, China Telecom	China
41	TD 2,500	2,496	2,690	2,496	2,690		
42	TD 3,500	3,400	3,600	3,400	3,600		
Interband 3DL CA including intraband noncontiguous CA							
1+3+5						SK Telecom	Korea
1	2,100	1,920	1,980	2,110	2,170		
3	1,800+	1,710	1,785	1,805	1,880		
5	850	824	849	869	894		
1+3+8						Korea Telecom	Korea
1	2,100	1,920	1,980	2,110	2,170		
3	1,800+	1,710	1,785	1,805	1,880		
8	900	880	915	925	960		
1+3+19						NTT DoCoMo	Japan
1	2,100	1,920	1,980	2,110	2,170		
3	1,800+	1,710	1,785	1,805	1,880		
19	800, upper	830	845	875	890		
1+3+20						Vodafone	Europe
1	2,100	1,920	1,980	2,110	2,170		
3	1,800+	1,710	1,785	1,805	1,880		
20 (RD)	800 dd	832	862	791	821		
1+5+7						LG U Plus	Korea
1	2,100	1,920	1,980	2,110	2,170		
5	850	824	849	869	894		
7	2,600	2,500	2,570	2,620	2,690		
1+7+20						Vodafone	Europe
1	2,100	1,920	1,980	2,110	2,170		
7	2,600	2,500	2,570	2,620	2,690		
20 (RD)	800 dd	832	862	791	821		
1+19+21						NTT DoCoMo	Japan
1	2,100	1,920	1,980	2,110	2,170		
19	800 upper	830	845	875	890		

Table 3.13 (continued)

Band	Name	Uplink		Downlink		Operator	Region
Interband 3DL CA including intraband noncontiguous CA							
21	1500 upper	1,447.9	1,462.9	1,495.9	1,510.9		
1+42+42						NTT DoCoMo	Japan
1	2,100	1,920	1,980	2,110	2,170		
42	TD 3,500	3,400	3,600	3400	3600		
42*	TD 3,500	3,400	3,600	3,400	3,600		
2+2+13						Verizon	US
2	1,900 PCS	1,850	1,910	1,930	1,990		
2*	1,900 PCS	1,850	1,910	1,930	1,990		
13 (RD)	700 c	777	787	746	756		
2+4+4						T Mobile	US
2	1,900 PCS	1,850	1,910	1,930	1,990		
4	AWS 1	1,710	1,755	2,110	2,155		
4*	AWS 1	1,710	1,755	2,110	2,155		
2+4+5						US Cellular	US
2	1,900 PCS	1,850	1,910	1,930	1,990		
4	AWS 1	1,710	1,755	2,110	2,155		
5	850	824	849	869	894		
2+4+13						Verizon Wireless	US
2	1,900 PCS	1,850	1,910	1,930	1,990		
4	AWS 1	1,710	1,755	2,110	2,155		
13 (RD)	700 c	777	787	746	756		
2+5+12						US Cellular	US
2	1,900 PCS	1,850	1,910	1,930	1,990		
5	850	824	849	869	894		
12	700 a	699	716	729	746		
2+5+30						AT&T	US
2	1,900 PCS	1,850	1,910	1,930	1,990		
5	850	824	849	869	894		
30	2,300 WCS	2,305	2,315	2,350	2,360		
2+12+12						AT&T	US
2	1,900 PCS	1,850	1,910	1,930	1,990		
12	700 a	699	716	729	746		
12*	700 a	699	716	729	746		
2+12+30						AT&T	US
2	1,900 PCS	1,850	1,910	1,930	1,990		
12	700 a	699	716	729	746		
30	2,300 WCS	2,305	2,315	2,350	2,360		
2+29+30						AT&T	US
2	1,900 PCS	1,850	1,910	1,930	1,990		
29	700 d			717	728		

Table 3.13 (continued)

Band	Name	Uplink		Downlink		Operator	Region
Interband 3DL CA including intraband noncontiguous CA							
30	2,300 WCS	2,305	2,315	2,350	2,360		
3+3+7						Telia Sonera	Europe
3	1,800+	1,710	1,785	1,805	1,880		
3*	1,800+	1,710	1,785	1,805	1,880		
7	2,600	2,500	2,570	2,620	2,690		
3+7+7						Orange, Deutsche Telekom	Europe
3	1,800+	1,710	1,785	1,805	1,880		
7	2,600	2,500	2,570	2,620	2,690		
7*	2,600	2,500	2,570	2,620	2,690		
3+7+20						Vodafone	Europe
3	1,800+	1,710	1,785	1,805	1,880		
7	2,600	2,500	2,570	2,620	2,690		
20 (RD)	800 dd	832	862	791	821		
4+4+12						T Mobile	US
4	AWS 1	1,710	1,755	2,110	2,155		
4*	AWS 1	1,710	1,755	2,110	2,155		
12	700 a	699	716	729	746		
4+4+13						Verizon	US
4	AWS 1	1,710	1,755	2,110	2,155		
4*	AWS 1	1,710	1,755	2,110	2,155		
13 (RD)	700 c	777	787	746	756		
4+5+12						US Cellular	
4	AWS 1	1,710	1,755	2,110	2,155		
5	850	824	849	869	894		
12	700 a	699	716	729	746		
4+5+30						AT&T	US
4	AWS 1	1,710	1,755	2,110	2,155		
5	850	824	849	869	894		
30	2,300 WCS	2,305	2,315	2,350	2,360		
4+12+12						AT&T	US
4	AWS 1	1,710	1,755	2,110	2,155		
12	700 a	699	716	729	746		
12*	700 a	699	716	729	746		
4+12+30						AT&T	US
4	AWS 1	1,710	1,755	2,110	2,155		
12	700 a	699	716	729	746		
30	2,300 WCS	2,305	2,315	2,350	2,360		
4+29+30						AT&T	US
4	AWS 1	1,710	1,755	2,110	2,155		

Table 3.13 (continued)

Band	Name	Uplink		Downlink		Operator	Region
Interband 3DL CA including intraband noncontiguous CA							
29	700 d			717	728		
30	2,300 WCS	2,305	2,315	2,350	2,360		
19+42+42						NTT DoCoMo	Japan
19	800 upper	830	845	875	890		
42	TD 3,500	3,400	3,600	3,400	3,600		
42*	TD 3,500	3,400	3,600	3,400	3,600		
Intraband Contiguous Aggregation							
3	2 downlinks+2 uplinks					China Unicom	China
		1,710	1,785	1,805	1,880		
7	2 downlinks+2 uplinks					Orange	Europe
		2,500	2,570	2,620	2,690		
23	2 downlinks+1 uplink					Dish	US
	S-band	2,000	2,020	2,180	2,200		
27	2 downlinks+1 uplink					NII Holdings, Mexico, Argentina, Brazil	CALA
	800 SMR	807	824	852	869		
39	2 downlinks+2 uplinks					China Mobile	China
	TD 1,900+	1,880	1,920	1,880	1,920		
40	3 downlinks+1 uplink					China Mobile	China
	TD 2,300	2,300	2,400	2,300	2,400		
42	2 downlinks+2 uplinks					China Mobile, NII Holdings, Bollore	China, CALA, Africa
	TD 3,500	3,400	3,600	3,400	3,600		
Intraband noncontiguous CA							
2	2 downlinks+1 uplink					Verizon	US
	1,900 PCS	1,850	1,910	1,930	1,990		
3	2 downlinks+1 uplink					SK Telecom	Korea
4	2 downlinks+1 uplink					T-Mobile	US
	AWS 1	1,710	1,755	2,110	2,155		
7	2 downlinks+1 uplink					Telecom Italia	Europe
	2,600	2,500	2,570	2,620	2,690		
23	2 downlinks+1 uplink					Dish	US
	S-band	2,000	2,020	2,180	2,200		
25	2 downlinks+1 uplink					Telus	US/ CALA
	1,900+	1,850	1,915	1,930	1,995		
42	2 downlinks+1 uplink					China Mobile, NII Holdings, Bollore	China, CALA, Africa
	TD 3,500	3,400	3,600	3,400	3,600		

Table 3.13 (continued)

Band	Name	Uplink		Downlink		Operator	Region
Intraband noncontiguous CA							
Interband TDD-FDD CA 2 downlinks+1 uplink							
8+40						Korea Telecom	Korea
8	900	880	915	925	960		
40	TD 2,300	2,300	2,400	2,300	2,400		
1+42						NTT DoCoMo	Japan
1	2,100	1,920	1,980	2,110	2,170		
42	TD 3,500	3,400	3,600	3,400	3,600		
19+42						NTT DoCoMo	Japan
19	800 upper	830	845	875	890		
42	TD 3,500	3,400	3,600	3,400	3,600		

*Interband noncontiguous carrier aggregation

gregation, intraband contiguous aggregation, and interband and intraband FDD/TDD combinations.

That is just Release 12 and does not include future options presently under review.

Sprint provides an example requirement for Tri-Band LTE mapped to Band 26 in low band intended for the Sprint Nextel Enhanced Specialized Mobile Radio market offer, Band 25 in mid-band, and Band 41 in high band with a nonstandard channel plan that varies across U.S. regional markets, a legacy of the Clearwire WiMAX TDD network rollout (Table 3.14). It does not align with the EMEA (Europe, Middle East, and Africa) Band 7/Band 38 band plan.

Table 3.14

Sprint Spark Tri-Band LTE

Low Band		Mid-Band		High Band	
Band 5	**Band 26**	**Band 2**	**Band 25**	**Band 7 FDD**	**Band 41 TDD**
824–849	814–849	1,850–1,910	1,850–1,915	2,500–2,570	2,496–2,690
869–894	859–894	1,930–1,990	1,930–1,995	2,620–2,690	
	Sprint 5 MHz, FDD LTE, ESMR/Nextel		Sprint 5 MHz, FDD LTE		Sprint 20 MHz, TDD LTE, Channel plan varies by U.S. region
	814–824		1,910–1,915	Band 38 TDD	
	859–869		1,990–1,995	2,570–2,620	

3.12 Potential Future Innovation: Full-Duplex Radios

The reference data in this section was provided by Bristol University. Frequency division duplexing (FDD) and time division duplexing (TDD), the two mechanisms used to manage self-interference, and introduce planning complexity and cost at the network level and loss of spectral efficiency. FDD and TDD introduce filter and/or switch path complexity into smart phones and user devices. The sources of self-interference in a handset include reflected reentrant energy coming back into the antenna, reflected energy caused by antenna mismatch, and duplexer leakage, also a function of mismatch. This is shown diagrammatically in Figure 3.3 [6].

An alternative is to provide transmit to receive isolation using a combination of analog and digital interference cancellation to allow the same radio channel to be used simultaneously for the uplink and the downlink, sometimes described as full duplex or in-band full duplex. The approach is being researched by some operators for use in 5G networks [7].

The particular challenge within the handset is the magnitude of self-interference. A 4G LTE handset has a theoretical transmit power of up to 23 dBm and sensitivity of −95 dB implying a need for 118 dB of self-interference suppression. In a base station it is possible to achieve 70 dB of isolation from antenna separation, but it is difficult to get more than 30 dB of separation from handset antennas working in the meter band due to volume and space constraints.

The additional wavelength separation available in the centimeter and millimeter bands should theoretically make it easier to increase the isolation available at the antenna, but in practice there will be a need to combine antenna-based isolation, RF domain analog cancellation, and digital baseband cancellation, irrespective of the band into which the devices are deployed.

Single-antenna, full-duplex architectures can use circulators to improve isolation, but performance is limited by antenna mismatch, which becomes worse and harder to compensate as bandwidth increases. A circulator works by summing and differencing the RF phase of two waves traveling in opposite directions around the circumference of a ferrite disk. It is a two-port device with constructive interference at one port and destructive interference at the second port. The bandwidth of the device will be significantly less than an octave, for example, 1–2 GHz, and the amount of available isolation will be of the order of 20 dB.

An alternative is to obtain isolation by exploiting balanced signals in hybrid transformers. These can realize about 45 dB of isolation across 20-MHz bandwidth, but the isolation is dependent on the accuracy with which the balancing impedance can mimic the antenna impedance. This is known as an electrical balance duplexer.

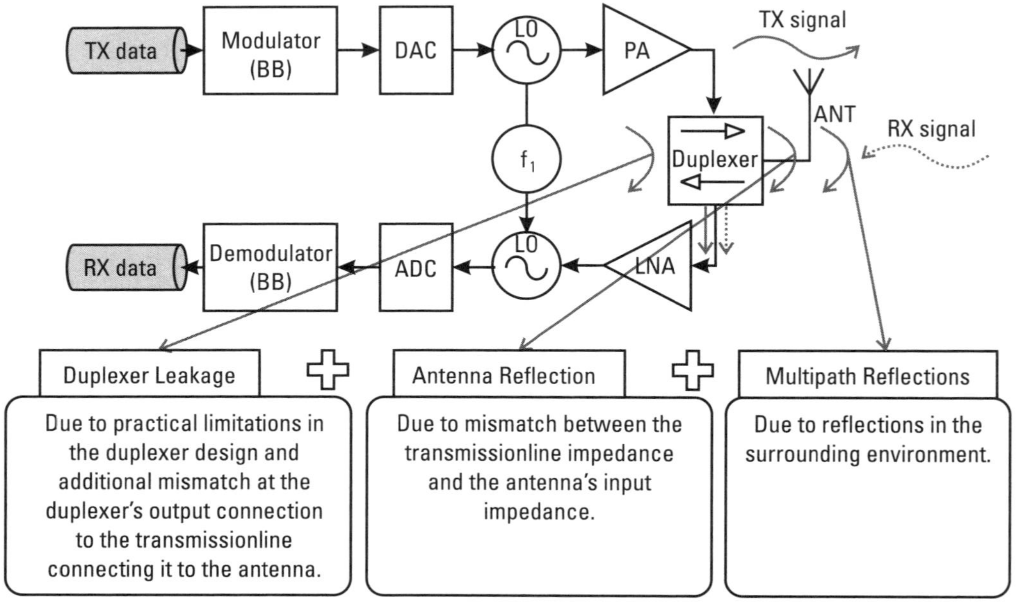

Figure 3.3 Sources of self-interference within a handset. (Courtesy of Interdigital Europe.)

In the RF domain cancellation can be either active or passive. In passive cancellation, a tapped portion of the RF transmit signal after the power amplifier is processed in the analog domain to replicate the interference, which is then subtracted at the input to the receiver. This helps to mitigate transmitter imperfections. This approach can only be subjected to low-order filtering.

Active cancellation uses an additional transmit chain to upconvert a digital baseband cancellation signal to cancel self-interference. High-order filtering can be applied to accurately model the self-interference. This does not cancel out transmitter imperfections, which means that cancellation levels are limited by the error vector magnitude (EVM) of the transmit chains. The combination of all these techniques together yields about 80 dB of TX to RX isolation and requires just one antenna which makes it potentially suitable for LTE smart phones. An example approach is shown in Figure 3.4.

Research at Stanford University references technical progress over the past 3 years with a claimed 110 dB of isolation now available coupled through a single antenna.

This approach is presently too power-hungry to implement in 4G handsets. It could be applicable to 5G handsets, but that will be dependent on the dynamic range and system channel bandwidth. Figure 3.5 shows the comparision of active and passie successive inteference cancellation (SIC) that shows potential improvements.

In the meter band, in-band full duplex would also need to coexist with legacy FDD and TDD systems, which would introduce new coexistence considerations. Earlier implementation in base stations could open up interesting opportunities including the ability to provide simultaneous access and backhaul in the same frequency band. This could also be of interest for longer-term 5G applications sharing existing backhaul radio channels in the centimeter and millimeter bands.

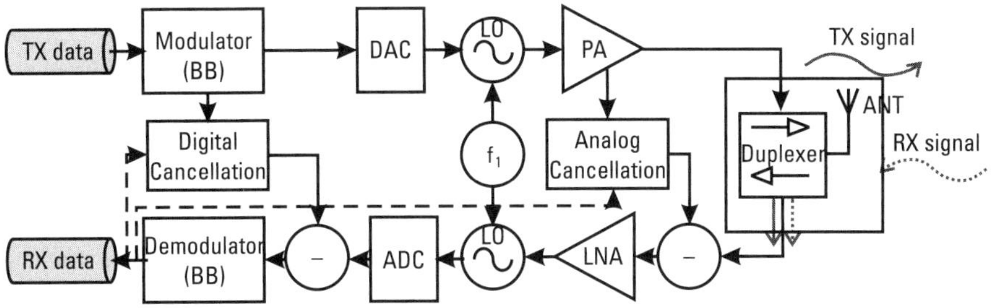

Figure 3.4 Isolation available from analog and digital cancellation and antenna separation. (Courtesy of Interdigital Europe.)

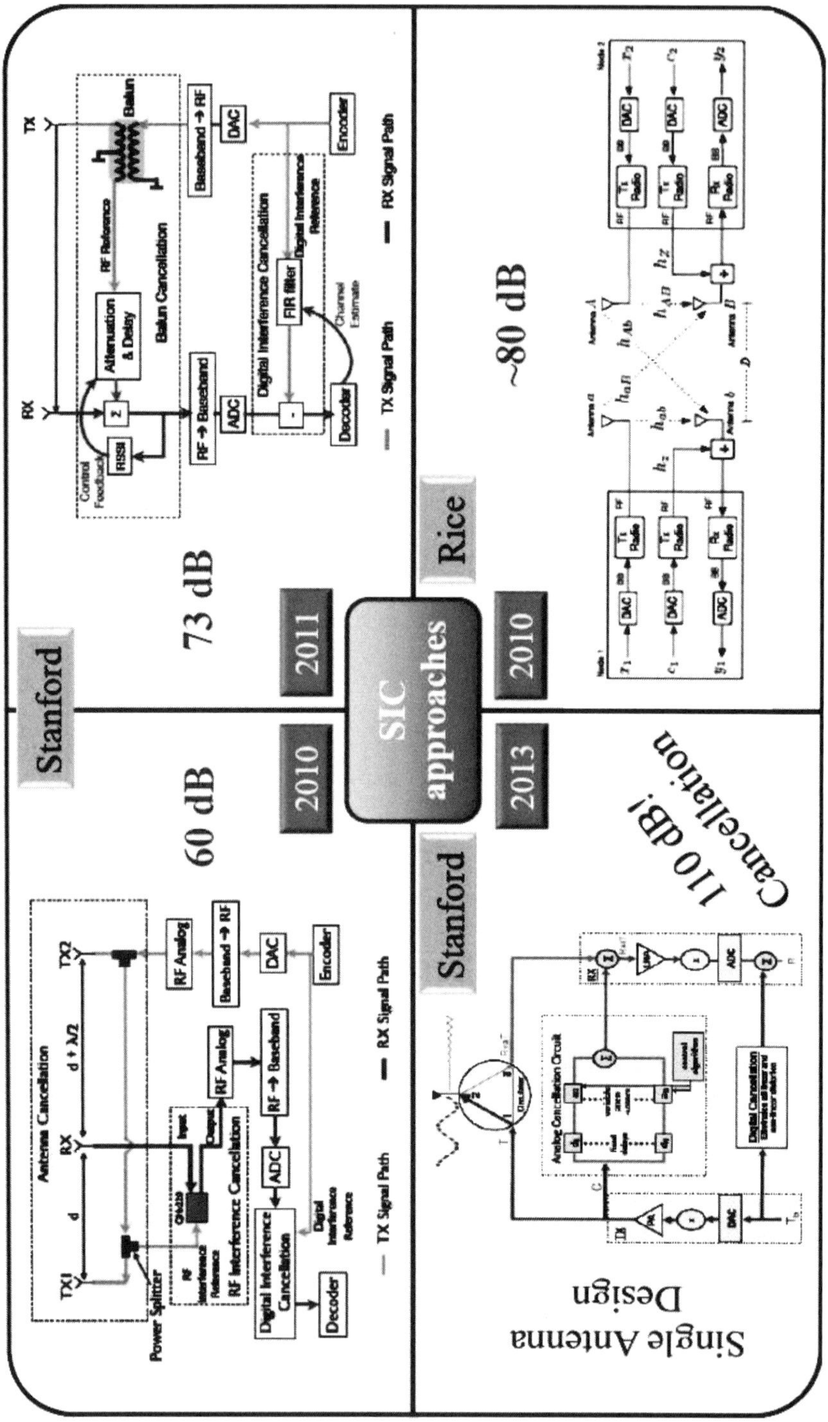

Figure 3.5 Progress to date. (Courtesy of Interdigital Europe.)

3.13 Wi-Fi Bands and LTE-U and LTE-LAA

Wi-Fi band allocations are significantly simpler than LTE and more regionally harmonized. Figure 3.6 shows the present allocations.

Some LTE vendors are promoting LTE-LAA (License Assisted Access) for higher-power outdoor Wi-Fi in the 5-GHz band using 1-W power output rather than the 200 mW for indoor Wi-Fi. Table 3.15 shows the proposed split of the band into three subbands, Bands A, B, and C.

LTE-LAA is proposed for implementation with a dedicated control channel. A separate proposal, LTU LTE Unlicensed, would use a polite protocol compatible with Wi-Fi contention protocols. Proposals are being developed within the IEEE for higher-power, wide-area Wi-Fi and automotive Wi-Fi at the top of the 5-GHz band, shown in Table 3.16.

3.14 Channel Aggregation in Wi-Fi

Some of the simplicity in the Wi-Fi global band plan is likely to disappear due to the present standards work including aggregation standards. This includes the 802.11ac standard, which extends the 40 MHz of aggregated bandwidth in 802.11 with 80 or 160 MHz of aggregated bandwidth at 5 GHz. Below the 5-GHz and 2.4-GHz bands, there are proposals to implement Wi-Fi in white-space VHF and UHF spectrum between 54 MHz and 790 MHz.

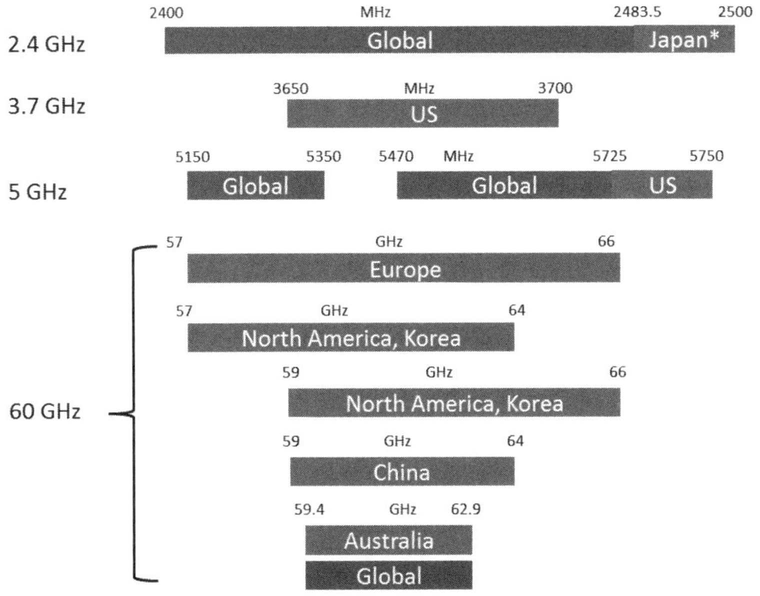

Figure 3.6 Present Wi-Fi global allocations. (Courtesy of Plum Consulting.)

Table 3.15
LTE-LAA in the 5-GHz Band

Band A		Band B		Band C	
5,150 MHz	5,350 MHz	5,470	5,725	5,725	5,850
200 MHz		255 MHz		125 MHz	
Indoor only		Outdoor		Outdoor	
30–200 mW		1W		1W	

Table 3.16
IEEE 802.11 P for Automotive and Wider Area Connectivity and Parallel ETSI Standard for High-Power Wi-Fi

United States	Europe
IEEE802.11p/1609x	CEN/ETSI EN302 663
5,850 5,925	5,855 5,925
7 × 10 MHz channels: two 20-MHz channels formed by combining 10-MHz channels	7 × 10 MHz channels
3–27 Mbps	3–27 Mbps
23–33 dBm (EIRP)	23–33 dBm (EIRP)

The standard is called 802.11af with a physical layer based on 802.11ac with frequency channels of 6 or 8 MHz to match the TV 6-MHz channel bandwidth in the United States, the 7-MHz channels in Australia, and 8 MHz in the rest of the world markets. Up to four channels can be bonded in one or two contiguous blocks. The four channels can support MIMO operation with up to four streams per channel delivering a claimed data rate of 26.7 Mbps for 6- and 7-MHz channels and 35.6 Mbps for 8-MHz channels implying a maximum data rate, albeit under perfect channel conditions of 426.7 Mbps for 6- and 7-MHz channels and 568.9 Mbps for 8-MHz channels.

There is also an 802.11aj standard for use in the 45-GHz band, specifically in China, and the 802.11ay standard, which will formalize the 60-GHz Wi-Fi band plan and functional extensions. The 802.11ay will support the bonding of two, three, or four channels, each with a channel bandwidth of 2.16 GHz producing a composite headline data rate of 100 Gbps.

3.15 Summary: LTE and Wi-Fi Device and Network Economics and Implications for 5G

The number of LTE bands has multiplied faster than our ability to support multiple bands in small form factor (thin and slim) smart phones. The constraints are technical and commercial and are cost and performance-related. The more bands that are supported, the higher the cost and the bigger the performance risk and performance cost.

Carrier aggregation adds complexity to multiband phones both in terms of RF front-end design and DSP design. LTE transceivers have to be capable of receiving multiple downlinks which can either be interband (low band, midband, or high band) and/or intraband contiguous or noncontiguous. Aggregated channel bandwidth can be up to 100 MHz. The combination of wide channel bandwidth and high data rates is particularly challenging for the DSP (covered in more detail in Chapter 11).

Operators are encouraged to deploy networks that support aggregation but need to factor in the cost and performance risks that this imposes on user devices. This cost and performance risk can compromise network economics. The same economic relationship is likely to apply to 5G networks. The cost and performance of the handset will be crucial to the viability of the network. There is presently no visibility to a solution for the multiband problem or at least a solution that is sufficiently power efficient to be compatible with a handheld device with an expectation of at least an 8-hour battery duty cycle.

The combination of LTE band and technology requirements and Wi-Fi band and technology requirements produces a to-do list for the RF and DSP design team, which gets longer every time there is either a standards meeting or a spectrum congress. The allocation of bands has moved significantly ahead of our ability to support the bands in a low-cost, low-power budget, small-form factor user device.

It is presently hard to see how 5G can add value to this already over complex process at least in the present meter band application domain between 300 MHz and 3 GHz. Complexity is a proxy for cost, and we have said that one of the aims of 5G is to achieve an order of magnitude reduction in the cost of delivered bandwidth. It is not going to do that by adding complexity. Given that the cost and performance of user devices is such an important part of this cost equation, it is going to be critical to define what a 5G user device is going to look like in terms of technical implementation, a task for the next chapters.

References

[1] http://niviuk.free.fr/lte_band.php.

[2] http://www.apt.int/.

[3] http://www.inmarsat.com/about-us/our-satellites/europasat/.

[4] www.craf.eu/iridium.htm.

[5] http://www.tranzeo.com/products/wimax.php.

[6] http://www.cambridgewireless.co.uk/crmapp/EventResource.aspx?objid=45190.

[7] http://www.sktelecom.com/en/press/detail.do?idx=1119.

4

The Cost of Coexistence

4.1 The Cost of Physical Proximity

In this chapter, we quantify the spectrum-related costs that are introduced by physical, spectral, and geographic proximity and explore the implications for 5G physical layer specification.

Physical proximity can be a few millimeters of separation within a handset, a few meters between users, tens of meters in a Wi-Fi network, hundreds of meters to 50 km or more in a cellular network, hundreds of kilometers in ultrahigh frequency (UHF) or very high frequency (VHF) terrestrial broadcast networks or satellite networks, or millions of kilometers in deep-space communication networks. In all of these radio systems, coexistence cost is introduced by the need to separate wanted signal energy from unwanted signal energy.

Within a handset, coexistence costs are introduced by radio signals interfering with each other. Signals need to be moved through the radio frequency (RF) receive-and-transmit chain with each RF component matched to the next RF component in the chain. As radio bandwidth increases, this matching process becomes increasingly imperfect.

RF amplifiers get hot and therefore become noisier and transfer at least some of that heat to SAW and FBAR filters and resonators and oscillators and MEMS or silicon on sapphire/silicon on insulator (SOS/SOI) switches. These devices can be mechanically compromised by aggressive heat cycling and will be subject to frequency drift. Digital devices create noise and are supported by digital control lines that can be running at megabit and gigabit data rates creating RF noise in all the wrong places.

Temperature drift can be accommodated by temperature compensation, for example, temperature compensated SAW filters or temperature compensated crystal oscillators, but this adds cost. Adding temperature compensation to a surface acoustic wave (SAW) filter improves the average Q of the device across its specified operating temperature range, for example, –40°C to +85°C, but will degrade the absolute Q of the device. This inability of components to coexist happily together explains why they often do not deliver the performance claimed in a specification data sheet. The performance difference is usually described as implementation loss.

Coexistence costs include the need to support multiple technologies within a handset either at different bands or within the same band. Coexistence challenges can be simply stated as the need to ensure that components, devices, and networks do not interfere with other components, devices, and networks. Coexistence costs are the cost of mitigation measures to ensure that this interference is avoided or minimized. Mitigation cost includes additional filtering in mobile broadband user devices and additional filtering in other victim receivers, for example, terrestrial TV sets or distribution amplifiers or satellite receivers.

Mitigation cost includes the need to back off power amplifiers in order to meet out-of-band emission requirements when supporting higher-order modulation or wide bandwidth full resource block Long-Term Evolution (LTE) channels, for example, 10-, 15-, or 20-MHz LTE or LTE channel aggregation. We have to consider whether 5G would increase or decrease mitigation costs if introduced in the meter band and how meter band coexistence costs are likely to compare with 5G implemented into the centimeter and millimeter bands.

4.2 The Impact of Wide Passbands Within Wide Channels: Out-of-Block and Out-of-Band Versus In-Band Performance

In the last chapter we identified a trend towards implementing wider passbands to support wider channel bandwidths (10 MHz and >10 MHz).

As passbands increase, filter rolloff reduces and filters become less well behaved. On the TX path, spectral emission masks and adjacent channel leakage ratios (ACLR) become harder to meet.

This coincides with a need to improve the Q (effectively the filter rolloff) of the RF front end of LTE smart phones. This is a consequence of the continued regulatory trend to improve spectral utilization while providing improved coexistence protection to politically, socially, and commercially influential spectrally and geographically proximate systems, for example, TV broadcast at 600 and 700 MHz, public safety radio at 700 MHz, Global Positioning System (GPS) in L-band, satellite broadcasting, and fixed and mobile satellite services in C-band and the centimeter and millimeter bands. Coexistence with fixed point-to-point systems including backhaul also needs to be managed.

Wider channel bandwidths combined with the need to improve front-end Q result in conflicting design and performance objectives. Coexistence issues are normally discussed in the context of out-of-band emissions. Out-of-band emission limits have a direct impact on the component specification and cost of mobile broadband user device filters and a related impact on performance. Generally stated any improvement (reduction) in out-of-band emissions translates into a loss of in-band performance.

Confusingly, both out-of-block and out-of-band emissions are described as out-of-band in the technical literature. The two functions are related but separate. Out-of-block emissions are the emissions from one operator's block of spectrum to an adjacent operator's block of spectrum. This includes frequency division duplex (FDD) to FDD coexistence and time division duplex (TDD) to FDD coexistence. An example is the 2.6-GHz band (Band 7 and Band 38) shown in Table 4.1.

In refarmed bands, for example, at 850, 900, 1,800, and 1,900 MHz, block-to-block interference may also include LTE to Global System Mobile (GSM) interference, LTE to General Packet Radio Services (GPRS) interference, LTE to HSPA interference and in the 900-MHz band in Europe, LTE to the GSM-based radio system for U.K. and European railways (GSM- R) interference. Different channel bandwidths, for example, GSM 200 kHz, W-CDMA 5 MHz, and LTE 5 or 10 MHz, introduce additional coexistence complexity.

Coexistence cost is therefore a consequence of the mix of bands and technologies.

4.3 Intersystem Coexistence: Impact on Conformance Specification

For new bands such as the 800-MHz band and 700-MHz band, coexistence with TV and other users including Programme Making and Special Events (PMSE), for example, wireless microphones, has to be managed.

Out-of-band emissions are the emissions from one radio system, for example, a mobile broadband network at 700 or 800 MHz into another radio system, for example, a TV receive channel. This includes the possibility of multiple sub-1-GHz LTE bands including LTE 900, LTE 850, LTE 800, LTE 700, LTE 450, and possibly LTE 600 and LTE 500.

LTE can manage coexistence by reducing user device resource block allocations and/or reducing user device TX transmitted radio power either by a fixed amount defined in the conformance specifications and described as maximum power relaxation (MPR) or by a variable amount in response to

Table 4.1
Block Allocations in Band 7 and Band 38*

70-MHz FDD Spectrum							50-MHz TDD Spectrum					70-MHz Paired FDD Upper Duplex						
2,500	2,510	2,520	2,530	2,540	2,550	2,560	2,570	2,580	2,590	2,600	2,610	2,620	2,630	2,640	2,650	2,660	2,670	2,680

* *Source:* [1].

reported interference conditions described as adaptive maximum power relaxation (A-MPR).

Adaptive maximum power relaxation can only be realized when feedback is available from the victim system. For example, it can be used to reduce TX to RX interference across the duplex gap of an FDD network (user-to-user and/or block-to-block interference mitigation). If different operators have channels on either side of the duplex gap, then network-to-network coordination is required.

In all other cases, power relaxation has to be done on an assumed worst-case basis. This can result in a 5- to 7-dB backoff, which has a direct impact on data reach and user-specific throughput. If there is interference in an LTE receive channel at the block edge, the scheduler will avoid the worst affected subcarriers, but this reduces throughput and network capacity. The performance impact of band configuration and coexistence is reflected in the conformance specification. Table 4.2 shows the differences between bands for 10-MHz bandwidth LTE.

The best conformance specification is –97 dBm for Band 1 at 2 GHz in Europe, the existing AWS Band 4 in the United States, Band 10, Band 18, Band 19, Band 21, and Band 23. Band 10 extends the passband of Band 4 from 45+45 MHz to 60+60 MHz. This is 3.4% of the center frequency, which is well within the limits of acoustic filters that are generally well behaved up to a boundary of 4% of the center frequency. The large duplex gap minimizes user to user interference. This therefore is an example of a widening of the passband without performance loss. Band 3 operational bandwidth by comparison is 4.29%, which is reflected in the –94-dBm sensitivity figure.

Band 2 loses a couple of decibels of sensitivity due to a narrow duplex gap (1.04% of the center frequency). Band 25 reduces the Band duplex gap by 5 MHz and increases the operational bandwidth/passband from 60+60 to 65+65 MHz. The sensitivity is reduced by 1.5 dB relative to Band 2 and 3.5 dB relative to the –97 dBm of Bands 1, 10, 18, 19, 21, and 23.

The –95 dBm of Band 5 is due to the lower operational frequency and assumes a loss of antenna efficiency. Band 26 also known as E850 (Extended 850) increases the operational bandwidth/passband from 25+25 MHz to 35+35 MHz and reduces the duplex gap from 20 MHz to 10 MHz, with a consequent loss of 0.5 dB of sensitivity compared to Band 5. The sensitivity of E850 and Band 8 ends up being more or less the same. This is not a surprise, as they have the same passband (35+35 MHz) and duplex gap (10 MHz) and operate at a similar wavelength.

Potentially there could be an Extended 850 band that consolidated Band 5, Band 26, and Band 27 creating a 42+42-MHz passband with a 3-MHz duplex gap, but with present technology this would be technically and commercially difficult (probably impossible) to implement.

Table 4.2
LTE 10-MHz Channel Bandwidth Receive Sensitivity*

LTE Band	Uplink/Downlink frequencies (MHz), Mob TX/ Mob RX	Band Name	Receive Sensitivity (dBm)	Factors Influencing Sensitivity
1	1,920–1,980/ 2,110–2,170	2 GHz	–97	Narrow operational bandwidth, large duplex gap
2	1,850–1,910/ 1,930–1,990	1,900 MHz	–95	Narrow operational bandwidth, small duplex gap (see also Band 25), –93.5
3	1,710–1,785/ 1,805–1,880	1,800 MHz	–94	Wide operational bandwidth, small duplex gap
4	1,710–1,755/ 2,110–2,155	AWS	–97	Narrow operational bandwidth, large duplex gap (see Band 10)
5	824–849/869–894	850 MHz	–95	Low operational frequency, moderate bandwidth (see also Bands 26 and 27)
6	830–840/875–885	850 MHz	–97	Obsolete band superseded by Band 19
7	2,500–2,570/ 2,620–2,690	2.6 GHz	–95	Narrow operational bandwidth, TDD in duplex gap
8	880–915/925–960	900 MHz	–94	Low operational frequency
9	1,750–1,785/ 1,845–1,880	1,800 MHz	–96	Narrow operational bandwidth, small duplex gap
10	1,710–1,770/ 2,110–2,170	AWS+	–97	Increases operational bandwidth of Band 4 AWS from 45+45 to 60+60 MHz but with large duplex gap
12	698–716/728–746	700 MHz	–94	Tight filtering required due to adjacency to U.S. TV channel 51 at 698 MHz
13	777–787/746–756	700 MHz	–94	Tight filtering required due to adjacency to Band 14
14	788–798/758–768	700 MHz	–94	Tight filtering required due to adjacency to Band 13
17	794–716/734–746	700 MHz	–94	AT&T-specific variant of Band 12
18	815–830/860–875	850 MHz	–97	Low operational frequency, narrow bandwidth
19	830–845/875–890	850 MHz	–97	Low operational frequency, narrow bandwidth
20	832–862/791–821	800 MHz	–94	Low operational frequency
21	1,448–1,463/ 1,496–1,511	1.5 GHz	–97	Narrow operational bandwidth
22	3,410–3,500/ 3,510–3,600	3.5 GHz	–94	Large operational bandwidth
23	2,000–2,020/ 2,180–2,200	AWS 4	–97	Narrow operational bandwidth, large duplex gap
24	1,625.5–1,660.5/ 1,525–1,559	L-band	–97	Narrow operational bandwidth, large duplex gap
25	1,850– 1,915/1,930–1,995	1,900 MHz+	–93.5	Small duplex gap

Table 4.2 (continued)

LTE Band	Uplink/Downlink frequencies (MHz), Mob TX/ Mob RX	Band Name	Receive Sensitivity (dBm)	Factors Influencing Sensitivity
26	814–849/859–894	850 MHz	–94.5	Low operational frequency, wide operational bandwidth
27	807–824/852–869	800 MHz	–95	Low operational frequency, narrow bandwidth
26+27	807–849/852–894	850++	?	42+42 MHz, 3-MHz duplex gap
28	703–748/758–803	700 MHz	–95.5	Requires two overlapping duplexers 2 × 30 MHz with 15-MHz overlap
29	717–728	700 MHz	–96	Downlink only, carrier aggregated

* *Source:* [2].

The U.S. 700-MHz bands are all specified at –94 MHz. Band 12 is the most challenging. The lower edge of the lower duplex at 698 MHz is immediately adjacent to the upper boundary of TV Channel 51. This means that there is no guard band between the Band 12 mobile transmit and a TV receiver.

Bands 13 and 14 similarly require tight filtering. Band 13 has to minimize out-of-band emission into adjacent Public Safety Radio (PSR) radio channels. Figure 4.1 shows what this means in terms of filter implementation to achieve the rolloff needed. A traditional LC filter as might be expected falls far short of the requirement. A standard SAW filter is not adequate and either a film bulk acoustic resonator (FBAR) filter or temperature compensated SAW is required.

Band 20 in Europe resolves the Digital Terrestrial Television (DTT) adjacency issue by being reverse duplex with mobile transmit in the upper duplex.

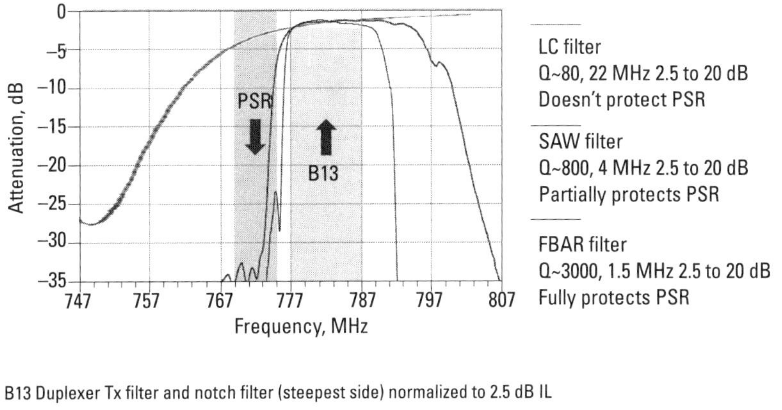

Figure 4.1 Filter rolloff requirement for Band 13.

Band 28 in Europe is more problematic. Band 28 is an interesting band for LTE. The 45+45 MHz passband potentially supports a mix of 5-MHz and 10-MHz LTE channels. Band 28 is known as the APT band as it was proposed by the Asia Pacific Telecommunity [3] at WRC 2012. It therefore has substantial market volume potential. The passband (42 MHz is ≥5% of the center frequency) is outside the operational bandwidth of acoustic filters (4% = 30 MHz at 730 MHz) so the front end is realized with two 30-MHz filters with a 15-MHz overlap. The lower filter pair is described as APT (a). The upper filter pair is described as APT (b). APT (b) overlaps Band 20 in Europe, but it still makes commercial sense to implement APT (a) for 30+30 MHz LTE with a 25-MHz duplex gap which could theoretically at least support TDD LTE.

However, Committee European Post and Telecommunications (CEPT) decided that the out-of-band protection to the DTT multiplex in Europe should be the same as the out-of-band protection specified for Band 20. The problem with this is that Band 28 is implemented as a standard duplex rather than reverse duplex. This implies a need to reduce the out-of-band limits by 20 dB. At the time of this writing, vendors had agreed to an additional 10 dB of out-of-band protection.

4.4 LTE to DTT Coexistence

The LTE to DTT coexistence challenge is similar to the requirement to achieve isolation between LTE transmit power and the GPS receive path in smart phones except that GPS is in L-band (at 1,575 MHz) several hundred megahertz away from many of the LTE signals, whereas the DTT receiver can be required to receive a signal with only a relatively narrow guard band (typically 7 or 9 MHz) to provide protection.

The degree of interference is a function of the transmitted power of the LTE signal, the frequency relationship between the LTE transmit frequency and DTT receive frequency, the spectral purity of the LTE signal, the filtering applied to the LTE signal, and (usually the most important factor) the free-space loss distance between the aggressor and victim device. (Propagation loss works on a fourth-power law, which means that for every doubling of distance, received signal energy reduces by a factor of 4.) The resilience to interference is a function of filtering, dynamic range, and the effectiveness and responsiveness of the gain control (automatic gain control) function in the RF front end (RFFE).

Because free-space loss is a function of distance and as the distance between devices can be anything from a few inches to several miles, it is evident that many assumptions have to be made in order to arrive at a statistically plausible assessment of how much interference will actually occur.

While the primary purpose of acoustic filters is to provide isolation between the LTE signals within the user or Internet of Things (IoT) device, they also help to attenuate the unwanted signal energy directed towards DTT receivers and are particularly effective at filtering out wideband noise, noise generated outside 1% of the transmit frequency, for example, 7 MHz at 700 MHz. Close-in interference is dominated by the output characteristics of the LTE power amplifier. The output characteristics are, in turn, influenced by the modulation applied to the transmission.

Since the auction of the 700-MHz band in the United States in 2007, the industry has accumulated experience on how to specify mobile cellular transceivers and broadcast receivers to minimize or mitigate mutual interference. Self-evidently, it is not desirable to have LTE signals swamped by high power (50 or 100 kW) high tower broadcast signals, nor is it desirable to have DTT receivers disturbed by transmissions from LTE user devices.

Historically regulatory bodies have concentrated on specifying transmit power (amount and purity) and have been relatively relaxed about receiver specification. Partly this was because television receivers and cellular phones used a receiver architecture known as the superhet (an abbreviation of "super heterodyne") where the incoming signal is mixed with a signal at another frequency to produce an image or intermediate frequency (IF), for example, 72 MHz in a TV receiver. This was because it was much easier to filter at the lower frequency (filtering could be achieved with relatively lower insertion loss). The disadvantage is that superhet receivers have to have two sets of filters at the receive frequency and intermediate frequency, which take up space, add cost, and introduce insertion loss.

This became an increasing problem for cellular phones supporting multiple bands and as a result there was a shift to using direct conversion where the incoming signal is mixed with its own frequency to produce a direct conversion to baseband. All contemporary multiband cell phones are now direct conversion. DTT receivers have also gone through the same transition, from traditional can tuners (superhet) to silicon receivers (direct conversion). Since 2011, all newly manufactured DTT receivers have been direct conversion [4].

This has meant that the specification of adjacent channel selectivity has become more important both for DTT receivers and for cellular phones. This is addressed in Europe by the Radio Equipment Directive [5].

In parallel, DTT receivers are transitioning from DVB-T to DVB-T2 (and in the longer term to T3). This transition improves spectral efficiency (the DTT multiplex can be upgraded from 20 Mbps to 40 Mbps), but the demodulator has to demodulate higher-order signals (in the longer-term 1,024 QAM).

This requires either a higher energy per symbol received or more extensive encoding and decoding. Increasing the complexity of the encode/decode process results in an artifact known as error extension where a disturbance produced by an interferer, for example, an LTE user device, propagates through the decoder trellis. From a DVB user perspective, this means that the picture breaks up for potentially several seconds.

Adjacent channel selectivity is only one of three factors influencing coexistence, with the other two being automatic gain control (AGC) and dynamic range. Automatic gain control in DTT receivers has historically been designed to adjust to the different levels of received signal from a remote or close high-power high tower. The difference in received power can be significant but does not change significantly over time apart from small changes in path loss caused for instance by weather effects, for example, temperature inversions. These changes happen slowly, over hours or days or years.

Interference from LTE user devices will vary within substantially shorter time scales but also will be bursty with some of the burstiness being periodic, a function of how channel sounding and signaling are implemented within the LTE physical layer.

This causes the AGC to reduce the gain in the front end of the receiver, which causes the picture to disappear. It is important to validate that these effects are not a significant factor when LTE user devices are in close physical proximity to DTT receivers. Note that this effect is substantially independent of the criteria used to establish acceptable out-of-band limits. This is because DTT receiver front ends are designed to receive the whole (or at least a large part) of the UHF receive band so the received signal (prior to direct conversion) is in band, not out of band.

Compression is a function of dynamic range and is related but separate to the AGC function described above. Dynamic range describes the ability of a receiver's range to handle large unwanted signals without swamping the wanted signal, which is typically a small signal just above the noise floor. In a situation of interference both conditions may exist simultaneously. The receiver must be able to discriminate wanted signal energy from unwanted signal energy.

Unfortunately a low noise amplifier (LNA) faced with a large unwanted signal can become nonlinear, which means that it distorts the signal that it is trying to amplify, which makes it harder or impossible for the demodulator to demodulate the signal.

This is not a major problem provided the receiver has been designed to accommodate high-level signals, although increasing the dynamic range increases the power drain, which is an issue in handheld devices. LTE transceivers now implement variable voltage rail front ends that increase the dynamic range only as and when required.

The extent to which this will be a practical problem still remains unproven though study work [6] has been done that suggests that this may be an issue with masthead amplifiers (shared antennas for apartment blocks), distribution amplifiers (home and in-building distribution) and launch amplifiers (DVB-T into cable distribution systems).

Most of these potential impairments would be mitigated by ensuring that the flux density of the received digital video broadcasting (DVB) signal was similar to the flux density of the received LTE signals. This could be accomplished by transmitting the DVB-T/DVB-T2 multiplex from cellular base stations.

There might be issues of space and wind loading to support wideband base station antennas on masts, although if cellular is being implemented at lower frequencies (>400, >500, >600 MHz), then this cost will be incurred irrespectively. So far, no commercial model has emerged to facilitate this integration.

4.5 LTE and DTT at 450 MHz

LTE to DTT coexistence issues may also occur further down in the UHF band. LTE Band 31 provides a contemporary example. This is presently an LTE Band allocation specific to Brazil with mobile TX between 451 and 458 MHz and mobile receive between 461 and 468 MHz with a duplex gap of 3 MHz. This is relatively narrow (0.6% of the center frequency) compared to, for example, Band 8 (1.08%). This is important because it will dictate a rapid rolloff to the duplex side, which will limit the rolloff capability to DTT spectrum of >451 MHz. The coexistence condition is eased to an extent due to the channel bandwidths being at the most 5 MHz and potentially 3 MHz or 1.4 MHz.

There is approximately a 10-dB difference (improvement in out-of-band) between a 5-MHz (25 resource block) and 10-MHz 50 resource block channel and a similar difference between a 10- and 20-MHz channel (the out-of-band emissions increase as channel bandwidth increases).

The other good news is that at least some of the applications in these lower bands, for instance, Band 31 at 450 MHz will be mobile rather than handheld, for example, LTE transceivers in cars, trucks, fire engines, ambulances, buses, and trains.

These transceivers can be larger, which means the antennas can be more efficient and they work off a 12- or 24-V power supply, so they can have higher dynamic range. Less onerous size and cost constraints also mean that ceramic filters or cavity resonators can be used rather than acoustic filters. These devices can deliver sharp rolloff characteristics but can also handle power levels that would be hard for acoustic filters to accommodate (the filters fail mechanically). This has enabled higher-power mobiles to be specified (+33-dBm output = +3W rather than+23 dBm = 250 mW), which are potentially more optimum for providing rural broadband geographic coverage or deep urban building penetration.

There is an industry consortium presently promoting the use of the 450-MHz band for LTE on a global basis [7].

4.6 Traffic Asymmetry and eMBMS

The impact of Band 29 (downlink only at 717 to 728 MHz) on U.S. digital terrestrial television also needs to be considered. The addition of downlink-only bandwidth is based on an assumption of increasing traffic asymmetry. For example, 1G and 2G networks in the 1980s assumed symmetric traffic (voice is essentially bandwidth balanced, although this does depend to whom you are talking). The 3G networks assumed a traffic asymmetry of 4 to 1 and 4G networks assume a traffic asymmetry of (more or less) 10 to 1.

In theory, this asymmetry can be partly supported by a higher link budget on the downlink, for example, 20-W base stations with sectored gain antennas supporting user devices with 250 mW of uplink power. However, there is also an assumed need for additional dedicated downlink channel bandwidth. This can either be a supplementary downlink channel with the uplink signaling carried on an aggregated duplex band or could be enhanced Multimedia Broadcast Multicast Service (eMBMS) (the LTE version of MBMS) transmitted over a multicast broadcast single-frequency network (MBSFN).

All user devices from Release 8 onwards are required to be able to demodulate an eMBMS sub frame. The MBSFN transmits on a 7.5-kHz frequency rather than a 15-kHz frequency subcarrier to allow a longer cyclic prefix, which, in turn, allows for larger cell deployments.

The narrower subcarrier spacing implies less resilience to frequency errors and phase noise and Doppler shift. This, in turn, determines which of the modulation levels (QPSK, 16 QAM, or 64 QAM) will be practical in real-life network conditions.

From a network perspective, eMBMS has well-thought-through features drawing on the lessons learned from the not-altogether-happy deployment experience of DVB-H. This includes time slicing to improve user device power consumption and E-MBMS counting, used to determine if there is a sufficient number of user devices interested in receiving a service to enable an operator to decide whether it would be efficient to deliver the service via MBSFN.

The supported functionality has been improved through Releases 9, 10, and 11. Release 11 user devices receiving MBMS transmission may also receive a unicast transmission in the same carrier time multiplexed onto different subframes and the terminal has the capability to inform the network about its MBMS interest and capabilities. A carrier aggregation capable terminal can receive MBMS on one carrier and unicast on another component carrier.

There is no present practical experience of how well or badly these user devices will perform in real-life network conditions and design teams may be motivated to pay more attention to improving performance in core LTE FDD bands (for Europe, 900, 1,800, and Band 1, Band 20, and Band 7).

4.7 Coexistence with DVB-T2, ISDB-T, and ATSC

The 4G or 5G coexistence also needs to consider coexistence issues with future terrestrial TV standards including Digital Video Broadcasting-Terrestrial 2 (DVB-T2) and possibly T3 in Region 1, Sistema Brasileiro de Televisão Digital (ISDB-T) in parts of Region 2, and Advanced Television Systems Committee (ATSC) in Region 3.

All three standards have road maps that develop support for handheld and mobile devices (ATSC3, for example) and all three standards talk about convergence (ATSC3 and DVB T3 as an example) and LTE integration. However, ATSC terminals have not as yet been introduced into the U.S. market and there is only marginal appetite for DVB- T2 in some markets despite the potential multiplex rate gain (40 Mbps over 20 Mbps).

Differences are also likely to continue to exist in channel and source coding including audio coding standards. The assumption is that these can be accommodated within standardized hardware, but in practice many advanced coding schemes require specifically optimized fast access memory and optimized parallel processing architectures. Field programmable gate array (FPGA) devices are not a solution as they still have practical clocking limits of the order of 100 MHz (at least in power constrained devices). Regional differences in carrier spacing (6, 7, or 8 MHz) also need to be accommodated.

Whether coexistence matters between terrestrial TV and 4G and 5G mobile broadband systems depends largely on how the incentive auction process proceeds in the United States. If it goes well, then terrestrial TV might become a historical curiosity and 200 MHz of spectrum comes available at 500 and 600 MHz. This will be more likely to happen initially in the United States and Latin America. Region 1 (Europe and Africa) will have to wait for the review of the UHF band now agreed as an agenda item for WRC2023.

4.8 Intersystem Coexistence Costs in L-Band

The principal coexistence issue in L-band over the past 3 to 4 years has been the proximity to the GPS receive band at 1,575 MHz. This derailed the plans by Light Squared to implement a hybrid terrestrial satellite LTE network, potentially an $8 billion investment in Band 24 (1,625.5–1,660.5/1,525–1,559 MHz).

Second-order and third-order intermodulation into the GPS band also needs to be considered.

Band 14 in the United States, allocated for Public Protection and Disaster Relief (PPDR) produces a second order product at 787.5 MHz, which falls directly into the GPS receive band ($2 \times 787.5 = 1{,}575$ MHz).

Protection also needs to be provided for the GPS L2 signal at 1,227.60 MHz and the L5 signal at 1,176.45 MHz. L2 is now supported in most consumer GPS receivers with L5 being added to improve resilience to jamming and to improve resolution and accuracy by correcting for atmospheric and ionospheric distortion [8].

These coexistence considerations will need to take into other GNSS systems as they become more widely available including Galileo in Europe and Beidou in China.

4.9 Intersystem Costs in S-Band

There have been some interference issues between Band 7/Band 41 at 2.6 GHz and aviation radar. These have been resolved with localized mitigation measures, although with some loss of capacity to the LTE networks in areas (close to airports) where capacity has high value.

4.10 LTE to LTE Interference: The Cost of Reducing RF Output Power in the User Device

All cellular systems to date have implemented power control, the process by which the base station measures the signal strength and signal quality on the uplink and downlink and then sends power-up or power-down instructions to the mobile to compensate for path loss and channel fading. The same process is used to manage handover decisions with interference as an added input to the decision process.

The transition to 3G systems made some things easier and some things harder. Channel spacing increased from the 25 or 30 kHz used in first generation systems to the 200 kHz used in second generation GSM to either 1.25 MHz (CDMA) or 5 MHz [Release 99 Universal Mobile Telephone System (UMTS)] for third generation networks.

This made frequency planning easier at the network level and to an extent relaxed the synthesizer performance and frequency reference requirements in user and base station equipment at least in terms of resolution though noise specifications became more stringent. Additionally, channel-to-channel selectivity and user-to-user selectivity had to be achieved using Walsh codes (CDMA) or OVSF codes (UMTS).

These provide an effective code domain mechanism for decorrelating wanted signal energy from a noise like channel, but the effectiveness of the process is dependent on closely managed power control to ensure that multiple users are received by the base station at a similar symbol level power, ideally within 1 dB of each other.

For an acceptable coverage area, this requires the mobile device to control its power over a 78-dB power range, nearly 100,000,000 to 1. By contrast, GSM/GPRS/ Enhanced Data Rates for GSM Evolution (EDGE) devices have 40 dB of mobile transmitter power control, less than 10,000 to 1. The power control loop is therefore more complex than prior systems and depending on the level of mobility in the user group, the associated signaling absorbs power and bandwidth.

This power control is there for a reason and works well in many present applications. It means that for much of the time user equipment is running at a fraction of its maximum output power, typically a few milliwatts or a few tens of milliwatts rather than the 125 or 250 mW potentially available. However, user equipment RF power amplifiers are not inherently efficient when run under these lightly loaded conditions and have to be built with switchable gain stages. These add cost and complexity. At cell edge, user equipment may need to be backed off in terms of output power to minimize cell-to-cell interference. This power may be necessary to maintain the call during handover to the next serving cell, avoiding a dropped call.

Even with uplink interference mitigation, the effect of this is that user equipment can be uplink limited more often than it needs to be. This translates into either a loss of range and data rate at cell edge, dropped sessions and or a loss of uplink capacity.

Close in to the cell, uplink offered traffic is limited by the noise rise of the base station receiver front end. The noise rise will initially shrink the size of the cell and ultimately approach pole capacity, the point at which user devices are instructed to increase their output power to overcome the noise rise at the base station. This then increases the noise rise.

As operators transition to mobile broadband data-dominant networks, these edge-of-cell and close-to-cell performance issues become more significant as the offered traffic, and by implication the offered traffic power requirements, vary substantially and rapidly both on a per-user and multiuser basis.

The overall aim is to achieve a sweet spot compromise between delivering data reach and data capacity irrespective of the cell geometry (the distribution of users within the cell). This means giving people acceptable data rates and data capacity at the edge of the serving cell. The dynamic range of the composite techniques used have to accommodate a wide range of operational conditions from dense urban to deep rural with users and IoT devices that are either stationary, moving slowly, or traveling at speed from cell to cell. These are wide-area high mobility networks, not static Wi-Fi networks.

Network efficiency in all data networks, ADSL being one example, is achieved by realizing multiplexing gain between users whose bandwidth requirements are continuously changing. This is different from circuit switching when a circuit is dedicated to one user for a specific voice or data or now a voice and data session.

In a mobile broadband wide-area network, admission control is more complex with a trade-off between user experience and network efficiency. User experience opinion scoring for edge of cell users will be improved when Round Robin scheduling is used. Round Robin scheduling allocates bandwidth to users, irrespective of their channel condition at any moment in time. This means that users at the edge of the cell will get either the same amount of bandwidth allocated to them as close in users or potentially additional bandwidth to compensate for the weaker link budget.

The alternative approach is to implement channel quality indication (CQI) based scheduling where users with the best instantaneous uplink and downlink channel quality will be given preferential access to radio channel bandwidth. Actually, it is a bit more complicated than that. CQI can be used to take advantage of short-term channel conditions (opportunistic scheduling on a time base of less than 100 ms) combined with multiuser quality of experience scheduling (on a time base of more than 100 ms). CQI is also used in the frequency domain in LTE.

If CQI is used as the only admission criteria, it will yield the highest data capacity per hertz of allocated bandwidth but edge of cell users will rarely get served. All schedulers therefore choose some intermediate position between the two extremes. This is known as proportional fair scheduling or fair throughput scheduling.

We have said that high-speed packet access (HSPA) devices may have to be power limited at the edge of cell to mitigate cell-to-cell interference given that most networks are implemented on a frequency reuse of one (the same channel allocated at all sites).

LTE (strictly speaking, LTE Advanced) introduces intercell interference coordination (ICIC). This manages intercell interference by ensuring that the user devices that are detected as being potential interferers do not use the same frequency subcarriers as users in the adjacent cell. Similarly problems with an insufficient guard band in the frequency domain can be accommodated by not using frequency subcarriers towards the edge of a 5-, 10-, or 20-MHz channel. If ICIC is not used, then the frequency-domain scheduler using CQI will do the job adequately and other mechanisms such as the use of the physical cell identifier help as well, at least within two or three sector deployments.

On the basis of using any or all of these intracell and intercell interference mitigation techniques, LTE devices can operate at the edge of the cell at full power. This increases data reach, the distance from the base station where acceptable data rates are still available.

Close in to the cell, user devices can be seen by the LTE base station at relatively different power levels, of the order of 20 dB or so. This is because selectivity is achieved in the base station in the time and frequency domains. This means that the user devices can be operated at maximum power with the power matched to the traffic requirement. This makes CQI scheduling more efficient close in to the cell, which means that more bandwidth can be made available to users at the cell edge. This increases data capacity and data reach.

The LTE admission control algorithm works on a time-domain resolution of anything between 10 ms and half a millisecond coupled with a decision as to how many frequency subcarriers are made available (the composite term for both of these together being physical resource blocks).

Admission decisions can be taken on the basis of multiple inputs (potentially 32 variables) but can also be beguilingly simple. As an example, for best-effort data the decision can be based on the buffer occupancy of the user's device. If the buffer starts to get full, the device sends "sad" bits to the network. If the buffer is relatively empty, the buffer sends "happy" bits to the network. The network then decides on an optimum physical resource block allocation for that single user taking into account the requirements of all other users in the cell and proximate cells.

This means that the allocation of channel bandwidth, and by implication, channel power can be done in the frequency and time domains. There is no need to power control the user's device, which, in turn, means that the channel signal energy previously absorbed by power control in the user's device is now available for user traffic (with associated user value) rather than signaling overhead. There are techniques in Release 7 onwards to reduce this overhead but not to eliminate it altogether.

Assuming that the mobiles can run at full power, the only constraint then becomes the dynamic range and selectivity available at the base station, the ability of the base station to handle the offered traffic power. Superficially, this seems odd. The base station might be transmitting 20W and the user devices are transmitting at most 250 mW, but then there could be hundreds of mobile devices firing in to the base station RF receiver front end.

Base stations with more dynamic range and selectivity on the receive path will therefore deliver a significant system efficiency gain. The difference in performance is determined by the architecture used. A low-cost base station, for example, might attempt to downconvert a whole band, which would mean as much as 70 MHz at Band 7 using one DSP (the state of the art is 60 MHz).

This will be a low-cost approach but will leave the RF front end and baseband DSP vulnerable to overload and nonlinear behavior.

Note that the constraints of a DSP are similar to RF component constraints. The ability of the DSP to handle dynamic range is a function of bit width. The ability to handle high-frequency signals across a given bandwidth is a function of bit width and clock cycle count. This is the same cause and effect but described in a different way.

4.11 Caveats: Fixed Power Control Overheads

There are caveats that need to be expressed. The reference symbols in LTE are fixed elements and regular (four per physical resource block) both on the uplink and downlink with the uplink channel quality being reported to the enhanced Node B (eNB) through the CQI symbols. This fixed part of the overhead is therefore inescapable, so you may as well get some benefit from it. In most cases LTE will be deployed with other legacy systems, so power control may be needed to manage intersystem interference both within the user device and within the network. While this is true, it is also true that if the signaling associated with the LTE power control loop can be minimized at the physical layer, then a link budget gain will be achieved.

4.11.1 User Equipment Sensitivity

There is no point in increasing the power output of a user's device if the sensitivity is compromised. This means that the isolation in the switch and filter paths needs to be carefully managed in both FDD and TDD systems. TDD systems are not immune to these effects. The techniques used to improve power amplifer (PA) linearity and efficiency in LTE user devices such as envelope tracking and predistortion and postdistortion produce digital noise, which can result in a loss of sensitivity.

4.11.2 ACLR and EVM Performance in Refarmed Spectrum

There is no point in increasing power output in a user's device if this causes problems in refarmed spectrum. This will require 5, 10, 15, or 20 MHz and potentially 100-MHz channels to be deployed in spectral and geographic proximity to 200-kHz GSM or EDGE channel bandwidth. This implies a need to optimize adjacent channel leakage ratio (ACLR) and error vector magnitude (EVM) performance.

4.11.3 Heat Gain

There is no point in increasing power output in a user's device if the heat rise in the device becomes unacceptable or if the duty cycle is reduced. The impact of all the above is reasonably profound. RF power amplifier manufacturers have become used to their customers asking for good efficiency at low output power and or across a wide range of output powers.

Even if low-power operation is not used in LTE, the LTE signal itself is a challenge to RF power amplifier designers. The selected signal has a peak power that exceeds the average (information-useful) power by 12 times. For the same average transmitter power (the same communication range) the LTE power amplifier must handle this peak power cleanly.

For 23-dBm (200-mW) average power, the PA must be designed to support 34 dBm (2.4W). Alternatively, a smaller and lower-cost power amplifier can be used, but the average output power must decrease, reducing the communication range. This is the reason for maximum power relaxation (MPR) provisions in the LTE (and HSPA) specifications. Furthermore, the complex LTE signal is not as tolerant of signal distortion as UMTS/HSPA signals, requiring the EVM specification to be tighter (raising costs).

4.12 Managing Intersystem Interference at the System Level

On balance and taking the above caveats into account, it can be generally stated that intersystem and intercell and intracell interference can and should be managed at the system level to allow LTE user devices to be run at or close to their maximum power level. This is because mobile networks and in particular mobile broadband networks are power-limited not bandwidth-limited. This begs the question as to whether operators should bid for new meter-band spectrum or concentrate on achieving better results from what they already have.

Improving the efficiency of existing spectrum will almost certainly result in a better return on investment (ROI) and Earnings Before Interest Tax and Depreciation (EBITDA) assuming the alternative is adding new bandwidth to an existing legacy band plan. On this basis the downlink and uplink performance of the user or IOT device makes a huge impact on the economic viability of the delivery network. Relaxing total isotropic sensitivity (TIS) and total radiated power (TRP) specifications to accommodate new bands, thereby reducing performance in existing bands, is therefore a bad idea.

If we take standard devices, by which we mean devices that support five or at most six bands, handset vendors are now suggesting that RF power amplifiers will be optimized to work somewhere between 19 and 23 dBm. Even this is a big spread with significant fiscal consequences for the operator community. A 3-dB backoff in user equipment peak power output results in an increase of

45% in the number of base stations required to cover a given geographical area with a given service level. Expressed in terms of constant base station numbers, this equates to a 32% reduction in coverage area.

Standards bodies are suggesting that maximum power reduction (MPR) should be considered where linearity requirements and or ACLR or EVM targets are hard to achieve. MPR is where a reduction of the UE maximum output power in the conformance specification is agreed. Additional MPR (A-MPR) allows for adaptive relaxation in certain operational conditions signaled by the network. As a consequence, user equipment designed to an MPR specification can be potentially 2.5 dB down on its originally specified output power and may be coupling with an antenna with a gain of the order of –7 or –8 dB.

This performance spread has to be considered not in terms of user experience expectations as they are today but in terms of user expectations in the future and the network and service value realizable from meeting those expectations. The LTE peak uplink data rate of 50 Mbps is specified to be an order of magnitude greater than the 5 Mbps promised by High Speed Uplink Packet Access (HSUPA). Expectations of available rates at cell edge may also be higher than presently assumed and may have a disproportionate impact on the quality of the user experience, both actual and perceived.

This is not dissociated with the concept that user value may be more uplink biased than presently assumed, that markets are cost-sensitive both in terms of user equipment costs and network costs and that many markets combine extreme urban density with wide open deep rural spaces, both of which may be subject to different but related uplink power constraints, for example, building penetration in dense urban environments and distance in rural applications.

4.13 The Economics of Increasing Network Density

One way of solving the coverage issue is to increase network density. One way of solving capacity issues that increase as coverage increases is to bid for more spectrum. Both increase capital and operational costs. These costs are directly a function of the RF link budget closely coupled to multiplexing and admission control techniques. Getting more performance from existing spectrum and existing networks improves ROI and EBITDA.

Buying more spectrum and or increasing network density reduces ROI and EBITDA. In terms of the impact of user equipment performance on network efficiency, the focus has generally been on receiver performance, determined by the selectivity, sensitivity, and dynamic range of the device. RF transmitter performance is at least as important. Every decibel lost on the transmit path translates into reduced data reach, reduced data capacity and lower user experience scores which in turn translate into revenue loss, churn and or higher retention budgets. The need to combine an ability to transmit at maximum power while maintaining ACLR and EVM performance is critical particularly when implementing LTE networks into refarmed spectrum.

We have possibly just moved the problem somewhere else. One consequence of reducing power control-related signaling load on the wide-area radio interface is that the eNode B base stations will need to do more interference coordination and as a result the backhaul signaling load may increase though if the scheduler is allowed to do the heavy lifting this should not be a major problem.

An additional counterargument is that a 3- or 4-dB gain in the link budget, irrespective of whether it is realized on the wide-area access transmit or receive path will reduce the number of point-to-point backhaul links, although not the bandwidth needed. Intuitively, given that reducing signaling load on the radio interface increases the ratio of income generating bits to overhead bits (what our Wi-Fi colleagues call "good put"), then our hunch is that this is probably where a useful net gain can be achieved.

4.14 The Impact of RF Efficiency on Network Economics

To summarize, the 4G physical layer is efficient at delivering in-band spectral efficiency, although block-to-block and band-to-band emission issues result in some of this efficiency being lost due to insertion loss. The 4G physical layer is not particularly power efficient. PA efficiency can be improved by backing off the PA. This has been formalized through the MPR and A-MPR work items that are now embedded in the conformance test process. This results in a reduction in data reach, particularly at the edge of a cell.

Improved intercell interference mitigation has meant that devices are able to transmit at higher power at the cell edge without causing significant loss of throughput in the adjacent cell. Running user device amplifiers close to their rated maximum power makes them more efficient and improves scheduling efficiency. The devices have to be kept linear in the presence of high peak to average power ratios.

Various techniques have been introduced to improve this compromise between output power and linearity including envelope tracking and predistortion and postdistortion. These techniques generate some digital noise which can result in a loss of receive sensitivity. This needs to be carefully managed.

The linearization and system efficiency gain techniques also tend to be bandwidth-limited and therefore work less well across wider channel bandwidths. The shift from voice traffic to low bit rate data (GPRS and EDGE) to bursty high data traffic and the physical layer evolution that has facilitated that shift has significantly increased the amount of power drawn by the RF power amplifier relative to other traditionally power hungry functions such as the screen and applications processor, the LTE transceiver and modem, and Wi-Fi. This shift in relative power requirement is shown in Figure 4.2.

Managing interference by reducing TX output power has an associated cost in terms of reduced data reach and edge of cell performance. This is particularly important in sparse rural areas but not unimportant in denser network applications.

High power requirements are hard to accommodate in small form factor devices both in terms of battery performance and heat rise. This suggests that improved power efficiency and heat management may be a critical requirement in 5G radios.

4.15 Summary

In all radio systems, reducing out-of-band and out-of-block emission will have a negative impact on in-band performance. This is particularly true when wideband channels are physically and spectrally proximate to narrower band channels.

In the meter band below 1 GHz, intersystem coexistence costs have been largely dominated by mobile broadband to terrestrial TV interference or the

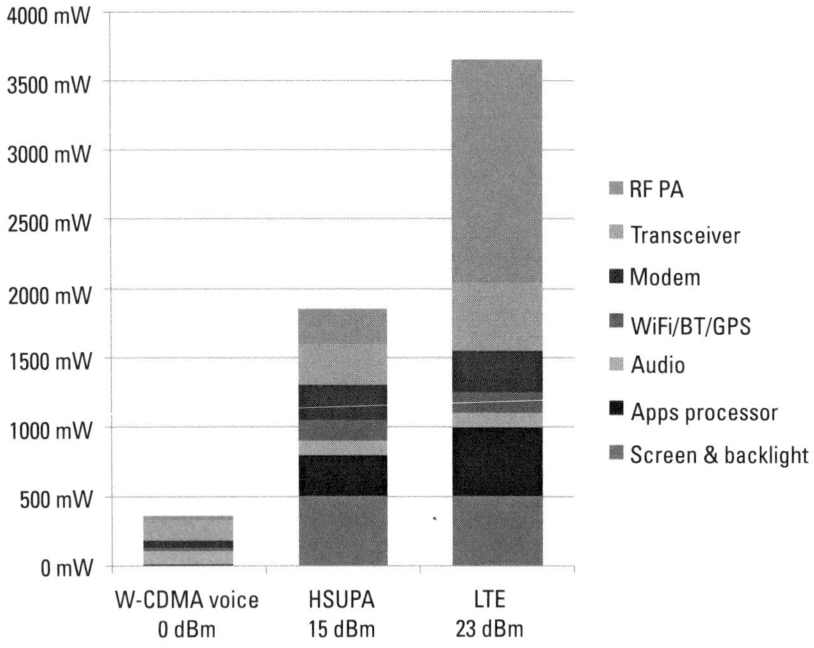

Figure 4.2 RF Power requirements for LTE relative to other functions. (Courtesy of Nujira.)

perceived risk of interference. Intrasystem coexistence costs have largely been defined by the need to manage operator-to-operator coexistence.

Between 1 and 2 GHz, operator-to-operator coexistence, for example, in the 1,800- and 1,900-MHz bands is technically and commercially important. Intersystem costs include the need to coexist with sensitive, easily jammed, receive-only, safety-critical signals from the GPS, Galileo, and Glonass and Beidou satellite constellations at 1,575 MHz. The addition into low-cost consumer devices of a second frequency (L2) at 1,227.6 MHz and a third frequency (L5) at 1,176.45 MHz to correct for ionospheric and atmospheric distortion potentially introduces additional coexistence issues. L-band satellite uplinks and downlinks also need to be accommodated.

Between 2 and 3 GHz, intrasystem coexistence cost includes LTE FDD/TDD coexistence, and intersystem costs include Wi-Fi/LTE coexistence, LTE, aviation and weather radar, and S-band satellite systems. Between 3 and 4 GHz intrasystem coexistence issues include FDD to TDD LTE and satellite broadcast downlinks particularly in Sub-Saharan Africa and parts of Asia.

Assuming that 5G is deployed into the centimeter and millimeter bands, the intersystem coexistence costs are likely to be determined by satellite and 5G spectral adjacency. Coexistence with automotive radar may also be an issue but could potentially be translated into a technology and commercial benefit. These issues are separately revisited in Chapter 8, 9, and 10.

References

[1] http://stakeholders.ofcom.org.uk/consultations/combined-award/update.

[2] http://www.3gpp.org/dynareport/36-series.htm.

[3] www.aptsec.org/.

[4] Digital Television Group.

[5] http://ec.europa.eu/enterprise/sectors/rtte/documents/legislation/review/index_en.htm.

[6] https://connect.innovateuk.org/web/spectrum.

[7] http://450alliance.org/.

[8] http://www.navipedia.net/index.php/GPS_Signal_Plan.

5

Allocation and Auction Economics: Theory and Practice

5.1 Forty Years of Cellular Spectrum Allocation

In 1977, the U.S. Federal Communication Commission released radio spectrum at 850 MHz for cellular trials. A working network was installed by AT&T (Bell Illinois) in Chicago. This became the basis for the American Advanced Mobile Phone System (AMPS) rollout in 1983. Networks were also deployed in Scandinavia, originally at 450 MHz, and in Japan.

In 1982, the U.K. government announced that two licenses would be offered to support mobile phone network services at 900 MHz. One license was awarded to the national operator British Telecom in partnership with Securicor. The second license was opened up for competitive bids and was won by Vodafone, a joint venture between Racal Electronics and Millicom. The bid process was structured to cover administrative costs calculated at £25,000. The primary purpose of the process was to introduce competition into the U.K. mobile phone market.

The original 15+15 MHz of spectrum was extended in 1986 with additional spectrum released by the U.K. Ministry of Defence [1]. Racal shareholders were at the time nervous about the buildout commitments. These had a major impact on the profitability of the company in 1985 and the share price. The subsequent financial success of Vodafone was at least partly due to a technically well-executed network, good marketing, and year-by-year improvements in user devices including size and weight, cost, and battery life, a trend that also helped other operators [2].

The use of radio spectrum as a way of introducing market competition is still a regulatory objective in many although not all sovereign countries, but the value now placed on spectrum is astronomically higher.

The U.S. AWS3 auction in 2014, as an example, attracted bids of $44 billion, with a significant obligation cost including a $5 billion budget to manage and mitigate interference with incumbent federal users [3]. Regulators are also increasing annual fees. Table 5.1 shows the latest fee structure in the United Kingdom [4]. The annual cost of the 900-MHz band is now £80.3 million and £119.3 million for the 1,800-MHz band.

The U.S. Federal Communications Commission (FCC) started auctioning spectrum in 1994, beginning a process that has been spectacularly successful at generating income for the U.S. government and sovereign nations around the world. The underlying theory is that the market pricing mechanism of a well-structured auction process guarantees that spectrum is valued efficiently and used efficiently.

The rest of the world has followed the U.S. market-led approach with varying degrees of enthusiasm. Japan, for example, simply issued 3G licenses to existing telecoms operators. Other governments have used "beauty contests" where licenses are awarded to interested parties on the basis of agreed-upon criteria determined by the government. In some cases the award process has not been transparent and the potential social and economic benefits have been less than fully realized.

5.2 Auctions and Bid Value

Auctions allocate licenses to the applicants who bid more than their competition. This means that the licenses end up with the bidders who place the highest value on the spectrum but this does not necessarily mean that the bidders have paid "true value" for the spectrum, as by definition, this is unknown at the time of the auction. Over the past 20 years this has resulted in what is known as "the winner's curse" where the liabilities of the spectrum exceed realizable value. At some point, this has to be reflected as an asset writedown.

Table 5.1
Current and Revised Annual License Fees for 900-MHz and 1,800-MHz Spectrum

£ million	Vodafone	Telefonica	EE	H3G	Total
Current	15.6	15.6	24.9	8.3	64.4
Proposed February 2015	62.6	62.6	77.3	25.8	228.3
Final decision	49.8	49.8	75.0	25.0	199.6

Source: [4].

The European 3G auctions between 2000 and 2002 provide an example. The value assumption was that 3G networks, deployed into new dedicated spectrum (Band 1 at 1.9 and 2.1 GHz), would increase average revenue per user and improve profit margins and return on investment.

Auction theorists were employed to design the auctions in order to maximize realized income. The five "winners" in the U.K. auction (April 2000) bid $35 billion, the six winners in Germany in August 2000 bid $46 billion, the Netherlands, Italy, Austria, and Switzerland netted another $14 billion, and with some rounding and bid expenses the total came to $100 billion. For the United Kingdom this equated to nearly $600 for every man, woman, and child. The equivalent in Switzerland was less than $15 per capita.

This huge variation is explained by local market conditions, auction timing, and the different ways in which the auctions were designed in terms of reserve pricing and bidding process. In the United Kingdom there was a guarantee that new bidders would win a license. The response from incumbents was to bid aggressively to ensure that those licenses would be relatively expensive. The Netherlands, in contrast, allowed joint bidding. As there were five licenses available and five incumbents bidding, it made sense for new entrants to partner with the stronger incumbents. There was only one new bidder. They did not last long and this helped to minimize price escalation.

Auctions can be structured as common value auctions, ascending price auctions, reserve auctions, incentive auctions, and/or reverse auctions. In a common value auction, the spectrum has the same value for all bidders, but the value is unknown. Operator bid teams estimate the true value but do not reveal their estimates. Some bid teams will overvalue the spectrum, and some will undervalue the spectrum. The winner's curse is most likely to occur in a sealed bid highest price common value auction or a second price common value auction in which the bidder who submitted the highest bid pays a price equal to the second highest bid amount. The two top bidders share the winner's curse.

However, spectrum auctions are not really common value because some operators will be better positioned to realize return on the spectral investment. For example, an incumbent operator with existing sites and backhaul and core network assets, existing customers, and billing and IT systems will be better placed than a new operator with minimal existing technical and commercial assets.

5.3 Auctions, Spectral Asset Value, and Obligation Costs

Most 3G auctions were ascending price auctions. In ascending price auctions the rival bids are known and the bidding continues until bidders drop out. The winner's curse still applies. The share price of the two largest European operators fell after they won bids in the German and U.K. auctions.

Partly this was due to the large increase in debt needed to pay for the spectrum, the network and new service offers. The European operator who won spectrum in both the U.K. and German auctions increased debt level by an order of magnitude between 1999 and 2000 and a downgrade from Moody's and S&P made servicing the debt more expensive. Return on assets also went negative; having to subsidize expensive handsets made things worse.

In 2003, one of the winners of the U.K. auction wrote down the asset value of the spectrum acquired 2 years before by 50%.

On November 19, 2007, the FCC invited bids for the 700-MHz band. There was nothing particularly new or unusual in the auction. Over 12 months, it raised $20 billion for the U.S. Treasury.

The band plan was complex and perilously close to the Advanced Television Systems Committee (ATSC) digital TV terrestrial multiplex at the lower end of the band. This has already been referenced in the earlier chapters on band fragmentation and coexistence, but as a reminder, the lower end of Band 12 is immediately adjacent to Channel 51. Band 17 was specified as a subband within Band 12 to allow for a guard band. AT&T bid and won a nationwide Band 17 footprint. Verizon similarly bid and won a nationwide Band 13 footprint. Band 13 has coexistence issues with Band 14.

The hidden cost of the band plan was the lack of availability of handsets for Band 12, which effectively invalidated the business model for the bidders who won this block of spectrum. The reason was that adding the filtering required due to the lack of a guard band added cost and performance loss to the handset and neither AT&T nor Verizon were inclined or motivated to tolerate this.

Also, the best way to optimize performance and minimize device cost in Band 17 and 13 was to support one or other of the bands in the handset. This meant that users were either locked into one of the operators or needed two phones. Thus, while the auction was structured to encourage competition, the auction process and band plan had the opposite effect. The winning bidders for Band 12 paid for the spectrum but 8 years later still had problems sourcing handsets.

This highlights an overall trend. As the realized value of auctioned spectrum has gone up over the past 20 years, the quality of the spectrum brought to market has gone down. The degradation in technical quality is a consequence of coexistence issues and/or spectral fragmentation.

This is not to say that the original spectrum allocations at 800 and 900 MHz did not have coexistence issues. They did and still do today (see Chapter 9), but after 30 years the issues have been largely resolved technically and commercially. Any new band allocation today needs to compete with 30 years of research and development and optimization investment in those original allocations.

Similarly, any new spectrum auctioned today needs to compete with 20 years of legacy-auctioned spectrum amortization and related research and development and network spending.

Europe is presently grappling with similar 700-MHz auction issues. The German regulator Bundesnetzagentur (B-Netz) [5] allowed bids from Telefonica Germany, Telekom Deutschland, and Vodafone for 2 × 30 MHz [the APT (a) spectrum] from 703 to 733 MHz and 758 to 788 MHz.

We have already documented the differences in out-of-band emission for this spectrum compared to devices developed for Asia Pacific Telecommunity markets. This introduces supply chain uncertainty about the cost and performance of user devices supporting this spectrum.

There are also technical and commercial coexistence risks associated with the repacking of TV channels into the lower end of the UHF band. This is predicated on a transition to DVB-T2, which may make TV receivers more vulnerable rather than less vulnerable to interference from LTE user devices.

The blocks were auctioned as a package with spectrum in the 900-MHz and 1,800-MHz and 1.5-GHz bands for fixed and mobile communications. After a slow start and after 16 days and 181 rounds of bidding, the auction raised 5.8 billion euros. Vodafone Germany was the biggest bidder at 2.1 billion euros. Deutsche Telekom spent 1.8 billion euros and Telefonica Deutschland spent 1.2 billion euros

Vodafone acquired 25+25 MHz in the 1,800-MHz band. This increased the price of the whole auction, which raised 16% more than the 800-MHz auction in 2010 but much less than the German 3G auction, which had raised over 50 billion euros. The 700-MHz spectrum will only become available as the TV licenses expire.

In France, the proposals of the French regulator Autorité de Régulation des Communications Électroniques et des Postes (ARCEP) [6] have been accepted as the basis for an auction process expected to last from the end of 2015 to July 2019 in line with the European Commission's *Lamy Report*, which recommended that 700-MHz spectrum allocation should take place across Europe by 2020, plus or minus 2 years [7].

The priorities of the auction were stated as "monetizing intangible state assets, regional development, promoting investment, and preserving fair and effective competition, to improve 4G coverage in rural areas and prepare for the potential development of 5G services in those bands, provide mobile data availability on board every day trains, high speed railways and underground lines." The roll-out requirements include an obligation to provide broadband coverage to at least 98% of households nationwide with an average transmission rate of at least 10 Mbps with mobile broadband coverage available along all national motorways.

The 30+30 MHz is divided into six blocks of 2 × 5 MHz. No single candidate can acquire more than three blocks in total and operators will not be able to hold more than 2 × 30 MHz of sub-1-GHz spectrum across 700, 800, and 900 MHz combined.

Free, the fourth entrant into the French 3G mobile market, petitioned ARCEP to reserve a block of frequencies to allow it to compete with the existing incumbents, Orange, Bouygues Telecom, and Numericable-SFR [8]. This was argued on the basis of resolving the difference in realizable value between the incumbents and the new player and followed a successful price-led marketing campaign by Free using the Orange network.

ARCEP said no, but it illustrates how the competitive landscape in Europe has changed. The risk is that too much competition lowers average revenue per user (ARPU) to a level at which spectral and network investment becomes unsustainable. In parallel, the European Commission has needed to respond to the regulatory implications of companies that offer mobile and fixed broadband, cable, TV, and telephone service bundled together. Numericable-SFR is just one of many new businesses formed by merging an incumbent with a relatively new player from another sector, in this case a cable provider. The assumption is that this will support a more comprehensive service offer at a lower price at a higher margin.

5.4 Spectrum for Public Protection and Disaster Relief

The 2 × 8 MHz of spectrum is also being reserved for public protection and disaster relief (PPDR). This brings us back to the United States. The problem with allocating spectrum for public safety radio is that you have to get different agencies to agree to what they want and need or think they want and need and then find enough money to build and operate the network.

In Europe the police, fire, and ambulance services are supported on either narrowband TETRA networks or (in France) narrower band Tetrapol networks either privately or publicly owned. The emergency services in the United States, or first responders as they are known, are supported on more than 10,000 separate private Land Mobile Radio systems, many of them incompatible with one another, or P25 narrowband analog or digital radios (more detail in Chapter 8). This incompatibility was highlighted as a problem during and after the terrorist attacks of 9/11 in September 2001 and substantially increased political awareness of the importance of a unified mobile communications infrastructure.

As part of the 2007 U.S. 700-MHz auction process, it was proposed that a public-private partnership would take on the task of building a nationwide interoperable public safety network based on a 5+5 MHz block of spectrum (763–768 and 793–798 MHz) with an adjacent 5+5 MHz D block allocation (758–763 and 788–793 MHz) for commercial auction with the winning bidder required to develop a shared wireless broadband network.

The auction was held in early 2008 and failed to produce a viable bid. In May 2010 the FCC adopted a waiver order granting 21 public safety jurisdictions to "pursue early deployment of state wide or regional public safety broadband networks." An additional waiver order determined that these networks should adopt 3GPP Release 8 (LTE) or higher as a common technology platform. This signaled a reverse from a previous policy of awarding spectrum on a technology-neutral basis. On February 22, 2012, the U.S. Congress enabled the Middle Class Tax Relief and Job Recovery Act, which allocated D Block spectrum as a 5 × 5 MHz passband within Band 14 [9].

The legislation directed the FCC to use the spectrum for a public safety nationwide broadband network to be run by a newly created entity known as the First Responder Network Authority (FirstNet).

There is a 6-MHz guard band separating the 5 × 5 MHz passband from narrowband spectrum allocated for 6.25-kHz channels. The problematic adjacency between the Band 13 passband and D Block is shown in Table 5.1 The passband edge of mobile transmit Band 13 at 787 MHz is immediately adjacent to the passband edge of Band 14 at 788 MHz and passband edge of Band 13 mobile receive at 756 MHz is only 2 MHz away from D block base station transmit.

5.5 Financing PPDR Spectrum and Network Rollout

Putting the technical challenges to one side, the commercial challenge was to produce some money to build the network. The 2012 Spectrum Act determined that $7 billion should be allocated to FirstNet to fund network construc-

Table 5.2
D Block Spectrum Allocation within Band 14

Band	Name MHz	Uplink		Downlink		Bandwidth MHz	Duplex Spacing	Guard Band	Area
FDD									
13 (RD)	700 c	777	787	746	756	10+10	–31	21	NAR
LTE	D Block	788	793	758	763		–31	6	NAR
Narrow Band		799	805	769	775				NAR
14 (RD)	700	788	798	758	768	10+10	–30	20	NAR
2007 public safety		793	798	763	768				

tion. This was to be raised from the auction of H Block of paired spectrum at 1,915 to 1,920 MHz and 1,995 to 2,000 MHz, the AWS 3 auction, and the 600-MHz incentive auction.

This proved remarkably easy to achieve with $1.5 billion raised from the H Block auction from Dish Networks who are pairing the spectrum with their AWS4 holdings at 2,000 to 2,020 MHz and 2,180 to 2,200 MHz and their unpaired lower 700-MHz E block spectrum (for LTE Broadcast). This was followed by the $44 billion raised from the AWS3 auction.

The money is there, but defining how the 60,000 public safety agencies are going to use the FirstNet network is going to take longer with a 46-step process defined for each of the 50 states and six territories involved including Puerto Rico, Guam, and the Virgin Islands. Given that the financing for the public safety network is already covered from the AWS 3 auction, the income from the 600-MHz incentive auction could be used for other purposes.

The incentive auction is now proposed to be held in 2016. It was originally proposed in 2010 in the National Broadband Plan [10]. It is structured as a reverse auction also known as a descending clock auction, which establishes the price at which the broadcasters are willing to relinquish their spectrum in exchange for a share of the proceeds from the auction of the new licenses sold in an ascending clock forward auction.

The assumption is that the broadcasters will be willing and able to repack their existing channels into the lower end of the ultrahigh frequency (UHF) band divided into Channels 14 to 36 between 470 and 608 MHz, Channel 37 between 608 and 614 MHz, and Channels 38 to 51 between 614 and 698 MHz. The repacking process is being overseen by the Office of Engineering and Technology based on study work on the impact of repacking on TV coverage [11].

The present FCC plan is to set aside 30 MHz of spectrum for mobile broadband in each market to carriers with less than 45 MHz of low-band spectrum. This would prevent AT&T and Verizon bidding on that spectrum in a number of major cities unless specific price thresholds fail to be achieved.

The spectrum will be allocated as frequency division duplex (FDD) spectrum with a duplex gap that could be used by TV stations and/or for unlicensed white-space radio including Wi-Fi or other unlicensed users or for licensed broadcast news microphones, the equivalent of PMSE in Europe. The overall intention is achieve a target of at least 84 MHz as the breakpoint for the auction. If this target is exceeded, this would provide more opportunity to support unlicensed usage.

5.6 Allowance for Spectrum Impairment

There is recognition within the process that at least 20% of the spectrum will be impaired by interference predominantly from broadcasters with an ongoing discussion as to whether the reserved spectrum should specifically be unimpaired.

The FCC is presently suggesting a dynamic reserve pricing system that arbitrates between what broadcasters think their spectrum is worth and what mobile broadband operators are willing to pay set against what spectrum is available for repacking on a market-by-market basis. Dynamic pricing will also need to comprehend the level of impairment per market.

The level of interference is going to be dependent on the number of mobile broadband users, which is dependent on pricing, which is dependent on the level of interference, so it is hard to see how this part of the process is going to work.

It is relatively easy to estimate interferences from and to base stations because you know what they are in terms of power output and antenna gain, where they are, and more or less what they will be doing in terms of daily traffic throughput. It is harder to estimate interferences from and to user devices because you do not know where they are going to be or what they are going to be doing. The safe option is therefore to plan on the basis of worst-case interference. This results in commercially unsustainable impairment costs.

The purpose of licensing and regulation is to manage this interference and to factor interference into spectral value calculations. The purpose of licensing and regulation is also to achieve a market price that provides an immediate return to the government but with a sufficient realizable margin for an operator to deploy a commercially sustainable network that complies with preagreed-upon coverage and service obligations.

This market price varies by country and by region within countries. In large countries such as the United States, there are significant differences in market condition from state to state and local regulations that have to be complied with.

Television broadcasting is based on a high-power, high-tower transmission model, hundreds of kilowatts from towers that can be hundreds of meters high [12]. Temperature inversions can create ducting conditions which allow TV signals to travel hundreds of kilometers. Regulations therefore have to be on a cross-border and/or interstate basis. The same requirements apply to satellite TV.

Worst-case interference modeling can therefore result in spectrum lying fallow that could be used most of the time without inflicting system interference to white-space spectrum. These cross-state interference issues are directly analogous to the cross-border interference, which occurs in Europe. Radio signals take no notice of national or state boundaries.

5.7 White-Space Spectrum: Licensed Versus Unlicensed Spectrum Asset Value

White-space spectrum exists technically wherever spectrum is unused or underused including spectrum that is only used at certain times in certain places or at random times at random places.

White-space spectrum includes guard bands, duplex-spaced bands, or existing allocated but underutilized radio system channels. These existing radio systems are already licensed or allocated for particular users, for example, military radio. White space is promoted by some agencies as a no-cost or low-cost option for providing Internet access in developing economies and is sometimes described as dynamic spectrum allocation or dynamic frequency allocation [13].

The deployment options are either to use interference detection, for example, "polite" devices that determine whether bandwidth is available prior to transmission or database-based allocation where a central database is continuously updated with channel availability or a combination of both.

The problem with device-based interference detection is that devices cannot always see potential victim receivers. This is known as the hidden node problem. The problem with databases is keeping them up to date and getting permission information from the database out to the devices in a network.

Even when well implemented, dynamic frequency allocation of white-space spectrum carries some residual risk of interference. The success of Wi-Fi technically and commercially proves that the risk is minimal and acceptable in low power local systems. The problem is how to scale white space to high-power wide-area radio.

From an auction theory perspective, it is particularly problematic to package an auction of licensed spectrum with an anticipated multibillion-dollar valuation with an obligation to allow "free" unlicensed access to the same spectrum or immediately adjacent spectrum, unlicensed white-space Wi-Fi in a guard band/or duplex gap being a specific example.

There are credible technical and market offers making progress in the use of white space and the sub-1-GHz ISM bands for low data rate low power budget Internet of Things (IoT) connectivity. We look at these in more detail in Chapter 8 [14].

5.8 Lightly Licensed Spectrum

Lightly licensed spectrum provides a third alternative. Light licensing is already extensively used in point-to-point radio. A company or individual wishing to install a point-to-point radio link applies to the regulator who checks that the link will not cause interference to other existing geographically proximate radio links. The cost of the license covers the administration cost of processing the request and is generally inconsequential relative to the value of the radio link to the user. The problem with light licensing is that it only works effectively with fixed radio systems.

For mobile broadband, including 5G mobile broadband, the debate therefore essentially revolves around how to manage the technical and commercial coexistence of licensed and unlicensed spectrum. This, in turn, is directly coupled with regulatory policy, which is directly coupled to competition policy, which is directly coupled to social, economic, and political policy. The complexity of this process becomes evident when you consider the range of industries and interests with which 4G and 5G mobile broadband need to work.

5.9 Competing Industry Interests

So far, we have referenced mobile broadband and terrestrial and satellite radio and TV, mobile broadband and wireless microphone and outside broadcasting, mobile broadband and public safety and mobile radio, mobile broadband and the defense community, mobile broadband and the LEO, MEO, and GSO satellite industry, mobile broadband and the subspace industry (drones and balloons), and mobile broadband and the radar industry (weather radar, aviation radar, automotive radar and military radar). We also need to add radio astronomy to the list.

All of these will be discussed as we work our way through Chapters 8 to 11, but the list in its own right highlights the regulatory challenge of industry and market convergence and an inherent uncertainty as to specific 5G regulatory challenges.

To date, mobile broadband spectral value has been, as you would expect, substantially determined by the willingness and ability of mobile operators to bid for spectrum. Mobile operators have always had cross industry shareholdings. Racal Electronics, the original Vodafone parent company, was a defense electronics business.

More recently, the biggest joint ventures have been cable companies and satellite TV companies merging with operators. This has been motivated by stock market sentiment. Shares in the U.S. T-Mobile business doubled in value after their merger with Metro PCS in 2013. The stock market value of Dish Networks went up by 86% over the same period.

AT&T acquired Direct TV, a major competitor of Dish Networks; Dish tried to buy Clearwire and Sprint but were outbid by Softbank of Japan and are currently developing their U.S. quad play (broadband, TV, home phone, and mobile phone) market offer with T-Mobile, which AT&T was attempting to acquire.

The spectrum allocation and auction process therefore has a profound negative or positive impact on the share values and debt ratios of mobile operators, cable operators, commercial TV and radio broadcasters, satellite operators, and defense contractors.

Conversely, the share value and debt ratios of mobile operators, cable operators, commercial TV and radio broadcasters, satellite operators, and defense contractors has a profound negative or positive impact on spectral value. Share values can rise or fall by large amounts very quickly; a 20% shift in a day can add or subtract billions of dollars of equity value. Spectral value in comparison is determined by the original price paid and any subsequent writedowns or revaluation. Revaluation can be realized through a merger or acquisition or outright transfer of spectrum if allowed in the license conditions.

5.10 Mobile Operators and Spectral and Market Asset Value

The positioning of existing mobile operators in the 5G bidding process, assuming such a process takes place, will be determined by how much money is made from existing 4G, 3G, and 2G networks over the next 3 to 5 years, their market position and regulatory policy including competition policy and antitrust legislation.

So far, the uptake of LTE has been encouraging when compared to 3G W-CDMA (see Figure 5.1) [15].

There are very large differences between regions and countries, with early adoption being particularly strong in China, Korea, and Japan (see Figures 5.2, 5.3, and 5.4).

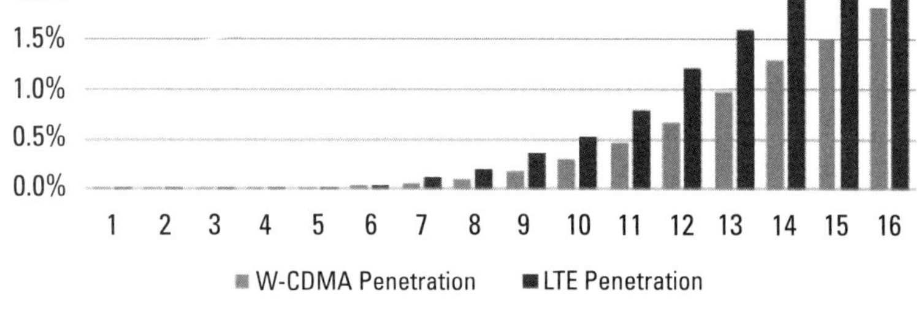

Figure 5.1 W-CDMA and LTE penetration, global market, first five years. (Courtesy of The Mobile World.)

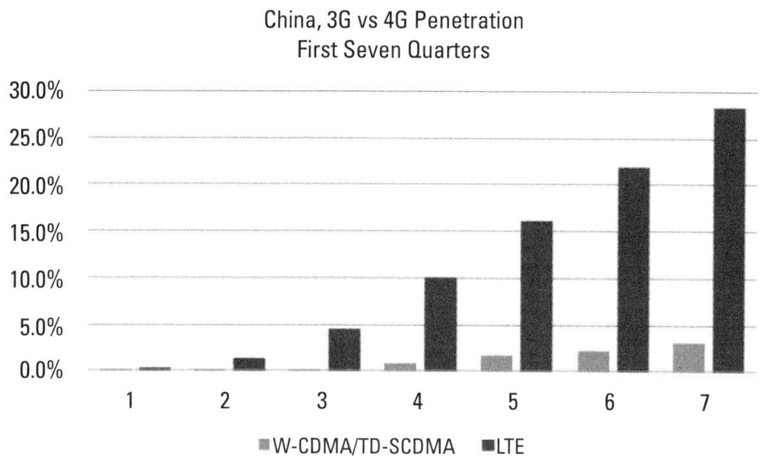

Figure 5.2 China, 3G and 4G penetration, first seven quarters.

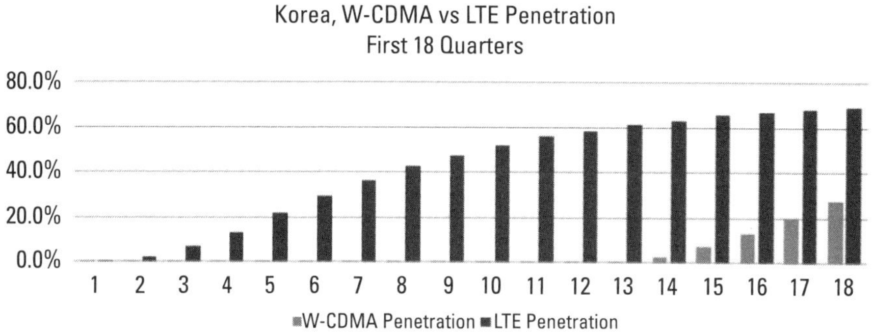

Figure 5.3 Korea LTE penetration, first 18 quarters.

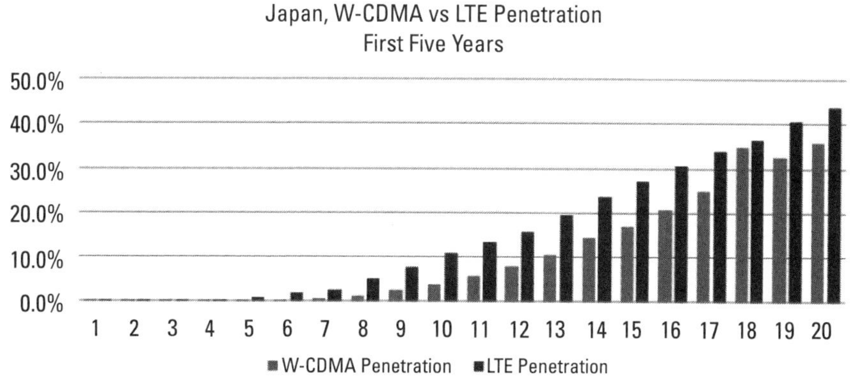

Figure 5.4 Japan, W-CDMA and LTE penetration, first five years.

5.11 Impact of Over-the-Top (OTT) Services

It might be expected that this relatively rapid transition would have a positive impact on stock market sentiment but there are two parallel offsetting trends. The first trend is the growth of over-the-top services where the added value of a service is captured and retained by a third party. Examples include Google, Facebook, and Amazon.

The second trend is the value that user device/smart phone vendors, particularly the two largest vendors, Apple and Samsung, have succeeded in capturing. This value has been realized from retail and operator subsidies including subsidized joint marketing campaigns. The cost of not range-listing the latest iPhone or Samsung Galaxy has made it hard to escape this subsidy model, which has resulted in Apple now having a cash mountain of over $200 billion.

Figure 5.5 shows how the capitalization of Apple and Samsung relative to mobile phone operators has grown since the Apple iPhone was introduced in 2007 and the impact this has had on the relative capitalization value.

This trend is made worse by the number of operators, over 600 globally, that have to compete with each other for subscriber revenue.

This extraordinary number of operators can be directly ascribed to regulatory and competition policy and auction policy, particularly the focus on auctions that are structured to result in at least five operators servicing a single market or country. This compares with a highly consolidated supply chain with only two dominant user device vendors, three scale-viable radio infrastructure vendors, probably at most two scale-viable baseband chip set vendors with one of those 10 times the size of the other and only four or five globally viable RF component vendors.

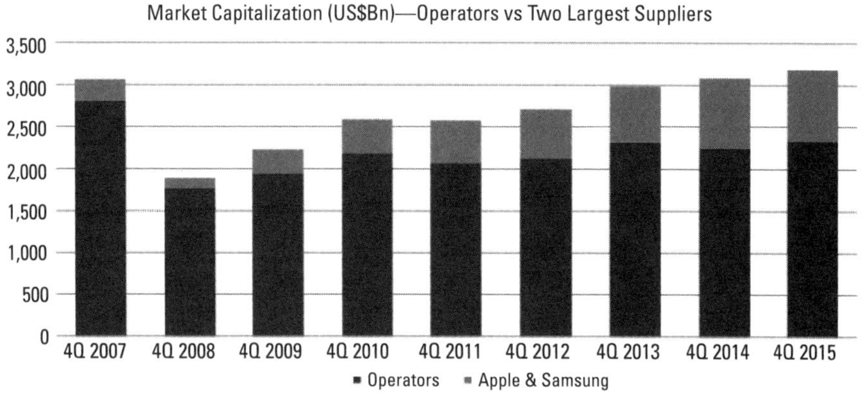

Figure 5.5 Operator market capitalization compared to the two largest suppliers, 2007 to 2015.

Figure 5.6 shows the top 20 operators worldwide by connection showing the numerical dominance of China Mobile. Many of the operators in the "others" segment are subscale and at best marginally profitable.

Figure 5.7 shows the country comparisons by connection showing the importance of the United States and China.

In terms of operator market share, market share by value shows a different picture with AT&T and Verizon at number 2 and 3 behind China Mobile. This is because the U.S. market represents 4% of the global market by volume but 10% by value, with the value figure remaining remarkably constant over

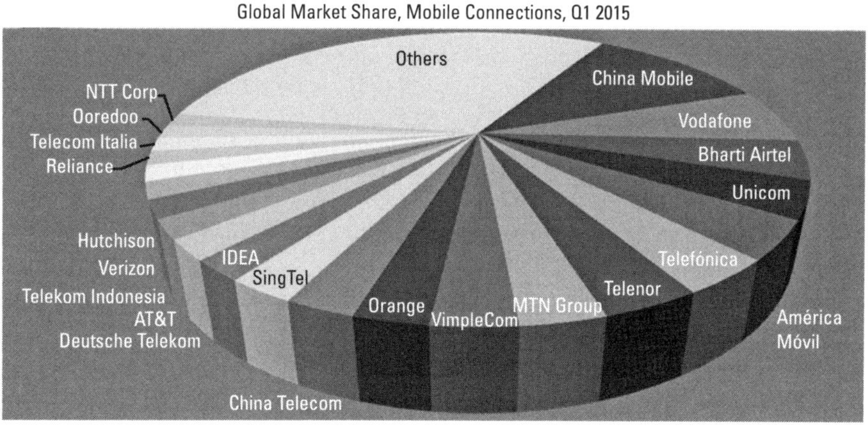

Figure 5.6 Global market share, mobile connections, Q2 2015.

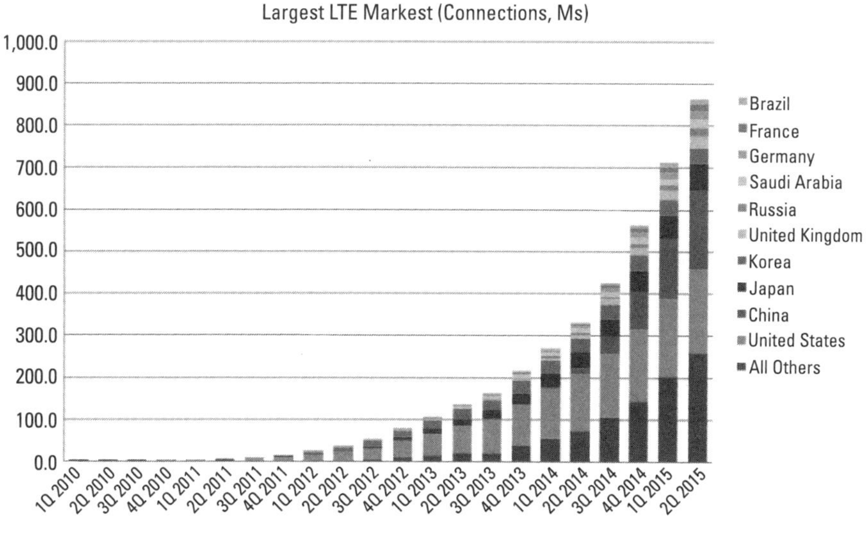

Figure 5.7 Largest LTE markets (connections, millions of subscribers).

the past 10 years. This is because the United States remains one of the highest ARPU markets in the world.

5.12 Commercially Efficient Spectrum

You might question how this could happen in a country where the auction process since 1994 has been ostensibly structured to encourage competition based on the market entry of new operators.

One answer we would suggest is that the two incumbent operators have managed to retain and/or acquire the most technically efficient and commercially efficient spectrum. This is both in terms of sub-1-GHz spectrum ownership and possibly even more crucially the quality of the blocks that they own or have acquired in each band. The most extreme example is the 700-MHz band where Bands 13 (Verizon) and 17 (AT&T) have been significantly easier than Band 12 from an RF component and user device price/performance perspective. The leverage of existing and dominant market volume on supply chain research and development allocation has also been crucial. Simply put, AT&T and Verizon always come top of most RF design team priority lists while their competitors suffer from the orphan band/orphan block orphan spectrum problem.

5.13 Orphan Spectrum and/Orphan Technologies

The doctrine of technology neutrality did not help either. The concept of technology neutrality was that new operators could buy new spectrum and then service that spectrum with a new radio technology, which would provide performance differentiation with existing operators.

In practice, it just added an orphan technology problem to the orphan spectrum problem. WiMAX is the most high-profile example. Developed though the IEEE standards process, WiMAX majored on TDD as being most suitable for per user maximum throughput with the reciprocal channel optimized for multiple input multiple output (MIMO) spatial multiplexing.

This works demonstrably well with Wi-Fi, but is challenging to implement in wider-area, higher-power radio systems.

The initial deployments were at 2.6 GHz, which gave the supply chain several problems. The U.S. band plan was different to the European Band 7 and Band 38 band plan and the technology was sufficiently different to make it difficult to repurpose 3G 3GPP RF components for WiMAX user devices. This proved problematic both for ClearWire and Sprint who acquired the ClearWire spectral and technology assets. The assets proved to be a liability.

In the United States at least it can therefore be observed that auction policy has followed the law of unintended consequences and unexpected outcomes.

Last but not least, there are significant differences in the way in which national operators have to manage regulatory and competition policy and the regulatory challenges faced by international operators such as Vodafone.

AT&T and Verizon can concentrate their regulatory management teams on their local market requirements. Vodafone has networks deployed in 30 countries and minority shareholdings or joint ventures with operators in a further 50 countries [16]. Every one of these 80 countries will have a measure of uniqueness in terms of regulatory policy, competition policy, auction policy, and existing and future band plan and technology combinations. The asset of geographic diversity can at times be a liability.

5.14 WRC 2015

The 2015 World Radio Congress ended at the time of this writing but did not produce any great surprises. There was no change in the UHF band, although footnotes were added for several Asian and American administrations, which should theoretically facilitate global harmonization.

The sub-700-MHz band is available for mobile in markets in the Americas. Several markets within the Indian subcontinent have now also announced their intention to use part of this band for mobile broadband. Spectrum between 614 and 698 MHz will be available for auction in North America, the Bahamas, Belize, Barbados, Canada, Columbia, and Mexico.

Spectrum from 470 to 698 MHz will be available for allocation and auction in Micronesia, the Solomon Islands, Tuvalu, and Vanuatu. Spectrum between 610 and 698 MHz will be available in Bangladesh, the Maldives, and New Zealand.

The U.S. incentive auction looks likely to be structured as paired spectrum with an 11-MHz duplex gap with 5-MHz channel spacing with potential use of 6 MHz for unlicensed operations.

The UHF band from 470 to 960 MHz in Region 1 (Europe) will be reviewed at WRC2023. There was agreement on a new globally harmonized L-band allocation at 1,427–1,518 MHz. Coprimary allocation of the 2.7–2.9-GHz band was rejected as was a global allocation of the 3.6–3.8-GHz band. Allocation of 3.7 GHz was agreed for the United States and 200 MHz of the C-band between 3.4 and 3.6 GHz was globally harmonized.

Spectrum for Unmanned Aircraft Systems (UAS) was provisionally agreed upon, but restricted to safe operation control and nonpayload communications (CNPC).

Allocation of spectrum for high-altitude platform stations was discussed but resulted in a decision that all "controversial" bands, including anything below 24.25 GHz, should be removed from 5G agenda items on spectrum above 6 GHz.

The FCC proposed the authorization of mobile operations in the 27.5–28.35-GHz band (28-GHz band) and the 38.6–40-GHz band (39-GHz band). The 37-GHz band (37–38.6 GHz) is proposed as a hybrid licensed scheme granting operating rights to property owners with geographic licenses based on counties for outdoor use. The assumption is that the licensing mechanism would facilitate advanced enterprise and industrial applications that would not be adequately served by unlicensed spectrum or public network services. There is a parallel proposal to grant mobile operating rights to existing fixed Local Multipoint Distribution Service (LMDS) operators and 39-GHz licensees.

The 64–71-GHz band is likely to be treated in a similar way to the 57–64-GHz band as unlicensed spectrum for Wi-Fi-like WiGIg operations.

It presently looks unlikely that this U.S.-specific approach to 5G spectrum will be replicated in Europe or Asia, which will make it harder to achieve scale economy both for 5G user and IoT devices and network RF hardware.

The candidate bands for IMT chosen for study for WRC 2019 are 24.25–27.5 GHz, 31.8–33.4 GHz, 37–40.5 GHz, 40.5–42.5 GHz, 42.5–43.5 GHz, 45.5–47 GHz, 47–47.2 GHz, 47.2–50.2 GHz, 50.4–52.6 GHz, 66–76 GHz, and 81–86 GHz. The 27.5–29.5-GHz band is excluded from study despite the likelihood that this is where 5G will be implemented first in Korea and the United States. All major satellite bands are excluded from consideration for 5G.

Preparations for WRC 2019 will now start with a series of Conference Preparatory Meetings (CPM) divided into six chapters. Chapter 1 will study land mobile and fixed service allocations including track to train railway and high altitude platforms (HAPS). Chapter 2 will address possible options for mobile broadband above 24 GHz (as per the above list and excluding 28 GHz). Chapters 3 to 6 will cover satellite and science services, maritime including Earth Stations on Mobile Platforms (ESOMPS), now renamed as Earth Stations in Motion (ESIM) but essentially broadband to ships, amateur radio, and other issues.

Coexistence issues will be studied by a new task group TG5/1, which replaces what was known as the JTG 4-7 process that filled up the 3 years between 2012 and 2015. The 3,000 delegates attending the Congress will spend a total of 792,000 man-hours in debate and discussion over the 4-week period which does not include the 4 years of prepreparation.

Note that the WRC does not determine band plans or deployment time scales or auction policy. These are the responsibility of national administrations. The ITU determines coexistence issues between countries. National administrations determine coexistence issues within a country.

The WRC process determines regulatory policy on newly emerging service platforms including high-altitude platforms and unmanned aerial vehicles though the process struggles to keep pace with the change taking place in these sectors. The regulation of nongeostationary (NGSO) spectrum is, for example, likely to require updating and will need to include nanosatellites and picosatellites deployed in almost any band including sub-1-GHz spectrum.

5.15 Summary

By default, most of this chapter has been about historic regulatory and spectrum allocation policy rather than 5G spectrum and regulatory policy. This is because 5G spectrum has only been discussed meaningfully at the November 2015 World Radio Congress with key allocation decisions deferred until the following Congress in 2019 or in some cases, for example, the UHF band in Region 1, until WRC 2023.

Apart from the intentions of the FCC referred to above, regulators do not presently have spectrum to allocate or auction for 5G. However, if we want to develop an opinion on how 5G spectrum and auction policy might change or should change over the next 15 years, then the past 15 years are a good starting point.

Spectrum policy, regulatory policy, and competition policy have defined and shaped today's mobile broadband industry. Regulatory policy and auction policy, particularly the use of auctions designed to ensure five operators per market, have meant that there are now over 600 operators worldwide.

Many of these operators are subscale; probably only the five largest operators worldwide could be described as being fully scale economic. Operators also rely on high debt ratios partly due to auction costs, partly due to network costs, and partly due to highly leveraged mergers and acquisitions, including recently mergers with cable operators who typically already have debt ratios at least as high as the operators with whom they are merging.

There are huge market-to-market and region-to-region differences. AT&T and Verizon are able to realize a 20% return on capital. European operators struggle to achieve a 2% return.

Licenses are typically issued on a 20- or 25-year basis. Some operators are only likely to show a return on these investments in the last 2 or 3 years of the license period. Many will never achieve a return and should probably have been far more aggressive in writing down past spectral equity value.

This highlights a peculiarity. Operators and their trade and industry associations have been efficient and effective in developing an advocacy position with regulators in which it is assumed that more spectrum is needed to support an exponentially growth in mobile broadband traffic. The regulators have responded by bringing new spectrum to market below 1 GHz and in L-band (1–2 GHz) and S-band (2–4 GHz). This growth in traffic is only going to happen if cost per delivered bit decreases or value per bit increases, but this seems presently unlikely.

There may be opportunities to use existing spectrum more efficiently, but most existing techniques come with an associated cost. Carrier aggregation, for example, produces a theoretical gain in multiplexing efficiency, which theoretically translates into capacity gain, which translates into lower cost per bit, but this is offset by a loss of RF efficiency in the user device.

Some of the new spectrum being auctioned is barely fit for purpose and has little chance of ever being scale-economic. This remarkably does not stop operators from bidding for it.

It can be observed that there is a safe absorption rate for spectrum. The industry supply chain can only manage to support additional spectrum at a relatively slow rate. This is because new bands and new band and technology combinations require research and development and band and technology-specific component investment. New band and technology combinations often reduce RF and DSP performance in the user device due to the need to support wider passbands and wider channel bandwidths within those passbands. Higher-order modulation requires more linearity. Dynamic range requirements increase. All of these factors degrade user device performance and user experience with a directly adverse effect on network economics.

The satellite industry is presently developing an advocacy campaign to convince the regulatory community that there is no need for additional spectrum for 5G in the centimeter band due to presently underutilized spectrum in the meter band. This can be positioned as a reasonably persuasive argument. There are TDD bands allocated and auctioned in the late 1990s that remain unused 15 years later, although the same could be said for some L- and S-band satellite spectrum.

The satellite industry is more than an order of magnitude smaller than mobile broadband in terms of service revenues but has sufficient strategic importance to guarantee that any changes to satellite spectrum in the centimeter band are likely to be minimal. WRC 2015, as an example, resulted in additional uplink spectrum being allocated to balance uplink bandwidth and downlink bandwidth at 14.5 to 14.8 GHz in 39 countries, although this excluded the United States, France, and Italy due to the current use of the spectrum by NATO.

Similarly, the mobile broadband community have been proactive in highlighting the opportunities to repack TV into the lower parts of the UHF band as part of the digital dividend but the broadcasting community in Region 1 led by the European Broadcasting Union have successfully delayed any meaningful discussions on this until 2023.

The satellite sector and broadcasters argue that mobile operators could coshare existing spectral resources more efficiently or use allocated and auctioned spectrum more quickly. The continued inability to use Band 12 Block A spectrum 8 years after the U.S. 700-MHz spectrum auction is a poster child for this argument. Similarly, L-band MSS spectrum in Europe has remained unused for at least as long.

Band 7 and Band 38 remain lightly loaded in Europe, particularly Band 38 TDD, and the 3.4–3.8-GHz FDD and TDD bands have yet to be auctioned and could have similar slow utilization time scales. As more mobile bandwidth has become available, the amount of unused bandwidth has increased. More critically, the ratio of unused to used bandwidth is increasing. This does not seem to square with increased auction activity and increased expectations of the value realizable from that spectrum.

As the quantity of spectrum available in the meter band has increased, the quality of the spectrum has decreased. The associated obligation costs, including rollout and coverage commitments and interference mitigation, have increased in parallel.

However, this hides a more fundamental problem. It is hard to see how 5G could add value to existing 4G networks in the meter band. Indeed, it is hard to see how 5G could be deployed without compromising the value of existing networks, many of which will not be amortized for many years to come. Some will never be fully amortized. The only exception would be if 5G could somehow deliver a step function decrease in delivered cost per bit.

This would imply a step function increase in market adoption. The only markets with that level of growth and uptake potential are the developing economy markets of "middle Earth" within the 48th parallel North to the 48th parallel South.

If high per user data rates are the main justification for 5G, then it is also hard to see how the meter band can deliver. Using discontinuous aggregated spectrum between 400 MHz and 4 GHz, for example, could theoretically yield very high data rates. These high data rate aggregated channels provide the basis for LTE Advanced demonstrations showing the throughput capability of various band and LTE Advanced technology combinations including higher-order modulation and higher-order MIMO.

There is no clearly visible path to supporting this sort of functionality in a small-form factor, low-cost user device. This suggests that high data rate 5G will have to be deployed into either the centimetre band or millimeter band. This at least partially overcomes the user device form factor issue. The millimeter band has the additional advantage of having less coexistence issues with the satellite industry.

The question is whether existing auction options would work for these shorter wavelength bands. Technically, there is no reason why it would not be possible to take two 5-GHz passbands at 71–76 and 81–86 GHz each subdivided into 1-GHz channels and construct an auction structured to produce five bidding entities, all of whom could then compete with each other to provide 5G services.

Technically, the spectrum could be FDD or TDD. Historically, better returns have been realized from FDD spectrum. China's present focus on TDD in the meter band between 2 GHz and 4 GHz might change this, but it is presently too early to be sure. Progress in baseband interference cancellation might mean that full duplex channels could be supported, although it is presently hard to imagine that an FDD separation would not deliver at least some useful performance gain in larger radius cells and as you can tell, we are arguing that large cell radius 5G is going to be an essential part of the 5G service offer.

In the final analysis, allocation policy, regulatory policy, and competition policy have to be defined and determined by network economics. The law of unintended consequences and unexpected outcomes applied to regulatory and competition and auction policy has resulted in mobile operators having collective debts of hundreds of billions of dollars. Conversely and some would say perversely, the same policy has resulted in the top two over-the-top providers, Google and Apple, having well over $200 billion of cash assets and a stock market valuation of which mobile operators can now only dream.

Why would Google and Apple or Facebook or Amazon (the GAFA Group) invest in 5G spectrum in any band including the centimeter and millimeter bands if they can continue to realize over-the-top revenue with no associated spectral or technology or rollout or return-on-investment risk? The answer we suggest is middle Earth, which brings us to Chapter 6.

References

[1] http://licensing.ofcom.org.uk/radiocommunication-licences/mobile-wireless-broadband/cellular-wireless-broadband/policy-and-background/history-of-cellular-services/.

[2] www.cntr.salford.ac.uk/comms/preservinghistory.php.

[3] https://www.fcc.gov/document/joint-public-notice-announcing-aws-3-coordination-details.

[4] Ofcom.

[5] http://www.bundesnetzagentur.de/DE/Home/home_node.htmlB-Netz.

[6] http://www.arcep.fr/.

[7] https://ec.europa.eu/digital-agenda/en/news/public-consultation-lamy-report-future-use-uhf-tv-broadcasting-band.

[8] http://www.numericable.be/fr/.

[9] https://www.fcc.gov/encyclopedia/700-mhz-spectrum.

[10] https://www.fcc.gov/national-broadband-plan.

[11] https://www.fcc.gov/document/oet-releases-final-tvstudy-information-related-incentive-auction.

[12] http://consumers.ofcom.org.uk/tv-radio/television/how-tv-transmitters-transformed-and-towered-over-the-uk/.

[13] http://www.dynamicspectrumalliance.org/.

[14] http://www.weightless.org/keyfeatures/spectrum-flexibility.

[15] http://www.themobileworld.com/.

[16] http://www.vodafone.com/content/index/aout/about-us.html.

6

Middle Earth

6.1 Mobile Broadband as a Facilitator of Economic Growth in Developing Economies and Developing Economies as a Facilitator of Mobile Broadband Market Growth

Ford Motor Company founder Henry Ford once said, "Any man who thinks he can be happy and prosperous by letting the government take care of him better take a closer look at the American Indian." He never had much confidence in government intervention to solve economic and social problems. However, he did have a policy of paying his workforce a wage that would be sufficient to allow them to buy the products that they were making.

This chapter looks at the potential demand for 5G radio systems that could be created by raising income levels in countries between the 48th parallel North and the 48th parallel South. Suppressing a natural sympathy with Mr. Ford's skepticism about government intervention, we also review the various initiatives that are presently under way to transform the economics of the developing world and specifically the use of 5G wireless system innovation to facilitate that transformation by increasing opportunities to earn and distribute individual income.

6.2 United Nations Development Programs

In July 2015 the United Nations (UN) issued a final report on the Millennial Development Program [1]. Established by the UN in September 2000, the

15-year program had eight defined goals covering poverty, education, gender equality, child mortality, maternal health, disease, the environment, and global partnership.

Progress has been made with the UN able to claim that 1 billion people have been lifted from extreme poverty. This still leaves 795 million people undernourished. Over 50% of the world's 1 billion extremely poor people live in five countries: India, Nigeria, China, Bangladesh, and the Democratic Republic of the Congo. The number of people displaced by war and regional conflicts is at its highest level since World War II. The Millennium Development Goals are being replaced by another 15-year program of Sustainable Development Goals grouped into six essential elements: dignity, people, planet, prosperity, justice, and partnership [2].

The ITU 2015 ICT statistics report [3] highlights the present reality that over 4 billion people remain offline. Fixed broadband penetration in Africa is less than 1%. Mobile broadband penetration in Africa is below 20% compared to more than 80% in Europe. There are more than 1 billion people living in Africa, a population that has doubled in size in less than 30 years [4]. It is geographically huge, over 30 million square kilometers, more than the combined geographic area of the United States, Argentina, India, Western Europe, and China. The rural delivery economics in Africa are therefore uniquely challenging.

The same challenge applies to the economics of other utilities considered essential in developed economies. Over 600 million people in Africa have no access to mains electricity [5]. This creates a related challenge for telecommunications connectivity including mobile connectivity and highlights a particular need for power-efficient RF base stations that can run off solar or diesel power and user devices that can run off AA batteries at least some of the time.

6.3 Dollar per Day Income Comparisons

Mobile broadband delivery cost is not just a question of size but of population density, population distribution, and wealth. Australia is a large country of 7.6 million square miles with a relatively small (33 million) but rich ($67,000 GDP per capita) population. Ethiopia, one of the economic success stories of Africa with growth rates of nearly 10% for the last 10 years, has an average gross domestic product (GDP) per capita of $315. Ethiopia and many other African countries remain vulnerable to drought, crop failure, and famine. Economic success in Africa has an underlying agrarian fragility.

Measured on the basis of purchasing power parity, what $1 would buy in the United States rather than local currency, over 13% of the people in China, 47.5% in Sub-Saharan Africa, 36% in South Asia, 14% in East Asia and the

Pacific, and 6.5% in Latin America and the Caribbean earn less than $1 per day, a total of 1.3 billion people. Table 6.1 shows the equivalent rankings for the percentage of the population living on less than $10 per day.

The problem is a combination of low income and low population density and a need for extended coverage into deep rural areas that are usually also the most deprived and economically vulnerable areas in a country.

6.4 The Need for Cost-Efficient Deep Rural Coverage: Australia as an Example

In previous chapters, we have documented that many modern mobile phones, including high-end LTE 4G phones, have less efficient antennas due to volume and space constraints and additional insertion loss and TX and RX efficiency loss due to the need to support multiple bands and technologies.

This does not matter particularly in countries that can afford dense networks, although there is still a capacity cost. It does matter in countries that need to support larger cells. Australia is one example, relatively rich but with large areas that are sparsely populated but dependent on being connected.

One Australian operator has found a partial solution to this problem with a Blue Tick scheme that identifies phones that have good RF performance, which means that they can provide service outside of the coverage maps [7]. Australia is also one of the most developed markets for relays and repeaters with window-mounted relays being used to provide connectivity to remote farms in the outback. Australian operators have historically developed variants of exist-

Table 6.1
Percentage of the Population Living on Less Than $10 per Day: Ten Poorest Countries Compared with Europe

Rank	Territory	Value	Rank	Territory	Value
1	Ethiopia	99.9	191	Czech Republic	0.072
2	Tanzania	99.8	192	Austria	0.039
3	Burundi	99.5	193	Germany	0.019
4	Yemen	99.5	194	Finland	0.004
5	Malawi	99.2	195	Belgium	0.003
6	Congo	99.2	196	Sweden	0.001
7	Rwanda	98.9	197	Denmark	0.001
8	Tajikistan	98.9	198	Japan	<0.001
9	Guinea-Bissau	98.5	199	Norway	<0.001
10	Madagascar	98.4	200	Luxembourg	<0.001

Source: [6].

ing technologies, which can support wider radius cells, Global System Mobile (GSM) with a range of up to 100 km being an example.

Technical solutions therefore exist to provide deep rural coverage both for voice (extended wide-area GSM) and data (LTE repeaters and relays). LTE transceivers with 2W (33 dBm) of transmit power will be available for the Band 14 LTE PPDR (Public Protection and Disaster Relief) market in the United States and could potentially be repurposed for enhanced rural broadband coverage. The issue is whether those solutions can scale from a high GDP market like Australia and or the United States to a low GDP market like Ethiopia.

GSM has the advantage of being fully amortized in terms of legacy research and development spending so potentially can be delivered at low cost, but there are some irreducible cost multipliers including a need to rely on diesel and solar power. Low-cost GSM does not address the data connectivity issue.

6.5 LTE Cost Escalation

The problem with LTE user devices is that they are designed to support high data rates close to a base station. This requires dense networks and dense networks cost more to build and run. This means that it is hard to build sparse networks that can meet the per-bit delivery cost requirements of the under $10 per day and under $1 per day markets.

The complexity required to support high data rates in user devices, for example, carrier aggregation, MIMO antennas, high-performance digital signal processing, and highly specified RF components, results in nonrecurring engineering costs that have increased by an order of magnitude every decade. Table 6.2 provides an indication of this engineering cost escalation.

This does not include the litigation costs associated with intellectual property disputes. To an extent, vendors might be able to take a view that subsiding LTE into low GDP markets, particularly low GDP markets that are growing relatively quickly, would be a worthwhile long-term investment.

Table 6.2

Nonrecurring Engineering Costs for Mobile Cellular Devices over the Past 30 Years

1G	2G	3G	4G	5G
1980	1990	2000	2010	2020
1990	2000	2010	2020	2030
$2–3 million	$20–30 million	$200–300 million	$2–3 billion	$20–30 billion?

*Includes RFIC NRE and directly associated radio frequency front end (RFFE) costs including system support.

6.6 Examples of Long-Term Market Development

In 1826, Guinness [8] started selling their Foreign Extra Strong Stout to Africa. The iconic Guinness brand logo, the toucan, is an African bird. There are a number of similarities between the drinks industry and telecommunications industry. Both industries are built on addiction and dependency. Both industries have relatively long return on investment (ROI) cycles; think of those 40-year-old malts maturing in Scottish distilleries. In 2007 Nigeria overtook Ireland as the world's number two market for Guinness behind Great Britain. The Guinness brand is now owned by Diageo, the world's largest drinks manufacturer. Over 40% of Diageo sales are from emerging countries with a target to increase to over 50% with Africa as presently one of the highest growth markets, not a bad return on 180 years of investment.

In 1965 the Chinese government offered to sponsor the construction of the railway line linking Tanzania, ruled by President Julius Nyerere, and Zambia. Fifty years on, Chinese vendors are strategically well placed to service fast-growth African telecom infrastructure, user device, and service provider markets with the side benefit of preferential access to Africa's mineral resources.

Julius Nyerere had a vision to create an egalitarian Socialist society based on cooperative agriculture. This involved collectivizing farmlands, a mass literacy campaign, and free and universal education with an emphasis on achieving economic self-reliance. The project was named *ujammaa*, the Swahili for family hood, and was based on economic cooperation and racial and tribal harmony.

Africa is not a country but a continent and there are large cultural differences from country to country and region to region. Tanzania has 120 tribes making the country intrinsically diverse. Kenya has three tribes making stability harder to achieve.

At the end of Nyerere's presidency in 1985, Tanzania was still one of the world's poorest countries, with a per capita income of US$250 and a third of the national budget reliant on foreign aid. Agriculture remained at subsistence level and industrial and transportation infrastructures were chronically underdeveloped. However, the country had one of the highest literacy rates in Africa [9] and was remarkably free of economic inequality and was, crucially and not coincidentally, politically stable. Tanzania compared to Kenya and many other African countries is a significantly peaceful place.

In terms of the original ambition, *ujammaa* in Tanzania failed but can be regarded as a good idea ahead of its time. In particular, 50 years on, Africa has potential access to Internet on Things (IoT) connectivity technologies that could revolutionize agriculture and access to communication technologies that could provide efficient mechanisms for delivering literacy and numeracy.

You cannot sell $500 smart phones and $50-per-month contracts to agricultural workers earning a dollar a day, but you can service low value but potentially high-volume markets provided the large differences in ROI time scale are understood and accommodated. ROI time frames of 50 years are not concordant with contemporary Wall Street or European investment return expectations although they are consistent with a working lifetime, a fact that most pension funds appear to have missed or ignored.

6.7 Mobile Micropayments Facilitate Economic Growth

There have also been notable mobile phone-related economic success stories with M-Pesa being a particular example, enabling low-cost micropayments and transactions to be done over the short message service (SMS). Twenty percent of Kenya's GDP is estimated to be directly facilitated by M-PESA based transactions [10]. Figure 6.1 show a mobile phone vendor stall in Cameroon, now an exhibit in the Science Museum Information Age gallery.

Transaction value by early 2015 exceeded $1 billion per month across Kenya, Tanzania, Uganda, and Rwanda. This is hardly surprising when you consider that for every bank account holder there are three mobile phone own-

Figure 6.1 Cameroon mobile phone vendor stall [11]. (Reproduced with permission of the Science Museum and Science Museum Picture Library.)

ers and hardly surprising given that the transaction costs are a fraction of the cost of processing through a traditional banking system [12].

However, this mitigates rather than solves the industry supply chain problem. In practice the opportunity cost of not servicing established high-value markets in order to realize possible long-term returns from low-value markets, particularly low-value markets with large rural lightly populated areas with low GDP (at most a few dollars per day) would imply too high a risk for most vendors.

6.8 Ultralow-Cost Handsets: An Expensive Risk?

This is corroborated by previous attempts to produce tailor made ultralow-cost devices that have not ended well. The GSM Trade Association [13] ran a competition in 2007 to produce an ultralow-cost (under $30) handset. The competition was won by Motorola and TI. This was the equivalent of the winner's curse that operators suffer after buying subscale spectrum with expensive obligations.

Delivering the cost target required an exceptionally high level of integration at the time which translated into $250 million of time to market risk. As a consequence, Motorola, once the world's largest most financially successful handset manufacturer, is no longer in the handset business and TI, the world's biggest supplier of GSM baseband in the 1990s, exited the market.

This highlights the reality that the supply chain is often just not able to respond to ultralow-price market requirements. It also suggests a need for regulatory intervention or at least a different regulatory approach for developing economies that more directly couples ITU policy with the UN 15-year Sustainable Development Goal (SDG) program.

6.9 The North to South to West to East Transition

The difficulty is that after 150 years it is quite hard for the ITU to change the way in which it has traditionally managed the spectrum and standards process. The ITU divides the world into three regions with boundaries defined by longitude (North to South) (see Figure 6.2).

For decades this has provided an adequate basis for realizing spectral and regulatory and standards policy for the cellular and broadcast and satellite industry. Historically the demography and GDP of countries north of the equator including South Korea and Japan in Region 3, the United States in Region 2 and Europe and the Middle East in Region 1 have had a gravitational effect on standards making and band allocation policy with countries close to or south of the equator following the technology lead of their northern neighbors. However the growth markets are better defined by latitude, west to east with the

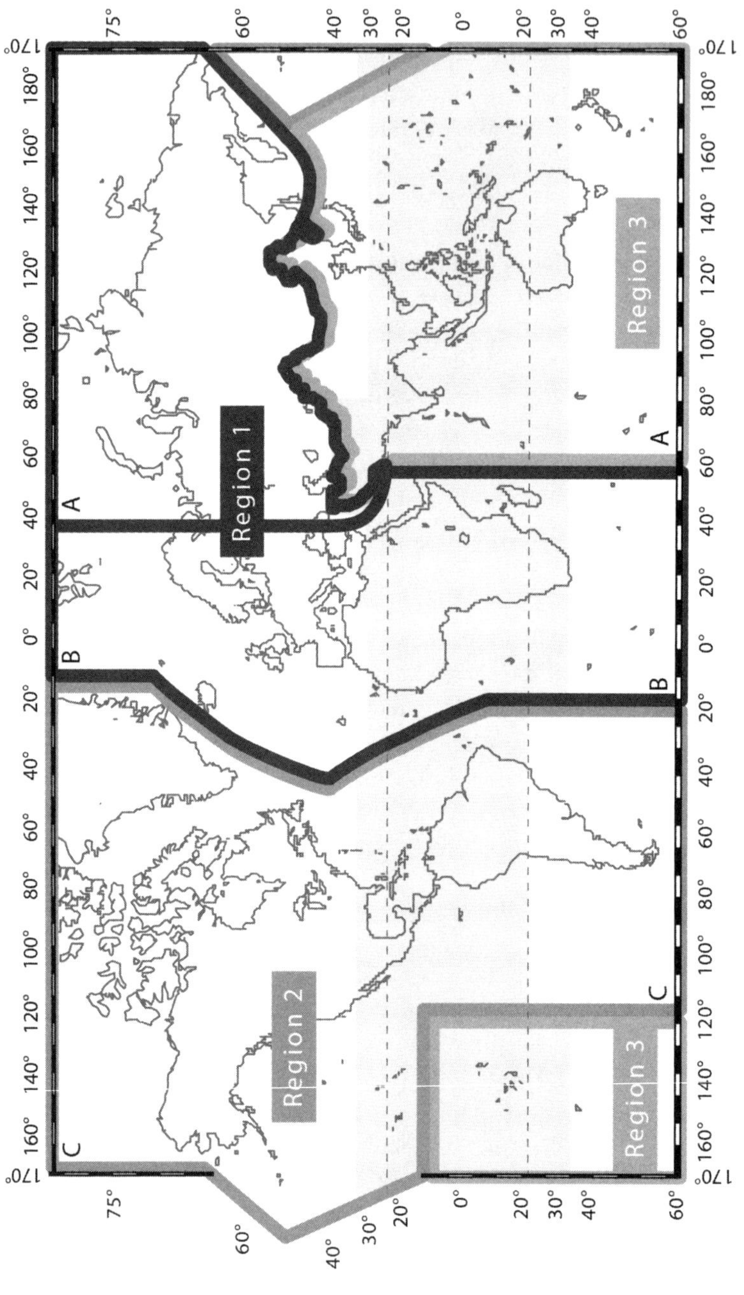

Figure 6.2 ITU regions.

developing economies largely concentrated between the 48th parallel North and the 48th parallel South.

This emerging West to East focus was evident at the WARC 2012 Congress with spectral allocations in the 700-MHz bands coupled to LTE technical specifications that look increasingly likely to become a de facto standard in Latin America, Africa, and Southeast Asia, a geographic focus defined expressly by latitude.

While WARC 2012 can be regarded as a turning point, the change started earlier with the formation of the African Telecommunications Union in Kinshasa in 1977 [14], and the Asia Pacific Telecommunity in Thailand in July 1979 [15]. The significance of WARC 2012 was that a significant part of Africa and Latin America expressed support for the APT 700 band, realizing a global 700-MHz FDD and TDD band plan (Band 28) that will be commonly deployed through Latin America, Africa, and Asia [16]. This was a surprise for some WARC 2012 attendees but in retrospect was understandable and predictable particularly as the United States and Europe had failed to coordinate a harmonized band plan in the prior 5 years from the U.S. 700-MHz band auction in 2007.

The southern latitude markets in Latin America, Africa, and Southeast Asia have traditionally been regarded as GDP constrained low penetration low average revenue per user (ARPU) markets. Low penetration is a market opportunity, particularly if GDP constraints are less severe than the statistics suggest.

First, GDP is an unreliable proxy for household income and it is household income that determines the money available to be spent on mobile communications. Second, this is not substitution spending; mobile phones are not being purchased as an alternative to wireline Internet and voice access because landline Internet and voice access is generally unavailable or prohibitively expensive.

Add together demographic factors such as large, fast-growing young populations, 75% of Kenyans are under 30 years old, for example, geographic factors (long distances making copper, cable, and fiber uneconomic), and rising GDP

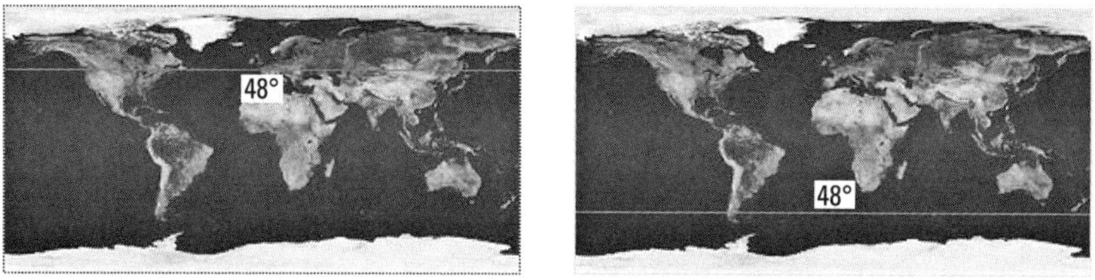

Figure 6.3 Middle Earth: the 48th parallel North to the 48th parallel South.

statistics with multiplier effects (household income and high relative economic value to each user) and compare these with saturated European and U.S. markets relatively better served by alternative fixed access options and it becomes clear that the North to South to West to East value shift has only just begun.

6.10 Loss of Market Efficiency in Developed Markets

The shift is potentially being accelerated by present U.S. and European competition and regulatory policy. This includes, at least in Europe, a progressively less sustainable belief that each national market needs four or five operators to be market-efficient.

If market efficiency is defined as the maximization of short-term spectral value, then this may be defendable, but if more broadly defined to include technical and commercial efficiency, then such a policy is manifestly indefensible. Technical and commercial inefficiency translates directly into higher per bit delivery cost. European-specific technical requirements that add dollars of cost to user devices and take several decibels off the link budget compounds the delivery cost problem. This translates into lower Earnings Before Interest Tax and Depreciation (EBITDA), which translates into a more adverse debt-to-equity ratio, a vicious rather than virtuous circle.

6.11 The Law of Unintended Consequences

It can be observed that Northern latitude regulatory and competition policy is obeying the tried and trusted law of unintended consequences and unexpected outcomes. Within the next 15 years, the time scale for the SDG program, the balance of global telecoms value, and the consequential influence that comes with that value will have shifted southwards but will be consolidated horizontally (West to East) across Latin America, Africa, and Southeast Asia. Even China will be small when compared to the size and power of these markets and the United States and Europe will be smaller.

The consequence of the southern hemisphere going horizontal is that the northern hemisphere will have to do the same in order to remain globally competitive. This means that European and U.S. standards and spectrum allocation and band planning processes will have to converge, presently an unlikely prospect.

6.12 White-Space Spectrum for Developing Economies: Wi-Fi and LTE-U

More fundamentally, there is an argument that licensed spectrum and the user devices designed to work on licensed spectrum are too costly for "middle Earth" markets. The Dynamic Frequency Alliance is an example of a lobby/interest group established to argue the case for greater use of unlicensed spectrum and greater use of opportunistic licensed spectrum access including white-space Wi-Fi in the 700-, 600-, and 500-MHz bands for deep rural coverage [17]. This includes standardization efforts to develop higher power wide-area Wi-Fi [18].

The point has already been made that previous attempts to scale Wi-Fi to wide-area via the WiMAX standards process have failed technically and commercially. Conversely, attempts to scale LTE to be cost-competitive with Wi-Fi in local area connectivity also seem unlikely to succeed. This includes present attempts to deploy LTE into the 2.4-GHz ISM band and 5-GHz U NII band (Unlicensed National Information Infrastructure Bands, also known as 5 GHz Wi-Fi).

Commercially, this is a sensible idea, mixing high-cost licensed spectrum with low-cost or no-cost unlicensed spectrum. Low cost in this context means public Wi-Fi hot spots where the operator still bears the network infrastructure cost but with no spectrum cost. No cost in this context means public access being provided from office or private Wi-Fi. This is sometimes described as an inside-out model. BT Open Zone in the United Kingdom is a contemporary example of this [19].

The assumption is that support for carrier aggregation and supplemental downlink only channels in user devices will make it relatively easy to support LTE at 2.4 GHz and 5 GHz with the same cost base as Wi-Fi. However, Wi-Fi will still exist and therefore coexistence with Wi-Fi has to be managed.

In Europe and Japan this requires Listen before Talk (LBT) to be implemented with sub-10-ms response times. This requires changes to be made to the LTE physical layer interface, which will be addressed in Release 13. The United States, Korea, and China do not require LBT to be implemented, so theoretically LTE-U could be rolled out in these markets.

The pragmatic alternative may be to improve LTE/Wi-Fi interworking, which is an ongoing but slow process. A Wi-Fi access point broadcasts a 48-bit base station system identity (BSSID) similar to an Ethernet MAC address. This identifies an access point with a second 32-bit string, the SSID identifying the network to which the access point belongs. Anyone can set up a network and choose an SSID, so it is hard to manage access rights, for example, in an automated Wi-Fi to LTE handover. The IEEE 802.11u standard aims to resolve this using Extensible Authentication Protocols to establish the authentication path, the basis for the Hotspot 2.0 Wi-Fi Alliance [20] and Next Generation

Hotspot (Wireless Broadband Alliance) [21] connectivity standards. The existence of parallel standards groups almost always slows and complicates market adoption.

6.13 Subspace High-Altitude Platforms

Wi-Fi from the sky is another option either using balloons or drones, known collectively as High Altitude (HAP) platforms. Authentication issues still apply, but coverage could theoretically be delivered at a relatively low cost.

The Google Loon (lighter than air) balloon project is an example [22]. Hot air, hydrogen, and helium are all possible choices for producing flying base stations, but helium is the most practical and safest option. Hydrogen is the lightest element on Earth, and helium is the second lightest. Helium has the significant advantage of not being flammable. On October 24, 2012, Felix Baumgartner flew in a helium balloon to 24 miles (39 km) and jumped out to break the sound barrier in a free-fall descent.

The concept of the Google Loon [23] project is to launch helium balloons into the stratosphere at around 20 miles (100,000 feet). The height of the balloons can be altered by pumping air in or out of the balloons to change their density. As their altitude changes, they catch different wind currents and can therefore be positioned to provide coverage as, where, and when required. The balloons take about 22 days to fly around the world, typically blown at wind speeds of 100 miles per hour, and should be able to stay up for 100 days.

Each balloon has solar panels to run the RF transceiver, the air pump, and a heater to stop the electronics from freezing. The present stated position is that the air interface will be 2.4-GHz and or 5-GHz Wi-Fi though they could feasibly use 5-GHz LTE-U. The LTE option would have the advantage of providing more closely controlled air to ground line-of-sight interference. This would include control of interference into other sky-facing radio systems including radio astronomy.

Helium balloons could be used at other heights. This could include the troposphere up to 36,000 feet using the jet streams that are generally moving West to East in the northern hemisphere and East to West in the southern hemisphere at up to 200 miles per hour.

Civil Aviation authorities would need to be reassured about how relatively low-speed balloons could coexist with higher-speed aircraft flying at a similar altitude. A regime similar to the present regulatory control of weather balloons would be less than adequate [24].

Higher altitudes above the stratosphere include the mesosphere at up to 150,000 feet (28 miles), the thermosphere at above 50 miles, and the exosphere, the boundary with space, at 300 miles. The highest that a gas balloon has flown to date is around 170,000 feet (32 miles), so thermospheric and exospheric altitude options are not presently feasible (and probably never will be).

At the time of this writing, FCC filings [25] showed Google wanting to use 71–76-GHz and 81–86-GHz spectrum for balloon-to-balloon communication and LTE to beam the Internet service back down to Earth.

6.14 Subspace and Space Communications Integration

By comparison, the orbital height of an Iridium satellite is 485 miles (780 km). Iridium satellites travel at 17,000 miles per hour in North to South polar 100-minute orbits. They are therefore different from balloons but share some of the same economic dynamics. For example, the present generation Iridium satellites have lasted longer than expected with a lifetime of well over twice the design expectation. Both present and next generation Iridium satellites successfully amortize delivery costs with other civilian and military payloads including sensing systems.

The delivery economics of Project Loon will be dependent on developing envelope materials that are sufficiently robust and resistant to damage including ultraviolet radiation in order to extend the flight time beyond present expectations, achieving further improvements in solar panel efficiency, and developing high-value hosted payloads to offset launch and flight and system costs.

Helium is a finite resource and one of the few elements that escapes gravity and leaks away into space. It is used in the electronics industry in the manufacture of silicon wafers, for superconductors and by deep sea divers. It will become more costly over time.

Facebook has a similar low-cost Internet access initiative to Google, although it is studying the use of drone technology rather than balloons [26]. These are quasi-stationary platforms (they go around in circles) flying at 65,000 feet to provide a city-sized coverage footprint with medium population density. This is close to the lowest altitude for unregulated air space with relatively stable weather conditions most, but not all, of the time (thunderclouds can reach 60,000 feet). Future generation solar-powered, high-altitude drones could stay on station for months or potentially years with their location precisely controlled. The communication system could be a conventional microwave transceiver or free-space optics, more power-efficient but susceptible to weather fading.

Project Loon and the Facebook Internet project are admirable, exciting, and potentially feasible options for delivering low-cost Internet access. It is not an either/or choice and would be more convincing if the two options were integrated together rather than being separately promoted. Google's recent acquisition of drone maker Titan Aerospace suggests that this might now happen [27].

A new generation of airships also suggests that alternative delivery platforms might become available [28]. However, the R101 disaster [29] was caused by an unexpectedly severe rainstorm over France that caused a fabric failure at the front end of the air ship. The weather and the resulting accident invalidated the economic business model for the commercial airship industry.

Eighty years later, we are able to forecast the weather with greater accuracy and foresight, but wind, rain, thunderstorms, typhoons, and hurricanes are elements that remain outside of our control. This is a definable risk statistically (ITU rain fading statistics for microwave links provide a starting point), but weather costs need to be fully factored in to Google's and Facebook's delivery cost calculations.

Line-of-site interference issues into terrestrial systems including upward-facing terrestrial systems such as radio astronomy also need to be addressed. Subspace systems are in principle and practice well placed to communicate downwards and upwards. A closer integration of space and subspace platforms would seem to be a sensible way forward.

6.15 Wi-Fi over Satellite Including New Generation Electric Satellites

Moving up into space provides a wide range of orbit options from low Earth, Iridium at 780 km being an example, through medium Earth orbit (MEO), GPS at 20,200 km being an example, to geostationary at 35,000 km. Typical round-trip latencies for low Earth orbit are 20 ms, 133 ms for an MEO, and 500 ms for a geostationary orbit (GSO) platform.

The delivery economics of the satellite sector are rocketing downwards, a function of commercial innovation including a wider choice of launch options, Elon Musk's Space X business being an example, shared payloads, and improved technical efficiency. Technical improvements include more efficient solar panel arrays that increase on orbit power output but also facilitate electric satellites [30]. These satellites use xenon ion thrusters that power the satellites from an interim orbit into final orbit. The thrusters are then used to maintain orbital position. Electric propulsion is not new and has been widely used for deep-space missions for several decades, but the thrusters are now more efficient and applied to Earth orbit platforms.

All-electric satellites weigh half as much as traditional chemically propelled satellites, so the launch cost goes down or the active payload can increase. They will also stay in orbit longer as they do not need to be de-orbited when the onboard fuel supply runs out. The only snag is that they take a few months to "sail" from interim to final orbit.

On-orbit lifetimes are already remarkable, with some of the original Inmarsat satellites lasting well beyond 20 years, but electric satellites should extend this significantly [31]. On-orbit servicing presently being researched by DARPA would further reduce operational costs.

In December 2013, the first Global Express Inmarsat satellite launched from the Baikonur Cosmodrome in Kazakhstan followed by a second satellite in February 2015, and a third and fourth satellite were being built, with the fourth satellite scheduled for launch by Space X in late 2016, a $1.6 billion space network investment. These are 6-ton satellites launched into geostationary orbit capable of generating 15 kW of RF power from multiple steerable antennas, maximizing the downlink and uplink link budget. This is enough to support data rates of up to 50 Mbps to fixes and portable devices. The satellite works in Ka-band and can switch between civilian bands (27.5–31-GHz uplink and 17.7–21.2-GHz downlink) and military bands (30–31-GHz uplink and 20.2–21.2-GHz downlink).

Inmarsat's integration of KA services with their L-band and potential S-band platforms should be a compulsory case study for all 4G and 5G product-planning teams [32]. The offer includes Wi-Fi access points [33] with backhaul on the L-band satellite downlink at 1,525–1,559 MHz and uplink at 1,626.5–1,660.5 MHz.

Also in L-band (1,616–1,626.5 MHz), Iridium has a parallel $3 billion investment program for their Next Constellation upgrade to their existing 66

Figure 6.4 Inmarsat Alphasat satellite. (Image courtesy of Astrium.)

low Earth orbit (LEO) satellites with launches scheduled through 2016 and 2017 [34].

6.16 Mixed-Use Models in the Satellite Sector

Inmarsat and Iridium and satellite operators in general are therefore able to amortize launch and operational cost across multiple civilian and military payloads including sensing and imaging applications. This is facilitated by brokerage arrangements that bring commercial users together with the satellite industry; the Hosted Payload Alliance is an example [35].

Iridium hosts a military payload that provides enhanced GPS performance for naval warfighters operating in areas that are RF signal restrictive due to topology (urban or rural canyons) or deliberate jamming [36].

The economics of these new satellite systems are therefore potentially attractive for rural coverage and or for delivering on-demand capacity as and where required, but pricing is presently determined by a customer base in aviation, maritime, enterprise, and government (including military customers). Access tariffs are therefore too high for low GDP markets. Wi-Fi/satellite access points supported by Inmarsat typically retail at over $2,500.

Satellites cannot compete presently with terrestrial systems in terms of flux density and it is hard to deliver acceptable indoor coverage without an external antenna and repeater or relay. It is possible to improve link budgets with user device dish antennas, particularly if accurately pointed, but this is only beneficial to portable or fixed devices. Satellite systems cannot therefore be considered as direct economic replacements for terrestrial networks supporting mobile users. The same caveats apply to balloon or plane/drone-based platforms, although it is easier to bring these back to Earth for repair.

Geostationary satellites are ideally placed in terms of their orbital position over the equator to service middle Earth markets. This, combined with commercial innovation (lower-cost launch options and mixed payloads), technical innovation (electric satellites, more efficient solar panels and spot beam antenna arrays) and longer-term market opportunity, suggests that their role in servicing low-value, low GDP middle Earth markets will increase over time.

6.17 LEO VHF Satellites: Orbcomm and Middle Earth Machine-to-Machine Connectivity

Closer to Earth (775 km) and further down the spectrum (in the VHF band between 137 MHz and 150 MHz), similar technology gains will be potentially realized from six new LEO satellites launched on a Space X Falcon rocket from Cape Canaveral in July 2015 upgrading an existing constellation of 29 satellites providing machine-to-machine and maritime monitoring [Automatic Identification Systems (AIS)] and messaging services [37]. One satellite had failed in orbit by August 2015, but another 11 constellation upgrade satellites were scheduled to be launched. The satellites individually weigh 172 kg and have a rated 400-W power output. Figure 6.5 shows a launch in progress.

The LEO orbits are optimized to provide coverage over North America, South America, Asia, Europe, Australia, and Africa and the sea in between.

All ships with a weight over 300 gross tons have a VHF transponder providing updates of position, speed, and navigational status via a VHF transmitter. The Orbcomm constellation uses onboard Spectrum Decollision Processing (SDP), which digitizes the VHF band and applies digital filtering to identify individual AIS signatures. The detection and refresh rate is a function of the number of satellite passes. For Brazil, at –5° latitude, the constellation upgrade increases the pass rate from 54 to 91 passes a day, which at 10 to 12 minutes per pass means the satellites are in view of a vessel for 5 rather than 9 hours.

Figure 6.5 Launch of six Orbcomm LEO satellites on a Space X Falcon rocket from Cape Canaveral, July 2015. (Courtesy of SpaceX.)

Argentina and Australia, at –35° latitude, increases the pass rate from 70 to 127 passes a day, increasing contact time from 12 to 21.7 hours per day.

Latency is a function of the number of satellites and number of ground stations. The constellation upgrade reduces Brazil latency (50% mean average) from 20 minutes to 3 minutes and Argentina from 20 minutes to 1 minute. Even considering the satellite count of the new constellation, this is a sparse network considering the area covered.

Ships are part of the Internet of Things (IoT) or rather the Internet of Slowly Moving Large Objects (IOSMLO?). This provides an opportunity to cross-subsidize or at least cross-amortize maritime and terrestial connectivity, a combination of technology and commercial innovation. Following WRC 2015, these platforms, formerly called Earth Stations on Mobile Platforms (ESOMPS), are now known as Earth Stations in Motion (ESIMS)

6.18 Macrosatellites, Microsatellites, Nanosatellites, and Picosatellites

The Orbcomm satellites with a launch weight of 172 kg are designated as microsatellites (10 to 500 kg). Macrosatellites are anything above 500 kg (Iridium and certainly Inmarsat). Nanosatellites are between 1 kg and 10 kg, and picosatellites are anything below 1 kg (see Table 6.3).

Communication may be direct or indirect via another satellite acting as a relay or repeater. The Hubble telescope as an example talks to Earth via five NASA Tracking and Data Relay Satellites (TDRS) in geosynchronous orbit at 35,000 km [38]. Hubble is orbiting the Earth every 97 minutes at a height of 569 km so it is officially an LEO satellite.

It might seem odd to communicate from the Hubble LEO to a GSO satellite and then back to Earth, but the result is a more efficient path link and a continuous link with a fast-moving object for the exchange of optical imaging and telemetry data.

The other notable LEO platform is the International Space Station, continuously occupied since November 2000, orbiting at a height of 248 miles (400 km), weighing 391,000 kg with solar panel arrays the size of a football field and with a living space and workspaces equivalent to a five-bedroom house. In the context of 5G and this chapter, the relevance of the Space Station is its role in fostering technical, political, social, and commercial cooperation between the

Table 6.3
Macrosatellites, Microsatellites, Nanosatellites, and Picosatellites

Macrosatellites	Microsatellites	Nanosatellites	Picosatellites
>500 kg	10–500 kg	<10 kg	<1 kg

five space agencies and 15 countries involved in the $100 billion project including NASA, the Russian Space Agency, European Space Agency, Canadian Space Agency, and Japan Aerospace Exploration Agency.

The ITU differentiates deep-space band allocations (greater than 2 million km from Earth) from near Earth applications (less than 2 million km from Earth). The Moon, a 2- or 3-day journey away at a distance of 384,400 km is definitely near Earth in radio terms (see Table 6.4).

The combination of spectrum availability and technology innovation supports the argument that the 5G physical layer including a low-cost IoT physical layer for remote area connectivity should at least comprehend the changes taking place in LEO, MEO, and GSO satellite delivery platforms.

6.19 Hybrid Terrestrial Satellites for Middle Earth

This brings us to a brief review of hybrid terrestrial satellite systems. In 2002 Teledesic conceded that their constellation of 288 LEO satellites, the Internet in the sky, was technically possible but commercially nonviable. Fourteen years on, the technology economics of the space sector have substantially changed and continue to change over time and it would be wrong to dismiss the potential of integrating terrestrial LTE with some combination of GSO/MEO/LEO service offer.

There is substantial negative investment sentiment due to high-profile failures such as Light Squared but this is offset by examples such as Thuraya. Thuraya, the Arabic name for the constellation of the Pleiades, has operated a commercially and technically successful satellite and terrestrial service. The offer has been restricted to voice and comparatively low bit rate data and targeted at a high-value demographic. The coverage footprint covers the 48th North to 48th South parallel very adequately [39] and the company has introduced innovative sleeve products that allow an iPhone or Android smart phone to have satellite connectivity [40]. Thuraya provides validation that a hybrid satellite/

Table 6.4
ITU Near-Space Satellite and Deep-Space Allocations

Band	Deep space >2 million km		Near space <2 million km	
	Uplink, Earth to space	Downlink, space to Earth	Uplink, Earth to space	Downlink, space to Earth
S-band	2,110–2,120	2,290–2,300	2,025–2,110	2,200–2,290
X-band	7,145–7,190	8,400–8,450	7,190–7,235	8,450–8,500
K-band				25,500–27,000
Ka-band	34,200–34,700	31,800–32,300		

Frequency allocations in MHz

terrestrial service offer can be technically and commercially viable. The challenge is to scale cost down to service low GDP markets and low-income users in developed and developing middle Earth markets.

6.20 Broadcasting to Middle Earth: Big Radio

6.20.1 Next Generation Broadcasting as a 5G Integration Opportunity

So far in this chapter, we have showed how technical and commercial innovation, for example, in the satellite sector, is allowing the industry to scale costs and amortize development to a point where supporting a lower-income demographic becomes commercially feasible. The same trend is happening in terrestrial radio and TV broadcasting. Radio and TV digital standards over the past 20 years have tended to concentrate on capacity rather than coverage. Generally, this has meant that the digital replacements for FM radio and analog TV have struggled at least initially to match the coverage provided by the older analog systems. This continues to be an issue for DAB in Europe, for example.

This is being at least partially addressed in second and third generation radio and TV standards. The second generation DVB-T standard in Europe and Asia known as T2 was published in 2009 with a 2011 update known as T2 Lite for mobile and portable reception. T2 receiver chip sets are now the same price as T1 chip sets, around $20 to $25, the benefit of global market scale. T2 services were launched in a limited way in the United Kingdom in March 2010 with subsequent launches in Italy, Sweden, Finland, Zambia, Namibia, Nigeria, Kenya, and Uganda.

The financial failure of first generation mobile TV has depressed investor sentiment in T2/LTE integration and competing standards such as eMBMS have made T2/LTE hybrids less likely.

T2 comes with the option of an 8K OFDM carrier, which means that the physical layer has a longer symbol duration that translates into a longer guard interval. The longer guard interval supports a longer cyclic prefix. In LTE, the cyclic prefix is used to compensate for the delay spread in the radio channel. In unidirectional/broadcast single-frequency networks, the cyclic prefix avoids intersymbol interference being imposed from adjacent transmitter sites.

In an eMBMS network implemented as a single-frequency network (an MBSFN), the LTE 15-kHz carrier subspacing is reduced to 7.5 kHz. This doubles the symbol length from 66.7 μs to 133.4 μs, which allows the cyclic prefix to be increased from 4.69 μs to 33.33 μs. The 33.3 μs is effectively a capacity cost but allows for larger radius cells. A capacity cost translates into a coverage gain. A 4.69-μs cyclic prefix allows a delay spread of 1.5 km. A 33-μs cyclic prefix allows a delay spread of 10 km.

The T2 multiplex cyclic prefix can be anything from 7 to 224 μs. The 224-μs guard band supports a single frequency network with 67-km cells. The DAB multiplex has a 246-μs guard band/cyclic prefix, which supports 74-km radius cells. The Digital Radio Mondiale (DRM) multiplex has a 2,660-μs guard band/cyclic prefix and supports 500-km radius cells.

6.20.2 Digital Radio Mondiale and Middle Earth Delivery Economics

If you want to discuss Big Radio, then a good place to start is the Rugby Radio Station. Transmitting at 16 kHz (18.7-km wavelength), the aerials of this transmitter were tuned by a 6-m-high tuning coil, which is now installed in the new Information Age gallery at the Science Museum.

Figure 6.6 The Rugby tuning coil. (Courtesy of the Science Museum and Science Museum Picture Library.)

From January 1, 1926, onwards, the Rugby transmitter sent messages to the British Empire from the Foreign Office, time signals from Greenwich, news, personal telegrams, and Christmas greetings. At the time, it was the most powerful transmitter in the world, producing 10 kW from 54 water-cooled thermionic valves producing an aerial power from the twelve 250-m masts of 350 kW. During the Cold War, the transmitter was used to communicate with submarines submerged at depths of up to 22m.

The radio was decommissioned in March 2003, 77 years after sending its first Morse code transmission. The transmitter is almost certainly one of the most long-lasting most fully amortized examples of cost-efficient Big Radio.

6.20.3 Digital Audio Broadcasting and Digital Radio Mondiale for Low ARPU Markets

Digital Radio Mondiale is to an extent a modern reinvention of the Rugby Radio though on a less grand scale. Digital Radio Mondiale has a relatively low profile in developed economies and is often perceived as having little or no relevance to present and future mobile broadband technologies and business models. DAB is often viewed as being of similar marginal interest.

However, 5G specifications include a remit to service rural low ARPU markets. This requires at least an order of magnitude decrease in delivered cost per bit. This can only be delivered by adopting an ultrasparse network topology. DRM and DAB are two examples of technically successful ultrasparse, low-cost digital radio deployment. It is therefore appropriate to include them in a comprehensive review of technologies relevant to 5G.

The DRM standards group [41] was formed in Guangzhou in China in 1997 with the objective of developing a digital radio standard for the AM broadcast bands at long, medium, and short wave.

The specification has been jointly developed by ETSI [42], the European Broadcasting Union (EBU), and Cenelec, which have responsibility for radio and TV receiver performance specification.

In 2005, it was decided to extend the specification with a DRM30 variant to cover the VHF audio broadcasting bands below 174 MHz (see Figure 6.7). This covers the Band 1 analog TV allocation (47–68 MHz), the OIRT [43] FM band from 65.8–74 MHz, the Japanese FM band from 76–90 MHz, and Band 2 from 87.5–107.9 MHz.

Long-wave DRM has been field tested at Orfordness/Erlangen at 1,296 kHz and 1,298 kHz (231-m wavelength), and medium-wave DRM has been field-tested at Rampisham/Bockhagen at 9505 kHz (31-m wavelength), Sines/Kotka at 17,740 kHz (17-m wavelength), and Sines/Limassol at 21,630 kHz (13-m wavelength).

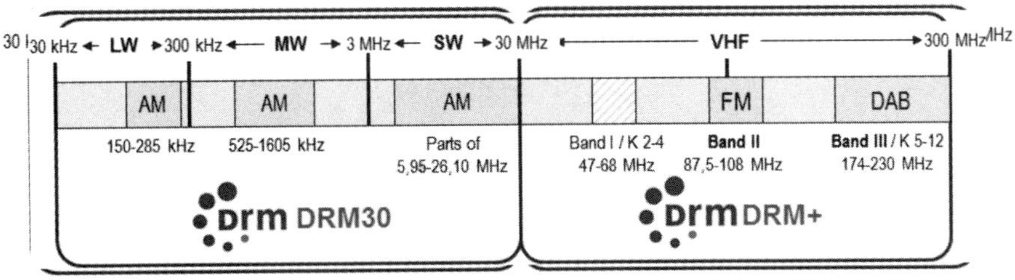

Figure 6.7 DRM30 and DRM + Digital Radio Broadcasting bands including DRM30.

All India Radio is transmitting long-wave DRM at 1,368 kHz (219m) and there are deployment plans in Africa across the 15 member states of the South African Development Community (SADC) including Angola, Botswana, Democratic Republic of Congo, Lesotho, Madagascar, Malawi, Mauritius, Mozambique, Namibia, Seychelles, South Africa, Swaziland, Tanzania, Zambia, Zimbabwe, and Brazil.

The specified frequency bands at long wave are 148.5 kHz to 283 kHz (2,000–1,000-m wavelength) in ITU Region 1 (Africa and Europe), the medium-band allocation is 526.5 kHz to 1,606 kHz (570m to 425m) in Regions 1 and 3 (Australia and Southern Asia), and 525 kHz to 705 kHz in Region 2 (Latin America and the United States), the short-wave allocation is 3 MHz to 27 MHz (100m to 11m) and is generally available worldwide.

The physical layer has to coexist with long distance interference from existing analog broadcast systems. In common with LTE, it has OFDM-based variable bandwidth channels. The channels are defined by signal bandwidth or transmission efficiency depending on required throughput (useful bit rate) and resilience to noise, multipath, and Doppler. The OFDM symbols are QAM-modulated to make dual-mode digital analog receivers easier to implement. Modulation levels are 4, 16, or 64.

A range of DRM radios and chip sets are available including the product shown in Figure 6.8 from Avion Electronics based in India [44]. The radio includes an emergency warning feature for events such as tsunamis and other natural disasters.

The channel spacing for AM radio under 30 MHz is 9 or 10 kHz. DRM is designed to be used within these nominal bandwidths or half-channel (4.5 or 5 kHz) bandwidths for simulcasting with analog radio or double-channel bandwidth (18 or 20 kHz) as and when allowed. The channel raster between 30 MHz and 174 MHz is 100 kHz.

The bit rate available for source coding for broadcast channels below 30 MHz is 8 Kbps for a half-channel, 20 Kbps for a standard channel, and 72 Kbps for a double channel. Channels from 30 MHz to 174 MHz support source coding from 35 Kbps to 185 Kbps. The source coders are MPEG-4 AAC

Figure 6.8 Avion Electronics DRM medium-wave, short-wave, and FM radio.

for stereo audio, MPEG-4 CELP (code excitation liner predictive codec), or HVXC (harmonic vector excitation codec) for voice with spectral band replication to support low data rate full audio bandwidth including Parametric Stereo and MPEG Surround Sound (MPEG/MPS).

There are 15 language description codes and 31 program codes including safety-critical information. Data services can be either synchronous or asynchronous. DRM networks can be deployed as single frequency (SFN) of multiple frequency (MFN). Table 6.5 provides a comparison of DRM symbol length and guard band and relative range available from DRM compared to DAB, DVB-T, and LTE.

6.21 DAB for Middle Earth

DAB suffers the same perception problem as DRM and, as far as this author knows, has not yet featured in any 5G technical discussions. As with DRM, there are potential positive technical and commercial touch-points with 5G that merit at least a brief summary.

The initial DAB standard [45] was produced in 1995 covering audio and data services. T-DMB was added in 2006 to address mobile TV and enhanced data streaming with a new codec introduced in 2007 (DAB+). In most countries it is deployable into the VHF TV band, Band 3 from 174 to 230 MHz (240 MHz in some countries). In Australia, a country with very large areas to cover, DAB and TV are multiplexed together.

Table 6.5
Symbol Duration and Guard Interval

	Broadcast Wide Area			4 G LTE	LTE Broadcast
System	**DRM**	**DAB**	**DVB-T**	**LTE**	**LTE eMBMS**
Frequency	>174 MHz	100 MHz, 220, and 1,500 MHz	470–700 MHz	450 MHz to 2.6 GHz	
Range	500 km	74 km	67 km	5, 30, 100 km	
Channel spacing	9 kHz	1.536 MHz	7.6 MHz	1.4, 3, 5, 10, 15, 20 MHz	
Max. data rate	72 kbps (double channel)	2.304 Mbps	5 to 31.7 Mbps	10–30 Mbps depending on cell size	
Modulation	QAM	QPSK	QPSK, 16 QAM, 64 QAM	QPSK, 16 QAM, 64 QAM	
Number of subcarriers	204	1536	1,705 (2K), 6,817 (4K)	300 across 4.5 MHz	600 across 4.5 MHz
Subcarrier spacing	41.66 Hz	1 kHz	4.46 kHz, 1.116 kHz	15 kHz	7.5 kHz
Symbol duration in microseconds	26,660	1,246	1,120	66.7	133.4
Guard interval in microseconds	2,660	246	7–224	4.69	33.33

DAB is also deployed to a limited extent in L-band between 1,452 and 1,479.5 MHz. The DAB Forum claims that over half a billion people are now within the DAB footprint with the standard adopted, although not necessarily implemented in 40 countries. The United States in particular is a DAB desert.

Penetration in developed economies is limited to an extent by automotive industry support, which varies from country to country. Penetration in developing economies is limited by the price of consumer devices. In 2015, entry-level DAB receivers were around $20 [46]. Mass adoption in these markets probably needs a price point closer to an FM transistor radio at around $2, an order of magnitude lower.

6.22 Radio, TV, and 5G

Terrestrial radio and TV are by far the lowest-cost option for delivering information and content. Transistor radios are by far the lowest-cost option for receiving content and information. It is therefore quixotic to dismiss terrestrial radio and TV as a delivery option for 4G and 5G and should contine to be considered as an option that could be more closely coupled with mobile and fixed wireless broadband.

Coexistence interefence has proved to be less problematic than broadcasters initially anticipated due to the tradition of always using worst-case assumptions in interference modelling. A transition to single-frequency networks over time will bring terrestrial TV planning closer to LTE single-frequency network planning.

Conversely, mobile broadband networks could deliver content significantly more efficiently over high-tower, high-power broadcast networks than over comparatively low-power, low-tower cellular networks that have been designed for two-way rather than one-way delivery .

Terrestrial radio and TV networks have additional physical assets that are useful to 4G and 5G systems including backhaul. These physical assets include DAB over L-band microwave, TV over C-band and satellite and terrestrial point-to-point links in Ku-band and Ka-band.

Radio and TV coverage is also reliant on a network of low-power repeaters and relays. LTE repeaters and relays are presently being standardized as part of 3GPP Release 10 through 12 [47].

Repeaters receive, amplify, and retransmit and do not demodulate and remodulate the channel. Low-cost HSPA repeaters are already available and include devices that use an SD card loaded with an operator-specific Absolute Radio Frequency Channel Number (UARFCN). This is used to program the synthesizer and phase lock loop in the transceiver to mix down the wanted 5-MHz channel from the designated passband, for example, a 5-MHz channel within the 35 by 35 MHz duplex passband of Band 8.

Relays, specifically advanced relays, decode before retransmitting. HSPA and LTE relays use an in band HSPA or LTE in band link to a host eNB (macro base station).

Repeaters apply front-end filtering to the passband and in a direct conversion transceiver translate the wanted modulated channel down to baseband. The signal to noise of the wanted channel will be directly related to the carrier-to-noise ratio but will also be affected by the quality of the downconversion process, so additional noise will be introduced by the mixing process and low oscillator (LO) injection.

Some window-mounted domestic repeaters then remix the signal on to a 5-GHz link to communicate with a second indoor unit, which then remixes back to baseband.

Repeaters lift the power of the carrier but do not improve the signal-to-noise ratio. They therefore depend on the noise floor in the passband being low (as a ratio to the low-power carrier). This is often the case in deep rural areas, so repeaters can be effective in providing reception a long way from a mast. This is why they are popular and effective in Australia. However, when the signal to noise is marginal, a repeater will often not help and may make things worse.

A relay does everything that a repeater does but demodulates and remodulates before retransmitting. The demodulator should therefore have a clean-up effect on the signal.

Existing HSPA repeaters for the domestic market typically support four bands, for example, Band 5 (850) and Band 2 (PCS 1,900) for the United States and Band 8 (900 MHz) and Band 1 (1.9/2.1 GHz) for Europe. They are single channel devices so, for example, will support a single 5 by 5-MHz channel within the Band 1 or Band 8 passband. They are therefore operator-specific, which reduces their market appeal particularly as most families and companies have smart phones that are registered with multiple operators.

Sharing radio, TV, and LTE repeater and relay infrastructure might help to resolve these issues or at a minimum help to reduce coverage cost. Backhaul can easily account for 30% of the CapEx and OpEx on a mobile broadband network with this ratio increasing as network density increases. We revisit this issue in Chapters 9 and 10.

6.23 Two-Way Radio: The DMR Standard

Last but not least, there is traditional two-way push to talk (do it yourself duplex) VHF and UHF radio used either to talk from device to device or via a repeater or relay. It is hard to beat the cost metrics of a walkie-talkie and there is a huge supply of specialist user devices including waterproof and explosion proof and ruggedized radios that you can drive a tank over. Do It Yourself Push to Talk duplex produces handsets with great sensitivity and a narrow RF channel bandwidth and passband help as well. Add this to the propagation gain from VHF and UHF when used with a decent antenna with a decent ground plane and an efficient RF amplifier due to the use of FM modulation and it is hard to see why you would want to reinvent this Cinderella sector of the radio economy. However, nothing, including common sense, ever stops technology progress, so there is now an ETSI digital mobile radio standard [48] implementing a TDMA air interface optimized for 12.5-kHz and/or 6.25-kHz channel bandwidth to support voice and low bandwidth data for use in unlicensed, licensed, or shared use (trunked) VHF or UHF spectrum.

The DMR standard has even less visibility within the 5G standards community than DAB and DRM, but it reinforces the point that technology and commercial innovation does not start or stop at 300 MHz.

If the 5G spectrum and standards process has a genuine ambition to produce a network of networks offer that can meet consumer expectations including consumers earning less than $1 or $10 a day, then it is important to at least study and understand the technologies and technology and commercial innovations that are being implemented at wavelengths of longer than a meter.

It is similarly important to study how other parts of the radio connectivity industry, including the MSS satellite community, are responding to changing user needs and exploiting new technical and commercial opportunities.

6.24 Summary

The middle Earth markets between the 48th parallel North to the 48th parallel South are potential high-growth markets for mobile and fixed wireless broadband in terms of volume, but require cost and price points at least an order of magnitude below existing mobile broadband radio systems.

A subscriber earning $100 a day is technically and commercially possible to service even in a remote rural area, for example, by combining mobile broadband with satellite access. A subscriber earning $10 a day is harder to service and a subscriber earning $1 a day is harder still.

Even at $1 a day, people find ways of either owning or using a mobile phone, a village sharing one phone, for example. This is because the economic benefits can be lifesaving.

Fully amortized network technologies such as GSM provide one option for wide-area coverage but incur high operating cost including the need to run base stations off diesel generators and solar power.

Attempts to produce ultralow-cost GSM handsets have not met with success, as high levels of integration are needed to meet the required cost floor. A selling price of $30 was considered an acceptable maximum in 2007, but today's target would be lower and would not meet the need for data. Other technologies such as DAB struggle to get below these $30 and $20 selling price targets which imply manufacturing costs significantly below $10.

Wi-Fi provides a lower-cost option for fixed or nomadic connectivity and can be implemented on a low-cost business model in which the spectrum is free and the real estate is free on the basis of users letting other users share their access points. Various initiatives are under way to facilitate and encourage these delivery models, the free Wi-Fi project in the Philippines being one example [49].

LTE is too expensive both in terms of network cost and user device cost, which is a function of component cost and a need to recover astronomically high nonrecurring engineering investment and arbitrate intellectual property value.

Wi-Fi over satellite or subspace high-altitude platforms such as balloons, drones, planes, or drones may be an alternative but whether these approaches can scale to the required cost point is uncertain.

The lowest-cost way of delivering information and content to people is terrestrial radio and TV. The lowest-cost way of receiving information is a transistor radio. Terrestrial radio and TV therefore remain important and many would argue critical to middle Earth markets. Technical innovation is also opening up new business model opportunities that have yet to be fully developed.

The WRC 2012 World Radio Congress signaled the emergence of middle Earth markets as agents of change both in terms of west to east spectrum allocation (APT a + b in the 700-MHz band) and technology (relatively relaxed but sensible out-of-band emission requirements). These markets may also be the most open to innovative allocation or auction policies and alternative delivery options including subspace systems.

By 2030, the end of the UN's sustainable development program, the world will be a different place with middle Earth as a market and technology driver. Present spectrum policy and standards policy and regulatory and competition policy has not fully comprehended the scale of this transition.

In the next five chapters, we explore what this means for 5G spectrum and standards evolution over the next 15 years and examine the possibility of producing a 5G standard that can deliver an order of magnitude decrease in per-bit delivery cost and user device cost.

References

[1] www.un.org/millenniumgoals/2014%20MDG%20report/MDG%202014.

[2] https://sustainabledevelopment.un.org/topics/sustainabledevelopmentgoals.

[3] http://www.itu.int/en/ITU-D/Statistics/Pages/facts/default.aspx.

[4] http://worldpopulationreview.com/continents/africa-population/.

[5] http://www.africaprogresspanel.org/publications/policy-papers/2015-africa-progress-report.

[6] www.un.org/esa/socdev/rwss/docs/2010/chapter2.pdf.

[7] https://www.telstra.com.au/coverage-networks/our-coverage.

[8] http://www.theglobalbrandonline.com/brand-success/brands-grow-over-time/guinness/.

[9] http://data.worldbank.org/region/SSA.

[10] http://www.vodafone.com/content/index/about/about-us/money_transfer.html.

[11] http://www.sciencemuseum.org.uk/online_science/explore_our_collections/information_age/cell.

[12] http://mobilemoneyafrica.com/details.php?post_id=462.

[13] http://www.gsma.com/.

[14] http://www.atu-uat.org/.

[15] http://www.aptsec.org/.

[16] http://www.gsacom.com/apt700/.

[17] http://www.dynamicspectrumalliance.org/.

[18] IEEE 802.22 standards for rural Wi-Fi.

[19] http://www.btwifi.co.uk/.

[20] http://www.wi-fi.org/discover-wi-fi/wi-fi-certified-passpoint.

[21] http://www.wballiance.com/key-activites/next-generation-hotspot/.

[22] http://www.google.com/loon/how/.

[23] http://www.google.com/loon/.

[24] http://ukhas.org.uk/guides:faq.

[25] Nov. 24 filing FCC, Google for 24 months beginning Jan. 1, 2016, Winnemucca City Council.

[26] http://internet.org/.

[27] http://titanaerospace.com/.

[28] http://steffanlewis.com/blog/worlds-longest-aircraft-the-airlander-hav304-airship/.

[29] http://www.airshipsonline.com/airships/r101/Crash/R101_Crash.htm.

[30] http://aviationweek.com/space/dawn-all-electric-satellite.

[31] http://www.satellitetoday.com/technology/2014/12/22/airbus-built-inmarsat-2-f2-satellite-breaks-world-record-for-mission-lifespan/.

[32] http://www.inmarsat.com/service/global-xpress/.

[33] http://www.groundcontrol.com/Hughes_9202_BGAN.htm.

[34] https://www.iridium.com/About.aspx.

[35] http://www.hostedpayloadalliance.org/.

[36] http://www.satellitetoday.com/publications/st/feature/2012/10/04/boeing-brings-iridium-capability-to-u-s-gps-system/.

[37] https://www.nasa.gov/content/tracking-and-data-relay-satellite-tdrs/#.Vb9levnNE9k.

[38] http://hubblesite.org/the_telescope/nuts_.and._bolts/spacecraft_systems/.

[39] http://www.thuraya.com/network-coverage.

[40] http://www.thuraya.com/SatSleeve.

[41] http://www.drm.org/.

[42] ETSI specification reference ES 201.980 v.3.11.

[43] Organisation Internationale de Radiodiffusion et de Television, also known as The International Radio and Television Organisation.

[44] http://www.avionelectronics.in/digital-wireless-radio-2248522.html.

[45] ETSI standard EN 300 401.

[46] https://www.worlddab.org/products.

[47] http://www.etsi.org/deliver/etsi_ts/125100_125199/125106/10.02.00_60/ts_125106v100200p.pdf.

[48] http://dmrassociation.org/the-dmr-standard.

[49] http://icto.dost.gov.ph/free-wi-fi-project-terms-of-reference/.

7

Three-Band 5G: The Wavelength Bands

7.1 Three-Band 5G

In Chapter 1, we argued that it is easier to discuss 5G design and performance parameters in the context of three wavelength-denominated bands rather than 300 LTE band combinations. The three bands of interest are the meter band (from 300 MHz to 3 GHz), as this is where present 4G and legacy systems are deployed, and the centimeter and millimeter bands, as these contain all present proposals for new 5G spectrum allocation. We should probably call the centimeter band the decimeter band, but centimeters are a more frequently used measurement base.

Although the terms are directly related and interchangeable, we argued that the use of wavelength rather than frequency provides a more useful way of defining Internet of Things (IoT)/user device and base station performance principally because wavelength relates to mechanical dimension. It is hard to design an effective quarter-wave or half-wave antenna at 1-m wavelength (300 MHz) in a space-limited smart phone or IoT modem or small-compact access point. It is easier to design an efficient antenna at the top end of the meter band at 0.1-m wavelength (3 GHz). In base stations and access points, antenna dimensions determined by wavelength directly influence site cost and wind loading for mast-mounted aerials.

Other industries use wavelength as a technical description. In the United Kingdom, we listen to the cricket commentary and shipping forecast and Parliament Today on Radio 4 Long Wave [1] at 198 kHz (1,500-m wavelength) or Radio 4 Medium Wave [2] at 720 kHz (416-m wavelength). Radio astronomers view the universe through radio telescopes denominated by radio and optical wavelength.

The three bands therefore are usefully viewed in the context of the overall electromagnetic spectrum, which includes longer wavelength spectrum [very high frequency (VHF), short wave, medium wave, long wave) and shorter wavelength spectrum, infrared, optical, ultraviolet, x-rays, and y-rays (gamma rays).

In Chapter 6 we identified some of the technical and commercial innovation taking place in longer wavelength spectrum that could be potentially relevant to 5G. This includes new, evolving radio broadcast technology standards for Digital Radio Mondiale at long-wave, medium-wave, short-wave, and VHF, DAB at UHF, and satellites at VHF for low-cost machine-to-machine (IoT) connectivity.

Long-wave, medium-wave, short-wave, and VHF radio is bandwidth-limited, but remains as the lowest-cost option for delivering voice and broadcast content to deep rural areas. Long-wave, medium-wave, and short-wave radio are the only parts of the spectrum that can be used to cover the world from one transmission site without the use of local repeater stations. Very long-wave radio is the only way to communicate with subsurface submarines. This is presently not discussed actively in the context of 5G research and development, but we argue that it should not be completely ignored.

Similarly, there is potentially relevant innovation at the shorter-wavelength end of the electromagnetic spectrum from the submillimeter band to optical wireless both from traditional wireless vendors and industries with a growing need for mobile broadband connectivity. This chapter and Chapters 8 through 12 explore the technical and commercial translation opportunities and research and development amortization opportunities, the positive touch-points, between these industries and the 5G community.

In terms of technical translation this includes the application or RF technology innovation and baseband processing and network topology innovation from radio and TV broadcasting, low Earth orbit (LEO), medium Earth orbit (MEO), and geostationary orbit (GSO) satellites, deep-space and near-space radio, radar, and radio astronomy. The technologies used in point-to-point backhaul in the centimeter and millimeter bands are also directly useful.

In terms of commercial innovation, translation opportunities include new service offers from radio and TV broadcasting and mixed payload models from the satellite industry.

We also document the potential negative touch-points determined by the coexistence issues introduced in Chapter 4. The dominant example for the past 10 years has been the TV industry and the battle for broadcast bandwidth in the ultrahigh frequency (UHF) band. The tension points between the mobile broadband industry and the TV industry have been exacerbated by an adversarial auction process. As the 2015 World Radio Congress demonstrated, the broadcasting industry particularly in Region 1 (Europe and Africa) remains protective of the sub-700-MHz UHF band (470 MHz–694 MHz, 0.63m–0.43m) and resistant to the idea of coprimary allocation [3].

7.2 Touch-Points and Tension Points

This battle is now likely to be extended both in terms of spectral scope and conflicting interest. This is because the new technologies being introduced in the backhaul industry, the satellite industry, near-space and deep-space radio, radar, and radio astronomy all require more bandwidth, wider channels within wider passbands. Table 7.1 shows these spectrum and technology touch-points and tension points.

7.3 Preferred Bands by Industry Sector and Industry-Specific Research Programs

Early in 2015, the U.K. regulator Ofcom asked for industry submissions from vendors and operators and other stakeholders on preferred 5G band allocations above 6 GHz, beginning at the top of the 5-GHz Wi-Fi band. Apart from infrastructure vendors, silicon vendors, and operators, submissions were received from COST IC 1004, ESOA, and ISG mWT.

COST is a trans-European Research initiative briefed to facilitate Cooperation in Science and Technology both within Europe and with scientific communities in emerging countries. IC1004 is a subgroup within COST studying Cooperative Radio Communications for Green Smart Environments [4].

ESOA [5] represents satellite operators in the EMEA (Europe, Middle East, and Africa) and Commonwealth Independent States including broadcasting, emergency communication, maritime and aviation communication, secure services for governments, and industrial process monitoring including energy and weather forecasting.

The submission from ESOA stressed the statutory oversight role of Ofcom and regulators in general to ensure that satellite operator business plan commitments including launch schedules are implemented in a timely manner and comply with national regulations and overall economic objectives. The submission argued that the consultation process should not include discussions about new spectrum allocation or sharing arrangements including preagreed-upon GSO orbit positions and associated frequencies on the basis that any such discussions would impact on investment sentiment in the sector. The global VSAT forum is advancing similar arguments on behalf of satellite TV broadcasting [6].

Table 7.1
5G Touch-Points and Tension Points

				Three-Band 5G								
Longer than a meter wavelength				**Meter Band**	**Centimeter Band**	**Millimeter Band**	**Submillimeter (Shorter than a millimeter wavelength)**					
30–300 kHz	300–3,000 kHz	3–30 MHz	30–300 MHz	300 MHz–3 GHz	3–30 GHz	30–300 GHz	300 GHz–3 THz	3–400 THz	400–900 THz	900 THz–300 PHz	300 PHz–10 EHz	<10 EHz
Kilometer	Meters	Meters	Meters	Meters	Centimeters	Millimeters	Millimeters	Millimeters	Nanometers	Nanometer	Subnanometer	
10–1	1,000–100	100–10	10–1	1–0.1	10–1	10–1	1–0.1	0.1 mm to 750 nm	750–350	350–1	1 nm to 30 picometers	<30 picometers
10-km band	Kilometer band	100-m band	10-m band									
Long wave	Medium wave	Short wave	VHF	Meter wave	Centimeter wave	Millimeter wave	Micrometer-wave terahertz radio	Infrared	Visible light	Ultraviolet	X-rays	Gamma rays Y rays
Digital Radio Mondiale			DRM+									
			DAB	DAB								
			Specialized Mobile Radio									
				Point-to-point backhaul								
				Point-to-multipoint backhaul								
				Satellites including GPS								
				Near-space radio								
				Deep-space radio								
			Radar									
			Radio astronomy									

ISG mWT is an ETSI-based Industry Specification Group on millimeter-wave transmission established in January 2015 with an initial brief to analyze present experience in the millimeter band, potential future applications, V-band and E-band worldwide regulations, V-band street-level interference, and the present and future status of the millimeter-wave semiconductor and component industry.

The initial study outputs of the ISG mWT highlight the growing need for additional millimeter band point-to-point connectivity identifying 57 to 66 GHz also known as V-band (where links are protected by oxygen resonance absorption) and 71–76 GHz and 81–86 GHz also known as E-band as preferred options implemented either for macro or micro cell backhaul, front haul (connection of a base station to remote radio heads), line of sight, near line of sight, or nonline of sight.

Each of these options is addressed by a European Framework 7 Programme [7] with three subsidiary programs. The E3Network Programme [8] is developing an E-band SiGe Bi-CMOS based transceiver for 10-Gbps, 1-km backhaul. The MiWaveS program [9] addresses backhaul access at 60 GHz, 71–76 GHz, and 81–86 GHz. The IPHOBAC-NG program [10] addresses the integration of millimeter-wave radio and photonics for backhaul and other applications described as photonic millimeter-wave radio (PMWR) to deliver 1–10-Gbps wireless access and 3-Gbps mobile backhaul. The project includes the use of coherent detection of dense wavelength multiplexed optical signals and optimized and integrated digital signal processing for RF and optical signals. The implicit ambition in these programs is to cross-amortize research and development investment across 5G backhaul and 5G mobile broadband hardware and software platforms.

A list of present group members is on the ETSI Web site Millimeter Wave portal [11].

The submissions also included inputs from EE based on initial study work by METIS (Mobile Communication Enablers for the Twenty-Twenty Information Society), the EU project tasked with setting out 5G ICT requirements for 2020 [12].

Satellite operators identify the millimeter-wavelength band as a priority and stress the likely negative impact on satellite sector investment that would be consequent on regulatory initiatives to change the primary status of centimeter band satellite spectrum. Changing this position is not impossible but would be critically dependent on constructing an incentive auction that placed a substantial premium on satellite sector share and asset value. The outcome of the 600-MHz TV to mobile broadband auction in the United States will be critical to establishing whether a similar process for satellite to mobile broadband could be plausible or economic.

The other inference from Table 7.2 is the lack of consensus across vendor and operator submissions on priority or preferred allocations. Each respondent has a particular view influenced either by their existing and perceived future commercial positioning or specific technology strengths. For example, a vendor with established technical capability in fixed point-to-point backhaul would potentially be able to leverage centimeter-band or millimeter-band point-to-point hardware into spectrally equivalent or spectrally proximate mobile broadband system design.

This is additionally only a U.K. submission process with at most a Europe-wide focus. The range of views and opinions globally is likely to be more diverse and could include identified requirements between 100 and 300 GHz.

7.4 Satellites in the VHF Band

In Chapter 6 we included a brief case study of the Orbcomm LEO satellite constellation between 137 MHz and 150 MHz and its role as a low-cost provider of wide-area IoT connectivity. Satellites in the VHF band are mainly between 137 and 138 MHz and include meteorological satellites transmitting data and low-resolution images and low data rate satellite downlinks with a matching uplink at 148 to 150 MHz [14]. Russian-manned spacecraft have historically used 121.5-MHz FM for voice communication and 143.625 MHz and 166 MHz.

The band 144 to 146 MHz is used for amateur satellites mainly in the upper half of the band between 145 and 146 MHz. The band 149.95 to 150.05 MHz is used by satellites providing positioning, timing, and frequency services for ionospheric research and for communicating with man-made objects in near-space orbits like the International Space Station [15]. The 240–270-MHz band is used for military satellite communication and lies within the 225–380-MHz passband for military aviation.

Although we are restricting our specific focus to above 300 MHz, we at least need to be aware that these systems exist and have the potential to be coupled either spectrally or commercially with future 5G systems.

7.5 Radio Astronomy and Space Radio: Relevance to 5G

This brings us to the very large, cosmological scale, radio astronomy industry. We include radio astronomy for a number of technical and commercial reasons. In terms of technical relevance, there are many technology innovations taking place in near-space and deep-space exploration, which have direct relevance to 5G component and system design. Radio astronomy is a growth industry and each new generation of radio telescope requires more bandwidth and is de-

Table 7.2
5G Spectrum in the Centimeter and Millimeter Bands: Vendor, Operator, and Stakeholder Submissions to Ofcom

	Centimeter Band	**Millimeter Band**
Wavelength	10–1 cm	10–1 mm
Frequency	3–30 GHz	30–300 GHz
Infrastructure Vendors		
Alcatel Lucent	Priority Bands: 27–29.5 GHz shared with microwave links Lower priority: 5.925–8.5 GHz, 15 and 18 GHz, 21.2–23.6 GHz, 25.35–27 GHz	Priority Bands: 36–37.5 GHz, 39.5–40.5 GHz, 42.5–52.6 GHz excluding 50.2–50.4 GHz, 55.78–66 GHz Lower priority: 36–40.5 GHz Bands for further study: 31.8–33.4 GHz, 40.4–43.5 GHz
Ericsson	10 GHz, 15 GHz	Above 30 GHz
	Prioritize bands not allocated to passive (receive only) services on a primary basis. Bands allocated to broadcasting services on a primary basis should be investigated to determine if they should be considered.	
Huawei	Priority Bands: 27.5–28.35 GHz, 29.1–29.25 GHz	Priority Bands: 37–38.6 GHz, 64–71 GHz, 71–76 GHz, 81–86 GHz Bands for Further Study 31.8–33.4 GHz
Silicon vendors		
Interdigital Europe		Priority Bands 55–71 GHz
Samsung Electronics UK	Priority Bands: 25–30 GHz focusing on 28 GHz	Priority Bands: 30–43.5 GHz Bands for further study: 40.4–42.5 GHz
Intel	Bands for further study: 24.25–24.45 GHz, 25.95–25.25 GHz	Bands for further study: 31–31.3 GHz, 42–42.5 GHz
Mobile Broadband Operators		
Vodafone	Priority Bands: 5.925–8.5 GHz	Priority Bands: 43.5–47 GHz, 51.4–52.6 GHz, 72–77 GHz, 81–86 GHz Bands for further study: 77–81 GHz
EE		Bands for further study: 31.8–33.4 GHz As per METIS submissions (https://www.metis2020.com/)
Confidential response	Priority Bands: 6–30 GHz for mobile	Priority Bands: Above 50 GHz for backhaul
Confidential response		Bands for further study: 37–39 GHz, 43.5–47 GHz, 57–64 GHz, 70–80 GHz
Research Groups		
COST IC 1004	Priority Bands: 25.25–29.5 GHz	Priority Bands: 36–40.5 GHz, 55.78–76 GHz, 81–86 GHz, 92–100 GHz
ESOA		Priority Bands: 37–39 GHz, 43.5–47 GHz, 57–64 GHz, 70–80 GHz
ISG mWT		Priority Bands: 31.8–33.4 GHz

Source: [13].

signed to work at ever higher levels of RF receive sensitivity. There are therefore coexistence issues that need to be considered.

The technologies and techniques used in radio astronomy span RF frequencies from VHF to 950 GHz. Channel bandwidths at shorter wavelengths can be greater than 500 MHz. The variable beamwidth, multiple frequency phase array antenna systems used in terrestrial antenna arrays require exquisite control of phase and amplitude combined with backhaul timing accuracy requirements over thousands of kilometers that are far in excess of any present 4G or proposed 5G mobile broadband network. The amount of raw data far exceeds present Internet traffic volumes and requires data mining and correlation techniques that are as yet untried and untested.

Radio astronomy is the science of using radio to study the stars and other extraterrestrial large and small (and compact) objects that emit radiation including the interstellar and inter galactic medium and the dust clouds of the Milky Way, the nursery for the formation of new stars and planets. We also use radar to examine planets in our solar system and the odd moon or two. It is a big subject getting bigger and faster all the time.

7.6 The Solar System and Our Galaxy: Our Local Backyard

7.6.1 Astronomical Units

This might seem indulgent but we are trying to make the point that if 5G aspires to be a final standard for mobile broadband connectivity, it needs to borrow techniques from the deep space radio industry, an industry that describes time in light years (9.5 trillion km) and distance in astronomical units (an Au, the distance from the Earth to the Sun, is 149.6 million km).

The distance from the Sun to the center of the galaxy is 26,000 light years. A light year is 9.5 trillion km. Our galaxy, the Milky Way, has 100 billion stars. The closest spiral galaxy, Andromeda, is 2.2 million light years away. It is approaching us at 1 million km per hour and is on course to collide with us in 5 billion years, an undesirable example of mobile connectivity. The universe has 100 billion galaxies and continues to expand at an ever-increasing rate. The way that we receive and process and analyze radio signals from man-made and natural objects in space including far, distant deep space and the bits in between has direct relevance to how we design terrestrial radio systems including 5G radio.

7.6.2 The ITU View of Space

The International Telecommunication Union (ITU) defines deep space as anything beyond 2 million km. The moon is 405,696 km from Earth, so is defined as near space. The deep-space communication bands are at S-band around 2

GHz, X-band at 7 and 8 GHz, K-band at 25 to 27 GHz, and Ka-band at 32 and 34 GHz. Frequencies down in the UHF and VHF bands are widely used for radio astronomy observation for objects observed with large red shifts.

7.6.3 Prewar and Postwar Radio Astronomy

Radio astronomy was invented by accident by Karl Jansky in the 1930s while investigating static on 30-MHz terrestrial radio links used by the Bell Telephone Company in the United States. After the World War II, a generation of radar engineers including Bernard Lovell in Manchester (Jodrell Bank Observatory) and Martin Ryle in Cambridge (The Mullard Radio Astronomy Observatory) repurposed radar antenna and receiver systems including the 8-m diameter German Wurzburg radar to look at signals from space.

This prompted a whole new generation of deep space radio observation techniques including Martin Ryle's use of multiple pairs of parabolic reflectors mounted on rails on an East-to-West axis, the One-Mile and 5-km arrays, large aperture antenna radio telescopes with resolution determined by the spacing distance. The East/West axis allowed the antennas to sweep across a segment of sky, hence the description of the device as an Earth rotation interferometer. The arrays were and are used to make high-resolution maps of radio galaxies (large-scale galaxies viewed at radio rather than optical wavelengths) and quasars (quasi-stellar, small-scale, compact objects of high radio brightness including neutron stars).

After an upgrade in the 1980s, the array was renamed the Ryle Telescope and is now used to help measure Cosmic Microwave Background radiation, the emission signature of the Big Bang 13.7 billion years ago and its immediate (370,000-year) aftermath (see Figure 7.1).

In 1968 one of Martin Ryle's colleagues, Antony Hewish, helped by a research student, Jocelyn Bell, completed an antenna consisting of 2,048 dipoles 3.7m in length spread across 4.5 acres of flat Cambridge countryside. This array working at a wavelength of 3.7m (81 MHz) detected the first pulsar.

Between the 1960s and today, the radio astronomy industry has produced radio telescopes with ever increasing resolution and sensitivity. The most recent telescope to be commissioned on the Cambridge site is the Arcminute Micro kelvin Imager (AMI) (see Figure 7.2). These are two separately correlated arrays of receivers operating at 12–18-GHz band with a small array of ten 3.7-m parabolic dishes in a compact configuration able to resolve angular scales of 2 to 16 arc-minutes linked to a large array formed by a compact configuration of eight of the 12.7-m Ryle Telescope dishes in the Ryle array including two offset dishes to create a North-South baseline to cover angular scales of 0.5–5 arc-minutes.

Figure 7.1 The Ryle Telescope now also known as the AMI telescope. (Courtesy of Cavendish Astrophysics, the MRAO and Stirling Essex.)

Figure 7.2 The AMI array. (Courtesy of Cavendish Astrophysics and the MRAO and Stirling Essex.)

The combination of the two systems works on the basis of the small array detecting shadows that galaxy clusters have imprinted on the cosmic radio background with the large array providing correction for contaminating radio sources. The overall bandwidth is 6 GHz divided into eight broadband 750-MHz channels. Independently of red shift, this combined array should be able to see clusters that are impossible or hard to detect optically, for example, galaxy clusters hidden behind dust clouds. This takes radio astronomy back to the period between 370,000 years and 1 billion years after the Big Bang, including pregalaxy structures coalescing under the influence of gravitational wave energy.

The AMI array demonstrates the complexity of present day radio telescopes in terms of their RF bandwidth, the RF phasing, and linearity required to preserve phase and amplitude information and the digital processing needed to perform multiple channel correlation. It also demonstrates the performance capability of present radio telescope systems when used with other systems.

As an example, the recently commissioned Atacama Large Millimeter Array (ALMA) in the Atacama Desert in Chile at an altitude of 5,000m cost $1 billion (the Hewish antenna cost £15,000) and has 66 steerable 12 and 17-m parabolic reflectors. The fiber connections for this array have tolerances of less than 10 microns. The array is considered to be ideal for studying the shifted spectral lines of water, carbon dioxide, oxygen, and nitrogen, the intellectual feedstock for a whole new generation of astrochemistry and for studying the radio emissions from black holes at wavelengths between 30 cm and 13 mm (1 GHz to 230 GHz).

The concept was developed from earlier schemes such as the Multi Element Radio Linked Interferometer Network (MERLIN), which linked the Lovell telescope at Jodrell Bank with the Ryle array in Cambridge 220 km away. These are known as long baseline interferometers. As the name implies, MERLIN originally used radio links between the antennas sites. These were replaced with fiber in 2011, which increased the bandwidth to 500 MHz centered at 1,500 MHz. The additional bandwidth allowed frequency diversity gain to be realized from each antenna pair and delivered an increase in sensitivity. The radio telescope operates between 151 MHz and 24 GHz. At a wavelength of 6 cm (5 GHz), MERLIN has a resolution of 50 milli-arcseconds, comparable to the Hubble Telescope at optical wavelengths. Very long base interferometers, by comparison, give a resolution of around 0.001 arc-second.

7.6.4 Radio Telescopes in Space

Longer baselines require a radio antenna in space coupled to a ground-based telescope. This was achieved by Japan with their HALCA satellite between 1997 and 2005 with an 8-m-diameter telescope coupled to an Earth-based telescope, a 21,000-km baseline, three times the possible distance between any pair of Earth-based telescopes.

The Russian Radioastron program launched in July 2011 had a 10-m diameter telescope in a highly elliptical orbit giving space/Earth baselines of 200,000 km using wavelengths of 1.3, 6, and 92 cm.

The Planck satellite [16] launched in 2012 spent 30 months observing cosmic background radiation from the First Lagrangian Point. The First Lagrangian Point is 1.5 million km inside the Earth's orbit and is the point at which the gravitational forces of the Earth and Sun are in balance, allowing the spacecraft-based telescope to hover in the sky.

The high-frequency Planck receiver measured radiation at wavelengths of 3, 2, 1.5, 0.9, 0.5, and 0.3 mm (100, 140, 220, 350, 550, and 850 GHz). The peak for observable cosmic background radiation is 160 GHz. The low-frequency receiver measured radiation at wavelengths of 10, 7, and 4 mm (30, 45, and 70 GHz). Measurements in space of cosmic background radiation (CMB) have provided accurate estimates of the age of the universe (13.7 billion years), the curvature of space (the flat universe) and a possible confirmation of inflation theory and the nucleosynthesis of helium. This is described by astro-scientists as precision cosmology. Planck produced enough data to estimate the relative contents of the universe as 4.9% baryonic matter (observable matter such as hydrogen and helium), 27% dark matter, and 68% dark energy. Radio astronomers are fond of pointing out that these are useful things to know and understand.

7.6.5 Even Bigger Terrestrial Telescopes

The largest ground based interferometer array is the very large baseline array in the United States with an 8,000-km baseline resolution giving a resolution of a thousandth of an arc-second. This array is able to measure pulsars at distances of over 7,700 light years with 10% accuracy. It has a primary beam, an area of sky covered by the individual elements, and a smaller synthesized beamwidth created by a combination of the elements in the array.

The Very Large Array radio telescope in Albuquerque, New Mexico, recently discovered an aurora on the exoplanet LSRJ 1835 18 light years away in the Lyra constellation [17]. The radio observations were correlated against images from the Hale optical telescope in Paloma in the United States and Keck Observatory in Hawaii. These are powerful machines made more powerful by their ability to work together.

A new low-frequency array (LOFAR) is being constructed by the Astron Astronomical Institute. There are 36 antenna clusters each containing a few hundred omnidirectional dipoles with the clusters distributed across the Netherlands, the rest of Europe, and Chilbolton Down in the United Kingdom.

LOFAR can survey the whole sky above the horizon or discrete parts of the sky or both simultaneously. It is a software telescope with no mechanical pointing. There are two antenna lengths, one covering low band 10 to 80 MHz and one covering high band 110 to 210 MHz missing the noisy FM band.

Amongst other tasks, LOFAR will be detecting hydrogen line radiation at 1,420 MHz which has shifted, by a large red factor into the VHF band and will help to tell us what happened between 370,000 years after the Big Bang and 1 billion years, the period when stars and planets started to coalesce.

The capabilities of ALMA and LOFAR will be combined in the next even more ambitious terrestrial project, the Square Kilometre Array (SKA), to be built in South Africa and Australia. SKA will cover the radio spectrum from the low frequencies of LOFAR through to the millimeter wavelengths of ALMA but primarily centered between 1 cm and 1m with three antennas spanning from 70 MHz to 10 GHz. There will be 3,000 high-frequency 15-m radius parabolic reflectors supplemented by lower-frequency dipoles. Like LOFAR, the dipoles are all beam-steerable based at dense centralized antennas farms with outliers along radial arms reaching out 3,000 km.

All sites will be interconnected by fiber with each channel carrying multiple frequency channels to allow fine resolution of spectral lines and rejection of narrowband man-made signals. Observations are planned to start in 2019 with performance improving in a second phase to be completed in 2025. SKA is therefore a project that is contemporary to terrestrial 5G deployment.

The SKA should provide exquisite resolution for resolving small compact radio objects including pulsars immediately adjacent to black holes. This will support research into gravitational waves, which, at the time of this writing, have just been measured for the first time at the Laser Inteferometer Gravitational Wave Observatory (LIGO) in the U.S. In parallel with LIGO, SKA will be able to do big sky searches efficiently and fast. The data handling requirements will be significantly higher than present day-to-day global Internet traffic and exceeds present super computer performance capabilities, but then that is the point of new technology on a cosmological scale.

This is the point at which the relevance of next generation radio astronomy to 5G radio systems should become apparent. The radio astronomy industry is producing radio systems that rely on electronically steerable antenna arrays capable of working from VHF to 950 GHz with channel bandwidths of 500 MHz or more at the shorter wavelengths. These arrays produce multiple beamwidths including adaptive electronically steerable narrow beams within wide beams with receive signals combined through ultralinear, ultralow, noise multiple receiver front ends. The wide-area timing accuracy required to maintain phase and amplitude information from thousands of antennas thousands of kilometers apart over fiber backhaul requires timing accuracy at least an order of magnitude better than present terrestrial radio systems.

7.6.6 Near-Space and Deep-Space Radio for Studying Planets and Exoplanets

Studying the planets in our solar system and exoplanets (the planets in other solar systems also involves extreme radio (and radar) techniques with relevance to 5G design and development. This includes the techniques used to communicate with spacecraft exploring the planets and the techniques used to manage and talk to Earth and space facing optical and radio telescopes and satellite systems.

The image processing and RF signal processing and data handling requirements of these systems provides a testing ground for future terrestrial radio and radio processing technologies. The failure of an RF or digital component in space compared with the launch cost of a spacecraft or satellite means that component and system reliability needs to be achieved almost regardless of cost. Therefore, although unit numbers are small, civilian and military space components and space systems command significant research and development spending.

Nuclear-powered deep-space missions are designed to be energy efficient, but solar-powered platforms including the new generation of "solar sail" deep-space exploration spacecraft have to manage every joule of precious generated energy. There are therefore a number of technical touch points between space radio communication and observation systems and 5G terrestrial radio.

In July 2015 the world started to receive high-resolution images from Pluto via the New Horizons [18] piano-sized nuclear powered spacecraft. Launched in 2006, the spacecraft has three optical instruments including a high-resolution camera, two plasma instruments for measuring charged particle emissions, a dust sensor and a radio receiver/radiometer to measure the radiant flux power of electromagnetic radiation. The payload consumes 28W [19].

The instruments are optimized for the low light levels and the cold of Pluto and the Kuiper Belt and are capable of measuring the geology of Pluto, the surface composition and temperature, atmospheric pressure, and the structure of the atmosphere, which extends 10,000 km out close to Pluto's moon, Charon.

Pictures of Pluto coming from the spacecraft have traveled 4.8 billion km and take 4.5 hours to arrive on Earth. Communications to and from the spacecraft are via three NASA 70-meter-diameter deep-space telescopes [20] spaced equidistantly 120° apart in longitude in California, Madrid in Spain and Canberra in Australia. These provide sufficient link budget to support the telemetry and communication needs of the mission, which at that distance means at best about 1 Kbps. The 1,024 pixel image from the camera is digitized as a 12-bit number. Lossless compression reduces the file size to 2.5 Mb, which means that it takes 42 minutes of transmission time to get one image back to Earth.

The spacecraft has two traveling wave tube amplifiers connected to a 2.4-m dish. The second amplifier was there to provide redundancy, but the two amplifiers can be made to work together with one signal with left-handed polarization and one signal with right-handed polarization. This increases the data rate by 1.9 times but doubles the power requirement. The nuclear power source is now 10 years old and cannot generate enough power to run both amplifiers simultaneously. This can only be done by shutting down the guidance system, which means putting the spacecraft into a spin to keep its pointing stable. This uses up the hydrazine needed for orbit or trajectory corrections and changes but can be worthwhile for short periods. It is going to be a while before we get videos from space, at least from the more distant planets.

7.6.7 Other Missions Supported from the Deep-Space Network

The deep-space network simultaneously needs to support at least 30 other spacecraft flying through space including Voyager 1 launched in 1977, which is now 40 billion km away, a 36-hour radio round trip. Other spacecraft include Rosetta chasing Comet 67P/Churyumov-Gerasimenko, 500 million km from earth. There is therefore substantial competition for a limited amount of deep space communications bandwidth.

7.7 Spectrum for Space

Most spacecraft use a portion of X-band at 8.4–8.5 GHz, which is set aside globally for deep-space communications. Because the signals coming back to Earth are weak, agencies such as NASA allocate dedicated frequency bands to avoid interference from terrestrial sources. Increasing amounts of terrestrial noise and a need to improve the sensitivity of near-space and deep-space receivers mean that space agencies are lobbying for higher frequencies around 32 GHz to be made available.

7.8 5G Spectrum and the Search for Extraterrestrial Life

There are now over 1,000 planets that have been separately identified outside of the solar system despite their low radio brightness when compared to their adjacent stars and over 4,000 candidate possible planets that need to be verified. Nine of these planets are similar in size to Earth and are in an orbit that is similar in terms of distance from the Sun. The challenge will be to find ways of identifying bio signature gases on the planets to provide an indication of potential life. This will require another generation of optical and radio telescopes.

The radio search for life is being partly financed by Yuri Milner, a theoretical physicist and internet entrepreneur. Mr. Milner has paid for thousands of hours of time on the Green Banks Radio Telescope in West Virginia and the

Parkes Telescope in New South Wales correlated against laser emission measurements (coherent emissions) from the Lick Observatory in California [21].

The combined measurements are calculated to be capable of detecting a 100-Watt signal 25 trillion miles away. The Breakthrough Listen project [22] has a $100 million budget. The radio measurements will be 50 times more sensitive and cover 10 times more sky than previous projects like SETI but will use some of the same analysis methods pioneered by SETI including the use of 9 million volunteer computers [23]. The search will cover the nearest 1,000 stars and will be the first project to scan the whole of the 1–10-GHz frequency band, the microwave window considered to be the most productive for studying planets. Particular frequencies of interest are hydrogen atoms at 1,420 MHz and hydroxyl molecules at four frequencies between 1,612 and 1,720 MHz. Collectively the frequency range between 1,420 and 1,720 MHz is called the Water Hole.

These narrow emission lines produced at characteristic frequencies by atoms and molecules need to be measured separately from continuum radiation. Radio continuum emission [24] is the broadband radiation emitted in the radio part of the spectrum by celestial objects. Its intensity (brightness temperature) varies relatively slowly as a function of wavelength.

Specific frequencies are assigned for spectral line narrowband observation and for continuum observation, solar wind observation, solar observation, pulsar observation, and very long baseline interferometers (VLBI). Narrowband frequency resonant spectral line observation is the intellectual feed stock for astrochemists. Most of the other observations address the mechanics of the universe. The spectral lines red shift over cosmological distances but can be recognized by their relative wavelength relationship.

Radio astronomy bands are designated to provide protection against radio interference from any unwanted source. The allocations do not mean that other wavelengths cannot be used, but they will not have regulatory protection. The frequency allocations in the European Union and their present uses are listed in Tables 7.3 through 7.10 [25].

These observations allow for precision dating and distance calculation and can be correlated with the shift of wideband cosmic background radiation, originally white light, shifted down to RF wavelengths.

7.9 Coexistence Issues: Terrestrial Radio Telescopes and Terrestrial Radio Systems

From the above it is clear that coexistence between terrestrial radio systems and terrestrial radio telescope receivers has to be managed both in the meter

Table 7.3
Sub-1-GHz Space Observation Bands

Frequency (MHz)	Band	Application
13.36–13.41 MHz	HF	
25.55–25.67 MHz	HF	
37.5–38.25 MHz	VHF	Continuum observations
73–74.6 MHz	VHF	Solar wind observations, continuum observations
150.05–153 MHz	VHF	Solar observations, continuum observations, pulsar observations
322–328.6 MHz	UHF	Continuum observations, VLBI
406.1–410 MHz	UHF	Continuum observations, pulsar observations
608–614 MHz	UHF	Continuum observations, VLBI

Table 7.4
L-Band Space Observation Bands

Frequency	Band	Application	Spectral Line
1,400–1,427 MHz	L-band	Spectral line observations	21-cm hydrogen line
1,660–1,660.5 MHz	L-band	VLBI	
1,660.5–1,668.4 MHz	L-band	VLBI, spectrum line observations, continuum observations	
1,668.4–1,670 MHz	L-band		
1,718–1,722.2 MHz	L-band		

Table 7.5
S-Band Space Observation Bands

Frequency (MHz)	Band	Application
2,655–2,690 MHz	S-band	Continuum observations
2,690–2,700 MHz	S-band	
3,260–3,267 MHz	S-band	
3,332–3,339 MHz	S-band	
3,345.8–3,352.5 MHz	S-band	

Table 7.6
C-Band Space Observation Bands

Frequency (MHz)	Band	Application
4,800–4,990 MHz	C-band	Continuum observations
4,990–5,000 MHz	C-band	Continuum observations, VLBI
5,000–5,030 MHz	C-band	VLBI

Table 7.7
X-Band Space Observation Bands

Frequency (GHz)	Band	Application
10.6–10.68 GHz	X-band	Continuum measurements, VLBI
10.68–10.7 GHz	X-band	Continuum measurements, VLBI

Table 7.8
Ku-Band Space Observation Bands

Frequency (GHz)	Band	Application
14.47–14.5 GHz	Ku-band	Spectral line observations, VLBI
15.2–15.35 GHz	Ku-band	VLBI
15.35–15.4 GHz	Ku-band	Continuum observations, VLBI

Table 7.9
Ka-Band Space Observation Bands

Frequency (GHz)	Band	Application	Spectral Line
22.01–22.21 GHz	Ka-band	Spectral line observations	Water line
22.21–22.5 GHz	Ka-band	Spectral line observations	Water line
22.91–22.86 GHz	Ka-band	Spectral line observations	Methyl formate, ammonia
23.07–23.12 GHz	Ka-band	Spectral line observations	
23.6–24 GHz	Ka-band	Spectral line observations, continuum observations	Ammonia line
31.2–31.3 GHz	Ka-band	Continuum observations	
31.3–31.5 GHz	Ka-band	Continuum observations	
31.5–31.8 GHz	Ka-band	Continuum observations	
36.43–36.5 GHz	Ka-band	Spectral line observations	Hydrogen cyanide, hydroxil

Table 7.10
Q-Band, V-Band, and W-Band Space Observation Bands

Frequency (GHz)	Band	Application	Spectral Line
42.5–43.5 GHz	Q-band	Spectral line observations	Silicon monoxide and other lines
48.94–49.04 GHz	Q-band	Spectral line observations	Carbon monosulfide
51.4–54.25 GHz	V-band		
58.2–59 GHz	V-band		
72.77–72.91 GHz	V-band	Spectral line observations	Formaldehyde
86–92 GHz	W-band	Spectral line observations, continuum observations	
92–94 GHz	W-band	Spectral line observations	Diazenylium and other lines
95–100 GHz	W-band	Spectral line observations, continuum observations	

band (300 MHz–3 GHz), centimeter band (3–30 GHz), and millimeter band (30–300 GHz).

Fortuitously and deliberately, radio telescopes tend to be situated in areas of low RF activity. Not a lot of people use their cell phones in the Atacama Desert in Northern Chile at 5,000m. However, some radio telescopes have to be in populated areas to meet required and specific aperture requirements. The deep-space network installation in Madrid is one example.

Space antennas face upwards most of the time but local RF power can still have a desensitization effect particularly at low pointing angles. Interference from Mobile Satellite Systems (MSS) also has to be managed and mitigated though MSS satellites know where they are and what they need to avoid and can turn spot beams on and off as required. The 5G terrestrial mobile user devices could be anywhere and cover a swath of spectrum directly adjacent to radio observation spectrum.

Each new generation of radio telescope is required to have improved sensitivity over a wider channel bandwidth over a wider passband. Contemporary radio telescopes have channel bandwidths > 500 MHz and passbands of many tens of gigahertz. Large amounts of money are spent on highly efficient RF front ends often cryogenically cooled to minimize noise floors. This increases the vulnerability of these systems to terrestrial and extraterrestrial interference.

In all three ITU regions there are submissions to allocate additional protected spectrum and higher protection ratios for existing spectrum for next generation near-space and deep-space radio observation systems.

These future requirements are being arbitrated within the ITU [26] and include designated frequency bands for radio astronomical measurement, protection of radio astronomy from adjacent channel interference and spurious emission, protection of radio astronomy services in frequency bands shared with other services, protection from unwanted emissions from wideband digital modulation, protection of radio astronomy measurements above 60 GHz from ground-based interference, radio quiet protection for the L2 Sun/Earth Lagrange point, sharing studies for frequencies between 10 THz and 1,000 THz, compatibility with nongeostationary satellite systems including MEO and LEO communication satellites, mutual planning between Earth exploration satellite services and radio astronomy in the 94-GHz and 130-GHz bands, and preferred bands for radio astronomy between 1 and 3 THz.

The spectral line frequencies, for example, the hydrogen line at 21 cm/1,400–1,427 MHz, need to accommodate the Doppler shift introduced when the spectral line is viewed from distant galaxies, a radial velocity shift of up to 100 km per second. The hydrogen line has been observed red-shifted to 500 MHz and some of the most abundant molecules have been detected in galaxies with velocities of up to 50,000 km per second, which translates into a 17% frequency reduction.

There are more than 3,000 spectral lines outside the allocated bands, which radio astronomers can observe as far as spectrum sharing and interference allows. The general point to make is that radio astronomy bandwidth requirements, including the requirements for a new generation of astrochemists and astrobiologists, are increasing over time and there is a growing appetite for higher protection ratios to support deep-space observation.

7.10 Radar and the Scaling of the Solar System

Radar started to be used after the World War II to scale the solar system. The first radar echo from the moon was achieved in 1946 by Zoltan Bay, a Hungarian scientist [27]. The combination of pulse delay and Doppler shift provided the basis for mapping the Moon. The Lovell radar achieved an echo from Venus in 1961 and provided the basis for measuring delay introduced by the atmosphere of the planet, the atmosphere of Earth and gravitational effects.

In 1988 the S-band (2,380 MHz/12.6 cm) radar transmitter and 305-m dish at the Arecibo Observatory in Puerto Rico mapped the Maxwell Montes region of Venus with a horizontal resolution of 2 km. The mountain is 11 km above the surrounding plain. Craters near the North Pole of Mercury were mapped using delay and Doppler at a resolution of about 15 km.

Terrestrially based radar beyond the solar system is not practical. However, radar on board spacecraft visiting planets and other objects can be used to examine the density of dust clouds and to calculate local distances and can be used for terrain mapping and surface and subsurface examination.

Radar is also used to detect Near Earth Objects including potentially hazardous objects. Near-Earth objects (NEOs) are comets and asteroids that have been deflected by the gravitational attraction of nearby planets into orbits that bring them close to Earth. The Near Earth Object (NEO) Program is a NASA program detecting potentially hazardous asteroids and comets that could approach the Earth. Ninety percent of the near-Earth objects larger than 1 km have been discovered (about 900 objects) and the hunt is now on for objects larger than 140m. A total of 1,600 objects are classified as potentially hazardous.

A meteor impact off the coast of the Yucatan Peninsula in Mexico is assumed to be the probable cause of the extinction of the dinosaurs 65 million years ago. A 150-m asteroid exploded 5 to 10 km above the Tunguska region in Russia in 1908 causing an air burst that flattened trees and killed animals over an area of several square kilometers. The 20-m meteor that landed in Russia in 2013 [28] caused over 1,000 injuries, including cuts from glass, concussion, retinal burns, and sunburn.

The NEO system is being upgraded to a 1-MW, 12.6-cm radar system with a back end that supports significantly more sophisticated signal processing than its predecessors. More bandwidth and an increase in receiver bandwidth and sensitivity together with a 10 MHz sampling rate and 20-MHz decoder should deliver an order of 40 times improvement in sensitivity [29].

7.11 Coexistence Challenges: Collaboration Opportunities

Over the past 60 years terrestrial and space-based radar has provided the basis for scaling the solar system, helped to map the terrain and surface and the below-surface structure of the nine planets in our solar system.

In parallel radio systems have evolved to bring us pictures from spacecraft either orbiting or landing on the planets in our solar system. These are supported from the NASA Deep Space network providing communications to spacecraft at distances up to 40 billion km from Earth.

Radio telescopes are also being used to search for extraterrestrial life. The quest for every more powerful wideband wide-channel bandwidth radio telescopes suggests that additional dedicated spectrum may be needed in the future in the meter, centimeter, and millimeter bands. This includes coexistence between terrestrial radio and wideband radio astronomy observations (continuum measurements and VLBI measurements), coexistence of terrestrial radio with narrowband spectral line detection and coexistence of wideband and narrowband astronomy with LEO and MEO mobile satellite (MSS) systems including L-band and S-band MSS. Coexistence issues in the submillimeter band from 300 GHz to 3 THz will also need to be addressed.

On the positive side, many of the technologies and techniques used in radar astronomy and space radar are potentially translatable to 5G component and system design. The enabling technologies of space based and terrestrially based radio astronomy include highly efficient RF centimeter-band and millimeter-band receivers coupled to multiple antennas with multiple low noise receive chains and associated digital signal processing and digital image processing techniques. Digital processing and correlation across multiple inputs and multiple frequencies provides the basis for interference and noise cancellation in both the optical and RF domain including the mitigation of narrowband RF interference.

Radio measurements are correlated across the whole electromagnetic spectrum including space-based gamma ray measurements used to research high-energy radiation including radiation from pulsars (due to gamma rays not being affected by the large-scale magnetic field of the galaxy). This requires data processing, correlation, and data analysis techniques that are significantly more complex than present 4G and proposed 5G mobile broadband network requirements.

7.12 Spacecraft and Spectrum

In 1957 the Mark 1 telescope, now known as the Lovell Telescope, was completed at Jodrell Bank just outside Manchester. With a diameter of 76.2m it was the largest steerable dish radio telescope in the world. Part of the gun turret mechanisms from the battleships HMS Revenge and Royal Sovereign were reused in the telescope's motor system. The telescope was finished just in time to track the launch of the world's first satellite, the Russian Sputnik 1 [30] just before midnight on October 12, 1957.

There are only two main windows in the electromagnetic spectrum that are open to space. One is the optical spectrum and is the reason we can see stars in space, the second is the radio spectrum with an optimum RF window (with some exceptions) from 30 MHz to 30 GHz though lower and higher frequencies (longer and shorter wavelengths) are useable.

Below 30 MHz, the ionosphere between 100 and 500 km absorbs and reflects radio waves (which is how long-wave signals propagate around the world). Above 30 GHz, the lower atmosphere or troposphere below 10 km absorbs radio signals due to oxygen (at 60 GHz) and water vapor. Even between 20 and 30 GHz, there are absorption bands that must be avoided including the first water vapor resonance peak at 23 GHz (used for weather radar).

The oxygen peak at 60 GHz produces an attenuation loss of about 15 dB per kilometer with a lower peak at just over 100 GHz producing attenuation of about 2 dB per kilometer. Water vapor losses peak just below 200 GHz with a loss of almost 40 dB per kilometer.

Sputnik carried two radio beacons at 20.005 MHz and 40.01 MHz. The Soviets continued to use frequencies around 20 MHz and 15 MHz for subsequent missions. The first satellite launched by the United States (Explorer 1) carried beacons on 108.00 and 108.03 MHz just above the terrestrial FM broadcast band (from 88 to 108 MHz) and just inside the civil aviation band from 108 to 136 MHz.

This frequency had been specified by an international committee for the International Geophysical Year (IGY, 1957/1958) as the one to be used for all scientific satellites launched in pursuit of IGY objectives. The Soviets had chosen to ignore this recommendation and use lower frequencies.

Sputnik was the first of many space probes that the Lovell telescope could and would track including the US Pioneer 5 between March 11, 1960, and June 12, 1960 [31]. The telescope was used to send commands to the probe including the instruction that separated the probe from its carrier rocket. The 43-kg probe set off towards Venus to explore interplanetary space and to test how far radio communications with a small (baseball-sized) object could be extended [32]. The solar-powered miniature spacecraft set a new record of 22.5 million miles before carrier wave contact was lost.

Pioneer 5 had two radio transmitters operating at 378 MHz, a low power 5-W transmitter used when close to Earth, and a 150-W transmitter that was turned on by a command from the Lovell telescope when the probe was 8 million miles away. The signals from the probe were analyzed to determine the Astronomical Unit (the average distance from the Earth to the Sun) used to express distances in the solar system. The cosmic ray flux density and flow of charged particles now known as the solar wind were measured to a distance of 17.7 million miles until telemetry encoding was lost. Data was collected that provided insight into magnetic fields in space. This set of instruments became the standard measurement kit for the next 10 years of deep-space exploration including four Pioneer spacecraft launched between 1965 and 1968.

The Lovell telescope tracked the U.S.S.R. unmanned Moon lander Lunar 9 in February 1966 and Lunar 15 in 1969. A new 15-m dish was constructed at Jodrell Bank in 1964, which was used to track the journey of Neil Armstrong and Buzz Aldrin to the Moon in Apollo 11 and their arrival on July 20, 1969.

7.13 NASA Deep-Space Network Bands

Today all U.S. spacecraft and many other spacecraft are supported from the three NASA Deep Space Network Ground Stations use S-band, X-band, and Ka-band for tracking and data/telemetry (see Table 7.11).

NASA spacecraft in the 1960s used S-band and then X-band in the 1990s and Ka-band from 2000 onwards. Most spacecraft have dual-frequency transceivers, initially S-band and X-band and more recently X-band and Ka-band. The shorter wavelengths provide better tracking ability but require more pointing accuracy.

If spacecraft become power-limited due to an accident or malfunction, more ground-based receivers are added to the receive array. The most famous example to date has been Apollo 13, which required the combined gain available from the 70-m DSN antennas and Australian Parkes Observatory radio telescope.

Table 7.11
NASA Deep-Space Bands

Band	Uplink Frequency (MHz)	Downlink Frequency (MHz)
S	2,110–2,120	2,290–2,300
X	7,145–7,190	8,400–8,450
Ka	34,200–34,700	31,800–32,300

7.14 Meter-Band, Centimeter-Band, and Millimeter-Band RF and Baseband Components in Space

Almost anywhere in space is an expensive place to have an equipment failure. Failure can be accommodated to an extent by redundancy. One of the two S-band transceivers on Hubble has failed for example. Low Earth orbit satellites like Hubble can be repaired in space. MEO satellites, for example, GPS at 20,000 km and GSO satellites at 35,000 km are far less accessible though there are studies to make servicing these more distant platforms in space technically and commercially feasible [33].

The reliability of RF systems and their supporting digital processing subsystems is therefore critically important. Part of the problem for space communications hardware is radiation damage. Digital components such as analog-to-digital and digital-to-analog converters, Digital signal processors and CMOS-based FPGAs can be particularly vulnerable and can inconveniently and unpredictable latch-up when exposed to high levels of radiation.

It is possible although expensive to produce radiation hardened DSP and FPGA chips that are resilient to high ionizing dosages, the cumulative effect of ionizing radiation on components on longer space missions described as total ionization dosage (TID) and single event effects, a random failure due to a charged particle arriving in the wrong place at the wrong time described as a single event upset (SEU). The TID and SEU rates differ by orders of magnitude depending on orbit trajectory, the Sun's solar cycle, and shield efficiency. The TID is relatively easy to calculate. SEU by its nature is more unpredictable.

The present approach is to build in hardware redundancy and manage failure through software resets. Given that we are due for another massive electromagnetic storm (on a 150-year rather than an 11-year Sun cycle), it might be useful and potentially profitable to translate this space sector experience into terrestrial component and subsystem design [34].

7.15 Time for Another Carrington Event?

A star is a large continuously exploding hydrogen bomb. Our own Sun has a 11-year cycle in which the intensity of this process both overall and locally on the Sun's surface, changes in intensity; the solar flares that coincide with high levels of radio interference from the Sun. Occasionally there is a coronal mass ejection emitting a sudden blast of x-rays, high-energy particles, and plasma (hot ionized gas). The biggest geomagnetic storm on record is the Carrington event in 1859 observed by Richard Carrington [34] through an optical telescope. In telegraph offices around the world, spark discharges shocked telegraph operators and set telegraph paper on fire. Even when batteries were disconnected, aurora-induced electric currents in the wires allowed messages to be transmitted.

Today the impact on a 5G radio system, electric utilities and aircraft, spacecraft, GNSS satellites, and communication and observation satellites would be close to catastrophic. These super storms occur every 150 years or so and we are just about due for one now. Space weather was added to the National Risk Register in the United Kingdom in 2011 [35]. The impact of space weather probably needs to be part of the longer-term 5G design and development brief.

7.16 Submillimeter Wavelengths in Space

The Hubble Telescope is an instrument of wondrous capability, which after a dodgy start has produced startling images of the known and previously unknown universe [36]. It is astonishing to realize that Hubble is now 25 years old and being replaced by a more capable optical telescope, the James Webb Telescope, named after the Apollo mission administrator and scheduled for launch in 2018 [37].

Unlike Hubble in a low Earth orbit, the Webb Telescope will be 1 million miles from Earth at Lagrange Point 2. The orbit has the advantage that the telescope is not rushing around the Earth but hanging at a stable point in space. The disadvantage is that there is no repair option if something does not work after launch or fails during the mission.

In addition to optical observations from the 6.5-m mirror almost three times the size of the Hubble mirror, the James Webb telescope will be looking at the universe at infrared frequencies. The infrared measurements will provide the basis for studying the universe from 200 million years after it was born (13.7 billion years ago) when the first stars and galaxies were beginning to coalesce. It will also study the planets around other stars and the planets in our host solar system.

This is part of a trend to extend space based optical telescopes either side of the visible wavelength bands including measurements below infrared in the terahertz band at 0.999-mm to 0.099-mm wavelength (300 GHz to 3 THz), also known as the submillimeter band.

Anything in the universe warmer than 10K (–263°C) emits terahertz radiation. Our bodies emit terahertz radiation and terahertz imaging systems are used for airport security. Submillimeter observations are also done from mountain-based observatories on Earth including the Caltech Observatory in Hawaii [38], the Atacama Observatory in the Atacama Desert in Chile (at 5,000m) [39], and the Heinrich Hertz Telescope in Arizona [40].

On the other side of visible light an increasing number of measurements are being made from Earth orbiting space-based telescopes and deep-space missions at x-ray and gamma ray (also known as y-ray) wavelengths. This is not absolute. Radio could be defined, for example, from 30 kHz (a wavelength of 10 km) to 3 THz (a wavelength of 100 μm), but longer wavelength measurements may also be important. Gravitational waves may have frequencies measured in days, months, or millions of years and we still need to understand magnetic fields in more detail.

To put the electromagnetic spectrum into perspective, gamma rays/y-rays at <0.01 nm are about the size of an atomic nucleus and are the result of nuclear reactions. They are emitted from pulsars, quasars, and black holes.

X-rays from 0.01 to 10 nm are about the size of an atom and are generated from exploding stars and quasars where temperatures are between 1 million and 10 million degrees. Ultraviolet radiation has wavelengths from 10 to 310 nm, about the size of a virus. Young energetic stars produce large amounts of ultraviolet light. Visible light from 400 to 700 nm has a wavelength equivalent to a molecule or protozoan (a single-celled microscopic animal). Conveniently, our Sun radiates most of its energy in the visible range. Infrared wavelengths from 710 nm to 0.1 mm (400 THz to 3 THz) are equivalent to the width of a pin point through to the size of a small seed plant. At 37°C our bodies emit infrared energy with a peak intensity at 900 nm; that is how all those infrared presence detectors work.

The infrared band could also loosely be described as the submicrometer band: 750 nm is 0.75 μm and 0.1 mm is 3 THz. The micrometer, also known as the micron, is used to scale biological cells, bacteria, and silicon chips.

7.17 Terahertz Radio

Below the infrared band, we are in to the top end of the radio band also described as the submillimeter band. The band from 300 GHz to 3 THz is also described as terahertz radio.

7.18 1 mm and longer < 300 GHz: Subterahertz Radio

Below the terahertz band, we have radio as we know it at wavelengths of 1 mm (300 GHz) to several kilometers (30 kHz = 10 km). Radio waves come from all parts of the universe including background radiation, the interstellar clouds, and the cool remains of supernova explosions, red shifted from visible to RF wavelengths.

The curiosity that drives us to discover more about the universe also drives us to invent new ways of measuring the whole electromagnetic spectrum. Some of this will be useful and directly relevant to future 5G research and development.

7.19 Quantum Telecom: The Science of the Very Small

It seems appropriate to finish a chapter on wavelength by looking at quantum computing, or rather, the transfer of the science of quantum computing to the telecoms industry and the related convergence of device and network physics.

Since 2008 the boundaries of the physical world have been explored under a mountain in Switzerland [41]. The research at CERN (the European Organization for Nuclear Research), has partly been about understanding the immediate aftermath of the Big Bang but is becoming increasingly relevant to device design as silicon scales to 22 nm and 14 nm. The physical oxide thickness needed for a 22 nm node is 0.5 nm, about twice the diameter of a silicon atom so an ability to harness rather than fight quantum physics at device level becomes progressively more important.

At the other end of the scale, mobile operators are beginning to invest in quantum computing [42]. The motivation is partly to gain visibility to next generation device performance but also to explore potential solutions to some of the timing and time distribution issues that are beginning to constrain high data rate wide-area mobile networks. The telecommunications industry has become progressively more dependent on accurate time, frequency, and phase references with ETSI producing synchronization standards specifying increasingly stringent requirements for jitter and wander at synchronization interfaces, for clock accuracy and stability and synchronization network architecture [43]. This includes a developing realization that network timing and time reference distribution is beginning to need to take into account relativistic effects.

Einstein was dismissive of the notion that quantum physics, in particular the property that entangled particles exhibit at a distance of apparently changing state instantaneously together (spooky action at a distance), could be harnessed for long distance communication. The flaw is that you need to know the state of the other entangled particle and you can only do that by conventional transmission. However, there are a new generation of super accurate quantum clocks using laser cooled atoms that could help reduce the vulnerabilities associated with GPS-based timing systems.

In 4G mobile broadband, localized and large-scale time coordination is made more complex by the need to manage interference through mechanisms such as intercell interference coordination (ICIC) and to manage link budgets through mechanisms such as coordinated multipoint (CoMP) transmission.

In 5G systems it is not unrealistic to assume that some applications will need end to end time coordination across significant global distances, the Australian brain surgeon working on an operation in the United States would be an example. Quantum telecom might be part of the answer [44].

References

[1] http://www.bbc.co.uk/radio4/programmes/schedules/lw.

[2] http://www.mediumwaveradio.com/bbc_national.php.

[3] http://www.gsma.com/newsroom/press-release/gsma-urges-governments-to-support-co-primary-allocation-of-sub-700mhz-spectrum-for-broadcast-and-mobile-at-wrc-15/.

[4] http://www.ic1004.org/.

[5] https://www.esoa.net/.

[6] https://gvf.org/.

[7] http://ec.europa.eu/research/fp7/index_en.cfm.

[8] http://www.ict-e3network.eu/-, Project reference FP7-ICT-317957.

[9] http://www.miwaves.eu/.

[10] http://www.ist-iphobac.org/ng/.

[11] https://portal.etsi.org/TBSiteMap/mWT/ListofmWTMembers.aspx.

[12] https://www.metis2020.com/.

[13] http://stakeholders.ofcom.org.uk/consultations/above-6ghz/.

[14] http://www.orbcomm.com/networks/satellite.

[15] Amateur Radio on the International Space Station, http://www.ariss.org/contact-the-iss.html.

[16] http://www.esa.int/Our_Activities/Space_Science/Planck/Planck_s_HFI_completes_its_survey_of_early_Universe.

[17] http://www.bbc.co.uk/news/science-environment-33711161.

[18] https://www.nasa.gov/mission_pages/newhorizons/main/index.html.

[19] http://pluto.jhuapl.edu/Mission/Spacecraft/Payload.php.

[20] http://deepspace.jpl.nasa.gov/.

[21] http://astronomynow.com/2015/07/21/lick-observatory-joins-search-for-intelligent-life-in-the-universe/.

[22] http://phys.org/news/2015-07-aliens-unveil-secrets-universe.html.

[23] http://www.seti.org/.

[24] http://www.hartrao.ac.za/continuum/.

[25] http://www.ukaranet.org.uk/basics/frequency_allocation.htm.

[26] https://www.itu.int/rec/R-REC-RA/en.

[27] http://history.nasa.gov/SP-4218/ch1.htm.

[28] http://www.bbc.co.uk/news/world-europe-21468116.

[29] http://echo.jpl.nasa.gov/~lance/radar.small.body.mission.targets.html.

[30] http://history.nasa.gov/sputnik/.

[31] http://nssdc.gsfc.nasa.gov/nmc/spacecraftDisplay.do?id=1960-001A.

[32] http://epizodsspace.airbase.ru/bibl/spaceflight/2005-5/pio5.html.

[33] http://www.space.com/27128-darpa-robotic-satellite-repair-droids.html.

[34] http://science.nasa.gov/science-news/science-at-nasa/2008/06may_carringtonflare/.

[35] http://www.metoffice.gov.uk/publicsector/emergencies/space-weather/met-office-role.

[36] http://news.nationalgeographic.com/2015/04/150423-hubble-anniversary-webb-telescope-space/.

[37] http://www.jwst.nasa.gov/.

[38] http://cso.caltech.edu/.

[39] http://www.almaobservatory.org/.

[40] http://aro.as.arizona.edu/smt_docs/smt_telescope_specs.htm.

[41] http://home.cern/topics/large-hadron-collider.

[42] http://exchange.telstra.com.au/2015/12/08/telstra-announces-plans-to-co-invest-with-federal-government-in-silicon-quantum-computing/.

[43] www.etsi.org/deliver/etsi_gts/05/0510/05.../gsmts_0510v050000p.pdf.

[44] https://www.sciencenews.org/blog/context/einstein-was-wrong-about-spooky-quantum-entanglement.

8

The Meter Band: 300 MHz–3 GHz

8.1 5G in the Meter Band

This chapter analyzes the merits and demerits of implementing 5G radio system at wavelengths of between 1m and 0.1m (300 MHz to 3 GHz).

In Chapter 7 we documented the work items within the European Framework 7 Programme (FP7) relevant to 5G in the millimeter band including ISG mWT [1]. Demonstrating an admirably broad vision but less admirable enthusiasm for tortuous acronyms, FP7 also has a 5GPP program known as FANTASTIC5G (Flexible Air Interface for Scalable Service Delivery within Wireless Communication Networks of the Fifth Generation) [2] with a brief to study a new 5G multiservice air interface below 6 GHz (meter band and the lower part of the centimeter band). The group membership includes satellite and space technology vendors [3].

To have a serious impact, European funded initiatives have to scale on a global basis. This was easier when Europe was relatively dominant in the mobile phone market [4]. The FP7 program is funded with the objective to secure standard essential patents for European-based industry. This complicates cooperation with countries and companies focused more specifically on U.S. and Asian markets.

The stated mission of the program is to identify the technology and commercial innovations needed to deliver a 1,000 times increase in capacity, dense deployment capability with capacity to support 7 trillion devices serving 7 billion people, an energy saving of up to 90% per mobile service, a reduction in service creation from 90 hours to 90 minutes, advanced user controlled privacy, and a 20% reduction in operational cost.

It is plausible that progress could be made towards delivering these objectives by improving interworking between existing meter-band mobile broadband networks and Wi-Fi at 2.4 and 5 GHz, but this work is already under way within existing 4G work streams. Companies like Google are introducing products that make the way that we use Wi-Fi easier and more efficient at the application level [5]. A new global standard is not necessarily needed to resolve basic user interface issues.

To be credible, any 5G initiative below 6 GHz has to add user and network value over and above the gains achievable from optimizing existing technologies. As the program only started in July 2015, there were, at the time of this writing, no outputs to document other than the general objectives. However, we can discuss possible physical layer options and assess their relative merits in the context of past industry experience.

8.2 Physical Layer Options for 5G

These options can be summarized as:

- A new physical layer wide band underlay beneath existing LTE and legacy narrowband radio channels;
- A new physical layer coupled to dynamic access to unused or underused spectrum in the frequency, time, or spatial domain, also known as the white-space spectrum;
- A new physical layer coupled to dynamic use of guard bands and duplex gaps in existing LTE and legacy spectrum;
- A new physical layer deployed into existing licensed or unlicensed spectrum below 1 GHz with a specific application focus, for example, an Internet of Things (IoT) and machine type communication (MTC) optimized physical layer;
- A new 5G physical layer deployed into new spectrum below 3 GHz for high-bandwidth, wide-area connectivity.

8.3 The Wideband Underlay Option

Although not presently included in any 5G proposals, it would be at least theoretically possible to deploy an ultrawideband underlay beneath existing LTE and legacy mobile broadband networks. Ultrawideband is defined by the U.S.

Federal Communications Commission (FCC) as a passband equal to 20% of fractional bandwidth, for example, 200 MHz at 1 GHz or 400 MHz at 2 GHz or 600 MHz at 3 GHz or any passband greater than 500 MHz. Short-pulse ultrawideband radio has been deployed for many years in the radar industry and is widely used today; automotive radar at 24 GHz is one example. It is also used in precision location systems.

A number of companies tried to convince the FCC in the late 1990s that ultrawideband technology could be deployed on a license-free basis under Part 15 of the FCC regulations for personal area, local area, and potentially wide-area wireless connectivity. In February 2002 the FCC issued the ultrawideband (UWB) rulings allowing technology commercialization and setting agreed radiation limits (Table 8.1).

In 2003, the IEEE published the 802.15 standard for high data rate (11–55 Mbps) personal area networks. There was then an attempt to establish an ultrawideband physical layer enhancement for imaging and multimedia, but this became derailed by two competing industry standards with one standard promoting multiband orthogonal frequency division multiplexing and the other promoting direct-sequence UWB. This standard (802.15.3a) was withdrawn in 2006 and replaced with (b) and (c) with the emphasis shifted to deployment in the unlicensed 57–64-GHz part of the millimeter band where 60-GHz Wi-Fi is now being deployed.

Two competing standards, with each standards option criticized by the other interest group, more or less guaranteed that UWB would fail commercially.

The European Commission has issued a mandate to ETSI to produce harmonized standards for UWB as part of the work stream for the Radio Equipment Directive being implemented in June 2016 [6]. This covers ground probing and wall probing radar, tank level probing radar, sensors, and in-building location but also includes communications applications. UWB therefore remains as a potential technical option with standards support. The likelihood

Table 8.1
FCC UWB Radiation Limits for Indoor and Outdoor Communication

	Indoor	**Outdoor**
Frequency in MHz	**EIRP in dBm**	**EIRP in dBm**
960–1,610	–75.3	–75.3
1,610–1,990	–53.3	–63.3
1,990–3,100	–51.3	–61.3
3,100–10,600	–41.3	–41.3
Above 10,600	–51.3	–61.3

of commercial adoption is open to question. Even at the low levels of allowed spectral density, existing incumbent users are likely to be less than enthusiastic about deployment, particularly if the end result is the introduction of new competition cosharing licensed expensive spectrum on an unlicensed (no cost) basis. In terms of potential L-band deployment, concerns about increasing levels of deliberate (malicious) and nondeliberate jamming at 1,575 MHz and the L2C frequency at 1,227 MHz and L5 frequency at 1,176.45 MHz would further frustrate deployment. The protection of interests of the Global Positioning System (GPS) industry and military and civilian users of GPS are coordinated through the GPS Innovation Alliance [7].

8.4 Operator and TV White-Space Spectrum: CDPD as an Early Example of Operator White-Space Spectrum

The second alternative is dynamic use of unused or underused spectrum in the frequency, time, or spatial domain, also known as white space spectrum [8]. White space can be operator white space or TV white space or any other unused or underused spectrum.

This is not a new concept. Cellular digital packet data (CDPD), introduced in 1995 in the United States, provides an example of a data service deployed by operators into cellular white space spectrum. The CDPD protocol supported the reuse of unused or underused 30-kHz Advanced Mobile Phone System (AMPS) channels. Gaussian minimum shift keying (GMSK) modulation delivered a data rate of 19.2 Kbps. If a channel was required for voice, then the CDPD session terminated within 40 ms. The 25–800-MHz passband had 833 × 30 kHz channels, so in a network planned on a 1 in 12 reuse ratio, there was theoretically plenty of unused bandwidth at any particular time at any particular place. Verizon and AT&T both supported and implemented CDPD initially in urban areas.

Constrained by a U.S.-only market, CDPD failed commercially due to a lack of consumer devices and uptake of GPRS which fulfilled a similar function in other parts of the world. A fully loaded CDPD network would also have projected a potential worst-case loss of 2 dB on the signal-to-noise ratio, which would have largely negated any potential long-term economic gain to the operator. The system was turned off in 2004.

8.5 TV White-Space Spectrum and White-Space Wi-Fi

Wi-Fi is not the only technology propositioned for TV white-space spectrum. Other options include proprietary technologies including generic IoT and MTC physical layer options (already covered in Chapter 3).

Variants of Wi-Fi produced within the IEEE standards process do have the potential to scale rather more easily than proprietary alternatives. The IEEE 802.11af standard [9], approved in 2014, also known as Super Wi-Fi or White Fi, provides a regulatory and technical framework for wireless local area network theoretically anywhere between 54 MHz in the ultrahigh-frequency (UHF) band through to 790 MHz at UHF implemented into 6-, 7-, or 8-MHz TV channels.

The af part of the standard adds in a geolocation database identifying available spectrum by place and time to manage coexistence with analog and digital TV, wireless microphones, and mobile operators at 700 MHz and potentially 600 MHz depending on the outcome of the incentive auction. The 802.11af complements 802.11ah, which is targeted at extended coverage Wi-Fi in the U.S. 900-MHz ISM band (902–928 MHz) and in other markets with sub-1-GHz ISM allocations (Table 8.2).

Whether deployed in TV white space or in any of the sub-1-GHz ISM bands, there is an inherent technical and commercial tension with mobile broadband operators understandably sensitive about their sub-1-GHz spectral and network investment. Similar concerns are voiced about 2.4-GHz ISM adjacency to LTE Band 30 (FDD 2,305–2,315, 2,350–2,360 MHz) and Band 40 (TDD 2,300–2,400).

An LTE mobile transmitting at +23 dBm in Band 30 within 7 meters of a Wi-Fi access point will reduce Wi-Fi throughput by 50% unless additional filtering is introduced into the Wi-Fi device. Conversely, there is the risk of Wi-Fi interference into the LTE mobile receive path, according to 2015 measurements undertaken by the Wireless Technology and Innovation Centre hosted by the Digital Television Group in Vauxhall. This is a tricky conundrum for regulators who have the conflicting objective of auctioning licensed spectrum to the highest bidder while facilitating low-cost or no-cost access to the Internet on spectrally and geographically adjacent spectrum.

Table 8.2
ISM Bands for IEEE 802.11ah

Country	Frequency MHz
United States	902–928
China	755–787
Europe	863–868
Japan	916.5–927.5
Korea	917.5–923.5
Singapore	866–869, 920–928

8.6 Guard-Band and Duplex-Gap White Space

The same tension points surface whenever the use of guard bands and duplex gaps are targeted as white-space opportunities. Proponents of guard band and duplex gap white space argue that frequency-domain guard bands and duplex gaps are dimensioned on a worst-case basis and can therefore be used dynamically when local conditions allow. Mobile broadband operators argue that guard bands are there for a reason and any additional use would increase edge of passband noise floors to an unacceptable level.

Proposed 5G multiple access schemes with inherently low adjacent channel leakage ratios such as filter bank multicarrier (FBMC) are at least partly motivated by the assumption that it will be easier to flexibly access small subsets of spectrum. Not having to use a cyclic prefix also increases spectral efficiency. While this may be true, there are counterarguments that any gains in spectral efficiency would be traded against higher-power consumption in user and IoT devices.

It could also be argued that the amount of available white space is reducing over time. This is partly due to increased deployment of single frequency networks both in TV broadcasting and LTE mobile broadband but also due to improved acoustic filters in LTE user devices. This has allowed for reduced guard bands and narrower duplex gaps, for example, in the U.S. E 850 band extension (Band 26 and Band 27 extensions of Band 5) and the U.S. PCS 1900 band (Band 25 extension of Band 2). Figures 8.1 and 8.2 quantify the

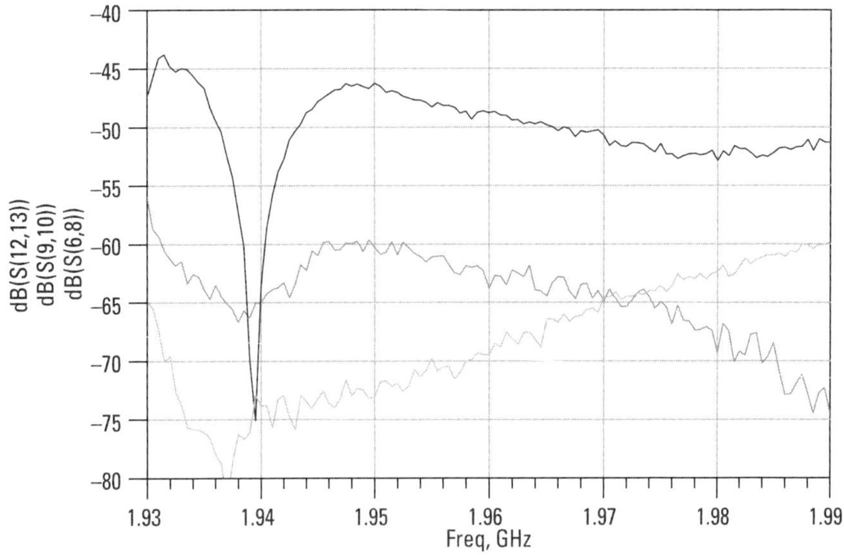

Figure 8.1 FBAR filter rolloff to the duplex gap of Band 25: performance improvements between 2000 and 2013. (Courtesy of Avago.)

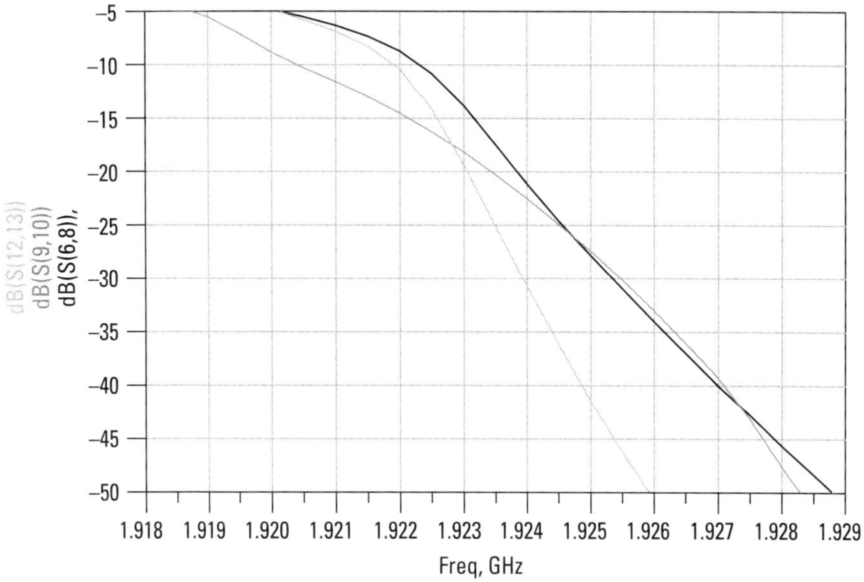

Figure 8.2 FBAR PCS 1900+ (Band 25) Isolation over time, 2000 to 2013. (Courtesy of Avago.)

performance improvements for FBAR filters for the 1,900-MHz band achieved by a combination of improvements in materials, processing, and packaging.

This increase in Q has helped resolve specific coexistence issues, for example, the proximity of Band 13 LTE in the United States to public safety radio. (See Figure 4.2 in Chapter 4). In parallel, vendors have achieved a steady reduction in device size. Combined with packaging innovations (filter banks with multiple matched filters on a single die), this has helped support high band count LTE user devices with (more or less) acceptable performance loss.

The challenge for any proposal to use spectrum differently in the meter band is therefore partly that LTE is getting more spectrally efficient. Note that spectral efficiency and economic efficiency are not directly linked. Additional filter performance generally comes with additional insertion loss which is compounded by other RF efficiency losses due to additional band support in small-form factor user devices. Spectral efficiency gain is therefore traded against a loss of range/coverage.

Also, at least some of the proposals, including white-space proposals, may project coexistence performance cost on existing users and existing networks. Even an apparently trivial loss of 1 dB results in an increase of 14% in network density (see Figure 8.4). If a mobile broadband operator is realizing an economic benefit from a white-space service, then this trade-off might be tolerable, but this is usually not the case.

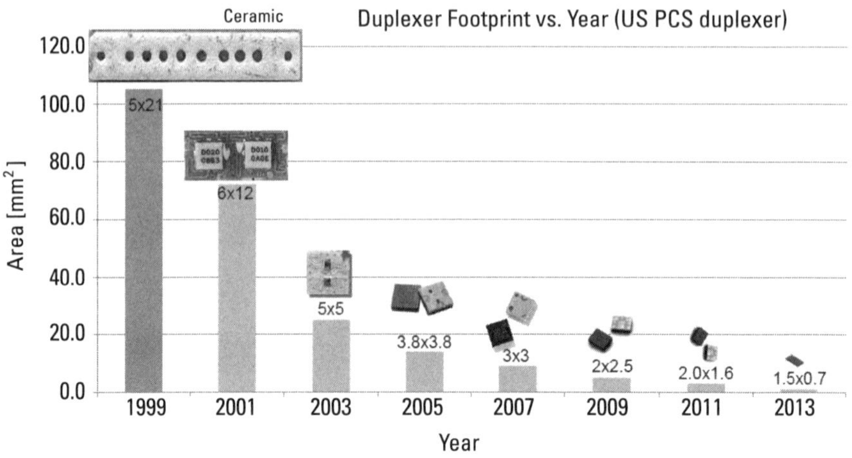

Figure 8.3 Film bulk acoustic resonator (FBAR) dimensions, 1999 to 2013. (Courtesy of Avago.)

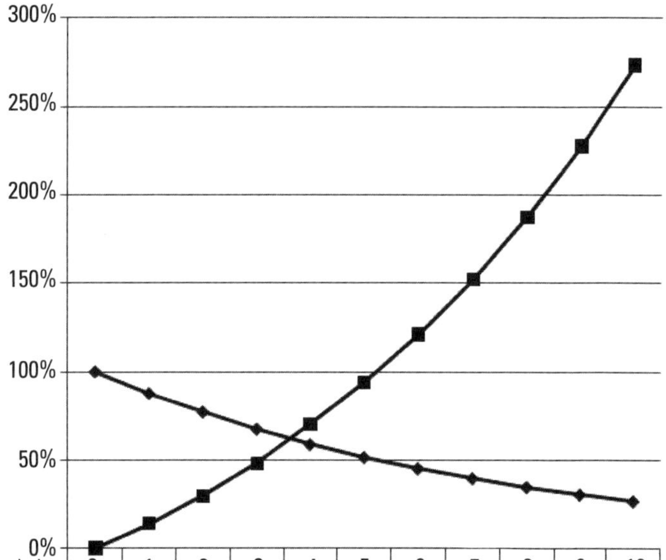

Pathloss [dB] degradation	0	−1	−2	−3	−4	−5	−6	−7	−8	−9	−10
Cell size area reduction	100%	88%	77%	68%	59%	52%	45%	40%	35%	31%	27%
Site Increase	0%	14%	30%	48%	70%	94%	121%	152%	187%	228%	273%

Figure 8.4 Impact of path loss on network density [10].

8.7 A New 5G Physical Layer for IoT Connectivity Below 1 GHz

An additional option is to introduce a new physical layer into existing licensed or unlicensed spectrum optimized for IoT connectivity with cost and power consumption as the two most important performance metrics [11]. This includes standby power consumption of the order of 100 μW compared to the 3-mW consumption of a 4G radio modem.

This is a back-to-the-future approach to 5G radio design using 12.5-kHz radio channels GMSK-modulated to maximize RF amplifier efficiency with a variable spreading factor of up to 64 in the uplink and up to 1,023 in the downlink supporting an adaptive data rate of between 200 bps and 100 Kbps with a 2-km urban range. Transmit power is up to 17 dBm.

The standard is proposed for use in the ISM license-exempt bands below 1 GHz, which, by the time you add all the various country-specific options, includes the bands in Table 8.3.

The relatively low output power and RF efficiency [no amplitude modulation (AM) in the modulation] means that power amplifiers can be integrated with a baseband chip. The 12.5-kHz channel spacing is argued to be an optimum compromise between wider band channels (higher front-end noise floors and more dynamic range increasing power drain) and ultranarrowband options (<12.5 kHz), which would require higher-cost temperature compensated crystal oscillators (TCXOs) to manage frequency offsets and drift. A 12.5-kHz channel can be implemented using a low-power, low-cost crystal oscillator or DCXOs (digitally controlled oscillator). Adjacent channel power is claimed to be better than −55 dBc. The end-point device cost point targets are below $5 U.S.

It could also be deployed in licensed spectrum, for example, legacy Global System Mobile (GSM) 200-kHz channels. One of the original companies promoting this standard [12] now focuses on operator licensed spectrum rather than unlicensed spectrum applications and has been acquired by Huawei.

The GSM Association has an initiative [13] to coordinate IoT connectivity based on an evolved LTE solution, an evolved GSM evolution or a clean slate solution. Note the GSM 900 band also includes GSM-R [14] for railway track to train communications, although there are no announced plans to use this for IoT connectivity. By December 2015, the two competing standards for IoT had been merged into a combined standard (see Chapter 2).

8.8 Support for Public Safety Radio Systems

Existing public safety radio systems including TETRAPOL, TETRA, and P25 [15] are typically narrowband 6.25-, 10-, 12.5-, or 25-kHz channel spaced systems within passbands of 3 to 5 MHz implemented at very high frequency

Table 8.3
ISM License-Exempt Bands Below 1 GHz

169 MHz	433 MHz	470 MHz	868 MHz	915 MHz	928 MHz

(VHF) and UHF up to 900 MHz. In Europe this includes TETRAPOL implemented in 12.5-kHz channel or 10-kHz channel spacing and TETRA implemented in 12.5-kHz channels spacing. In the United Kingdom the TETRA system used by public safety agencies (fire, police, ambulance) is deployed at 380–385 MHz (mobile TX) and 390–395 MHz (mobile receive).

Some European emergency service allocations have a 3 MHz rather than 5-MHz passband (380–383 MHz/390–393 MHz) and are supplemented by civil system allocations. Table 8.4 shows the allocated bands below 400 MHz, below 500 MHz including adjacency to Band 31 LTE and at 900 MHz including adjacency to GSM R and E GSM/900-MHz Band 8 LTE.

In the United States, analog and digital radios for the public safety sector include 12.5-kHz and 6.25-kHz channel spacing with implementation at UHF and in the 700-MHz and 800-MHz bands. The technology standards for these radios are coordinated by P25 with user needs coordinated through the Association of Public Communication Officers [16]. Table 8.5 shows the 700-MHz U.S. public safety allocation (Band 14) and adjacency to consumer mobile broadband LTE (the outcome of the 2008 U.S. 700-MHz auction). Band 17 is effectively a subband of Band 12.

The U.S. 800-MHz allocations including the overlap with Band 26 LTE (extended U.S. 850 band 5) (see Table 8.6).

Evolved variants such as Release 2 TETRA scale to 150-kHz channel spacing and use 64 QAM to deliver 500-Kbps data rates, but 150-kHz channel

Table 8.4

TETRA, LTE, and GSM R and E GSM LTE Allocations in Europe

TETRA emergency services <400 MHz		**TETRA civil systems < 500 MHz**		**LTE Band 31 <500 MHz**		**TETRA civil systems <900 MHz**		**GSM R**		**E GSM/LTE**	
Mob TX	Mob RX	Mob TX	Mob RX	Mob TX	Mob RX	Mob TX	Mob RX	Mob TX	Mob RX	Mob TX	Mob RX
MHz	MHz	MHz	MHz	MHz	MHz	MHz	MHz	MHz	MHz	MHz	MHz
380–385	390–395	410–430	450–470	452–457	462–467	870–876	915–921	876–880	921–925	880–915	925–960

Table 8.5

U.S. Public Safety Band Allocations at 700 MHz

Band 12		**Band 13 (Verizon)**		**Band 14**		**Band 17**	
Block (a)		**Block (c)**		**Public safety**		**Block (b)**	
		Reverse duplex		**Reverse duplex**			
Mob TX	Mob RX	Mob TX	Mob RX	Mob TX	Mob RX	Mob TX	Mob RX
699–716	729–746	777–787	746–756	788–798	758–768	704–716	734–756

Table 8.6
U.S. 800-MHz Public Safety Bands

Band 26 LTE Extended 850 band 850+		**Band 27 E SMR**	
Mob TX	Mob RX	Mob TX	Mob RX
814–849	859–894	807–824	852–869

spacing is still relatively narrow band compared to LTE. Release 2 TETRA increases the physical layer range from 58 km to 83 km to facilitate air-to-ground communication, but this also reflects the public safety sector need for large cell coverage for voice and data at a level not presently available from LTE networks. Partly this is due to economics, but the commercial challenge of delivering low-cost, long-distance LTE is also due to the increased channel bandwidth. This increases capacity due to the additional multiplexing gain but also increases the noise floor. Potentially, there could be an optimized version of LTE Advanced similar to the proposed IoT physical layer implemented in 1.4-MHz channel spacing.

The 5G IOT connectivity includes mobile IoT (objects that move). Public safety radio also has objects that move, for example, fire engines, police cars, and ambulances. This potential opportunity remains largely unexplored.

8.9 A New 5G Physical Layer Deployed into New Spectrum Below 3 GHz for High-Bandwidth, Wide-Area Connectivity

A final option would be to deploy a new 5G physical layer in the meter band for high-bandwidth, wide-area connectivity. The question is where would it go and how would it add value to existing 4G networks particularly given that these existing networks are a long way from being amortized.

Table 8.7 shows present sub-1-GHz FDD band allocations by region and the lack of global harmonization and the potential new bandwidth below 700 MHz.

National regulatory authorities such as Ofcom are required to work with national governments and international institutions with spectrum responsibility which includes the International Telecommunication Union (ITU), the European Conference of Postal Telecommunications Association (CEPT) and spectrum committees of the European Union.

Within the European Union, the Radio Spectrum Policy Program (RSPP) has a set of policy objectives which includes an obligation to identify 1,200 MHz of spectrum for consumer mobile broadband. The European Radio Spectrum

Table 8.7
Sub-1-GHz FDD Mobile Band Allocations by ITU Region

Region	**>900 MHz**			**>800 MHz**			**>700 MHz**			**>600?**	**>500?**	**>400 MHz**			**VHF**
	Band	Mo TX	Mo RX	Band	Mo TX	Mo RX	Band	Mo TX	Mo RX			Band	Mo TX	Mo RX	FM
Region 1															
Europe	8	880–915	925–960	20	832–862	791–821	APT(a)	703–733	758–788						
Africa	8	880–915	925–960	20	832–862	791–821	APT(a)	703–733	758–788						
							28	703–748	758–803						
Region 2															
US				5	824–849	869–894	12	699–716	729–746						
				26 UE	814–849	859–894	17	704–716	734–746						
				27 LE	807–824	852–869	13	777–787	746–756						
							14	788–798	758–768						
							29		717–728						
Latin America	8	880–915	925–960	20	832–862	791–821	APT(a)	703–733	758–788			31 Brazil	451 458	461 468	
				5	824–849	869–894	28	703–748	758–803						
				26?	814–849	859–894									
				27?	807–824	852–869									
Region 3															
Australia	8	880–915	925–960	5	824–849	869–894	28	703–748	758–803						
Asia				18	815–830	860–875	28	703–748	758–803						
				19	830–845	875–890									
				20	832–862	791–821									
Japan				6	820–840	875–885									

Policy Group (RSPG), the agency providing strategic advice to the European Union, is also tasked with reviewing spectrum above 400 MHz up to 6 GHz.

Ofcom's Public Safety Release Program is working on a U.K. government ambition to release 500 MHz of public sector spectrum for consumer and civilian use by 2020. The CEPT view of candidate bands for WRC 2015 shown in Table 8.8 included 470–694 MHz, L-band allocations at 1,350–1,375 MHz, 1,375–1,400 MHz, 1,427–1,452 MHz, 1,452–1,492 MHz, and 1,492–1,518 MHz.

Bands subject to further consideration for sharing and compatibility are shown in Table 8.9.

The Ofcom Mobile Data Strategy Document issued in October 2014 provides an illustration of the total potential mobile broadband downlink spectrum that could be released between now and 2028 or shared between now and 2028 if suitable technical and commercial arrangements could be made with existing users.

L-band allocations and the CEPT candidate bands are shown in Table 8.11.

Some caution needs to be exercised about L-band mobile broadband deployments due to the evolving requirements of global navigation satellite systems including the introduction of the second (L2) civilian GPS signal at 1,227 MHz. L2 and L5 allow for ionospheric and atmospheric correction to improve resolution and accuracy. All satellites launched since 2005 support L2 transmission. Combined with increased satellite power, users get faster signal acquisition and better coverage under trees and indoors or at least close to windows indoors. The system is also harder to jam, which given our increasing reliance on GPS in modern transport systems is probably quite important.

One option, as implied by the Ofcom study, is to make mobile broadband in L-band downlink only. This may mitigate coexistence issues, but periodic noise introduced from any source, in this case an LTE or 5G base station transmitter, could compromise the GPS correlator in user devices. The Light

Table 8.8

CEPT Bands for Mobile Broadband at WRC 2015

L-Band CEPT Bands	
1,427–1,452 MHz	1,452–1,492 MHz

Table 8.9

UHF and L-Band Spectrum for Study for Cosharing and Compatibility

UHF	L-Band		
470–694 MHz	1,350–1,375 MHz	1,375–1,400 MHz	1,492–1,518 MHz

Table 8.10
Downlink Spectrum Availability, 2012 to 2028: Ofcom Illustrative Example

Meter Band + Lower end of the centimeter band									**Total MHz**
2012 Bands available for mobile data downlink (MHz)									
900	1,800	2,100							162
2014 Bands available for mobile data downlink (MHz)									
900	1,800	2,100	2,600						290
2016 Bands available for mobile data downlink? (MHz)									
900	1,800	2,100	2,600	1,452–1,492	2,300				
Plus the 3.4-GHz TDD band = 100-MHz downlink in the centimeter band=									491
2022 Bands available for mobile data? (MHz)									
900	1,800	2,100	2,600	1,452–1,492, 1427–1452	2,300	700	2-GHz MSS		
Plus the 3.6-GHz FDD band = 100-MHz downlink in the centimeter band =									671
2028 Bands available for mobile data? (MHz)									
900	1,800	2,100	2,600	1,452, 1,492, 1,492, 1,518	2,300	700	2-GHz MSS	2,700, 2,900	
Plus an extended 3.6–4.2-MHz FDD band									941

Table 8.11
L-Band Spectrum Including CEPT Candidate Bands

L-Band									
LTE Band 11, Japan, Mob TX	LTE Band 21, Japan, Mob TX	Band 11, Japan, Mob TX	LTE Band 21, Japan, Mob RX	Band 24 MSS + Light Squared	GPS	Iridium	Band 24 MSS	Inmarsat 4	Inmarsat 4
1,428–1,453		1,476–1,501							
	1,448–1,463		1,496–1,511						
				1,525–1,559	1,559–1,610	1,616–1,626	1,625–1,660	1,518–1,559	1,626–1,675
WRC CEPT candidate bands – L-band									
1,427–1,452		1,452–1,492							
For further consideration at WRC 2015									
1,300–1,350	1,375–1,400		1,492–1,518						

Squared experience should make operators conscious that extensive engineering due diligence will be needed before bidding for L-band spectrum.

8.10 S-Band Spectrum

The same caveats apply to S-band spectrum, although with coexistence issues determined by satellite services and deep-space exploration rather than GPS. Table 8.12 shows the S-band MSS allocations that Inmarsat are now proposing to use to provide Internet connectivity to aviation passengers and the spectral proximity of Earth exploration satellites and deep-space radio to LTE Band 1 and LTE Band 40.

The band 2.025 to 2.3 GHz includes the Unified S-Band (USB), which has been used for many years for spacecraft, notably for the Apollo Lunar missions. These provided an early example (1965) of multiplexed imaging, data, and voice. A Universal Serial Bus (USB) antenna could transmit and receive simultaneously. Voice, telemetry, and television were all received together with slow-scan television frequency modulated on the carrier. Telemetry was phase modulated on the subcarriers. The system also allowed for accurate ranging to determine the distance of the spacecraft from Earth [17]. Today the band is used for military space links, military meteorology, and Earth resource sensing.

The 2.52–2.67-GHz band overlaps with LTE Band 7, 2,500–2,570 MHz, 2,620–2,690 MHz, and LTE TDD band 38 (2,570–2,620 MHz) and 41 (2,496–2,690 MHz). Space use is now limited to fixed point-to-point com-

Table 8.12
S-Band Spectrum

S-Band												
LTE Band 39, TDD	LTE Band 33, TDD	LTE Band 1, FDD, Mob TX	Inmarsat and Solaris, S-Band, Mob TX	LTE Band 1, FDD, Mob RX	Inmarsat and Solaris, S-band, Mob RX	Earth Exploration Satellite	Deep Space	LTE Band 40, TDD	Wi-Fi	LTE Band 7, FDD, Mob TX	LTE Band 38, TDD	LTE Band 7, FDD, Mob RX
	1,900–1,920	1,920–1,980	1,980–2,110	2,110–2,170	2,170–2,200	2,200–2,290	2,290–2,300	2,300–2,400	2,400–2,480	2,500–2,570	2,570–2,620	2,570–2,620

Table 8.13
LTE Band 31

Band	Name MHz	Uplink		Downlink		Bandwidth MHz	Duplex Spacing	Guard band	Area
31	450	452.5	457.5	462.5	467.5	5+5	10	5	CALA

munication and space-to-Earth broadcast links in parts of Asia and the Middle East.

8.11 Satellites in Other Parts of the Meter Band

Satellites are more spectrally ubiquitous than you might expect. Generally, co-existence issues between Earth-facing satellites and terrestrial radio can be effectively managed and only generally become problematic if something is changed, for instance, a regulatory change to shared use from primary use or a technology change, which creates potential system to system interference.

On the positive side and going with the interpretation of 5G as among other things a network of networks, we should at least be aware of what services are implemented where and how they might evolve in the future.

8.12 Satellites in the UHF Band

The 399.9–403-MHz band includes timing and frequency standards, navigation and positioning. The band has also been used for two-way radio. The 432–438-MHz band is an amateur satellite and Earth resources satellite band. The 460–470-MHz band supports meteorological and environmental satellites including uplink frequencies for remote environmental sensors. It overlaps with the downlink of Band 31 LTE [18] being deployed in Brazil for rural broadband and the focus for some vendor interest groups developing LTE sparse network service propositions [19].

The Russian International Space Station (ISS) uses 628–632 MHz.

8.13 Satellites in L-Band

These satellites include GNSS (referenced earlier) and MSS satellites, for example, Iridium and Inmarsat. At the time of this writing, there was an ongoing initiative by the LoRa Alliance [20] to integrate wide-area low-cost IoT connectivity with the Iridium low Earth orbit satellite constellation.

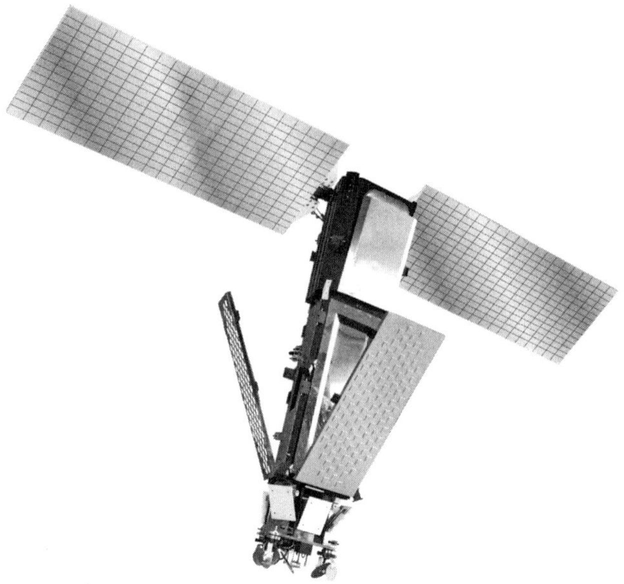

Figure 8.5 Iridium next constellation satellite. (Courtesy of Iridium Communications.)

The Iridium constellation is presently being upgraded with IoT connectivity as one part of a mixed civilian and military payload. The North-to-South orbits provide a global coverage footprint.

The band of 1.67 to 1.71 GHz is one of the primary bands for high-resolution meteorological satellite downlinks for data and imagery. Terrestrial broadcasting has bandwidth allocated at 1.4 GHz. In the United Kingdom this is used for Digital Audio Broadcasting (DAB) point-to-point backhaul and utility point-to-point links.

8.14 S-Band, C-Band, and X-Band Coexistence with Weather and Aviation Radar

S-band terrestrial radio systems also need to coexist with ground and aviation weather radar systems. The choice of band for weather radar is determined by the size of the raindrop to be measured, and heavier rain produces bigger raindrops. Raindrops should correctly be described as hydrometeors.

The 10-cm wavelength (longer wavelength) S-band weather radar has good, heavy, rain cloud penetration and is best for long-range weather radar up to 300 km. It is a critical civilian and military asset. Proposals to increase the bandwidth of existing Band 7 (FDD) and Band 38 (TDD) spectrum is therefore likely to meet with a stormy reception.

Table 8.14
Weather Radar Bands in the Meter Band

Band	Wavelength (cm)	Frequency (MHz)	Range	Application
S-band	11.7–10.33 (10 cm)	2,700–2,900	300 km	Heavy rain

8.15 5G in the Meter Band

So where would 5G go in the meter band, what would it do, and what would it do differently to 4G and more to the point how would it add value to existing 4G networks? Specifically, it would need to deliver a value greater than the writedown cost of underamortized 4G investment. We do not have a satisfactory answer to these questions.

A 5G IoT narrowband physical layer could be deployed into licensed 200-kHz GSM channels or indeed any legacy narrow band bandwidth including 12.5-kHz two-way radio systems and any of the unlicensed ISM bands, but the present proposals for this are not presently supported within 3GPP or 5GPPP standards processes and competing standards will almost certainly slow market uptake. It is presently hard to see what value a 5G work stream can bring to existing physical layer standardization efforts.

A wideband (high peak data rate), wide-area 5G physical layer would need to be differentiated technically and commercially from LTE Advanced including channel aggregated (100-MHz) channel bandwidth LTE.

Filter bank multicarrier modulation and filtering might produce a more spectrally efficient physical layer for accessing noncontiguous slices of spectrum of variable bandwidth, but the power consumption in user devices and IoT devices needs to be equal or lower than equivalent LTE devices and legacy GPRS devices. Irrespective of the preshaping applied to modulated spectrum there will be a perceived risk of spectral regrowth through the RF transmission chain. This should make operators wary of any proposed relaxation of protection ratios or use of guard bands or duplex gaps to support services from competitive providers from within or outside the industry.

The 5G could be positioned as an improved interworking protocol with 2.4-GHz Wi-Fi, but this is already being addressed within existing 4G workstreams.

Last but not least, would it be remotely plausible to have a three-band 5G network with bandwidth in the millimeter, centimeter, and meter bands? The answer to this is probably yes and there could be a basis for developing a physical layer with, for example, a 1-GHz channel bandwidth in the millimeter band, 500 MHz in the centimeter band, and 100 MHz in the meter band with the meter band option made to be directly compatible with 100-MHz LTE Advanced. Having said that, there is present limited visibility to how 100-MHz carrier-aggregated LTE will be realized in small form factor user devices without an unacceptable loss of RF and DSP efficiency.

Tables 8.15 to 8.18 show present LTE bands sorted by size of passband in ascending order. Table 8.15 lists the FDD bands below 1 GHz with Band 41 at the top with a 5 by 5 MHz passband at 450 MHz and Band 28 at the bottom with 45+45 MHz. Because acoustic filters are limited to 4% of center frequency, the Band 28 passband requires two 30-MHz filters overlapping by 15 MHz.

Given that these bands typically support between three and five operators with their own channels within these passbands, it is obvious that the opportunities for contiguous carrier aggregation are limited. Noncontiguous carrier aggregation is possible, but the transmit and receive chains will be passing through multiple nonharmonically related individually filtered and individually matched passbands.

Assuming that the RF signals survive the inward and outward journey without phase distortion, they then have to go through either a wideband analog-to-digital conversion or more practically, multiple parallel conversions.

Data can be moved quickly and session times are short. However, the noise floor will be higher and more dynamic range will be required out of the

Table 8.15

FDD <1 GHz

Band	Name MHz	Uplink		Downlink		Bandwidth MHz	Duplex Spacing	Area
31	450	452.5	457.5	462.5	467.5	5+5	10	CALA
6	UMTS only	830	840	875	885	10+10	45	APAC
13 RD	700 c	777	787	746	756	10+10	–31	NAR
14 RD	700 ps	788	798	758	768	10+10	–30	NAR
29	700 d	Downlink only		717	728	11		NAR
17	700 b	704	716	734	746	12+12	30	NAR
18	800 lower	815	830	860	875	15+15	45	Japan
19	800 upper	830	845	875	890	15+15	45	Japan
12	700 a	699	716	729	746	17+17	30	NAR
27	800 SMR	807	824	852	869	17+17	45	NAR
5	850	824	849	869	894	25+25	45	NAR
20 RD	800 DD	832	862	791	821	30+30	–41	EMEA
8	900	880	915	925	960	35+35	45	All
26	850+	814	849	859	894	35+35	45	NAR
28	700 APT	703	748	758	803	45+45	55	APAC

*PS = Public Safety, SMR = Specialized Mobile Radio, APT = Asia Pacific Telecommunity.

Table 8.16
FDD 1–2 GHz

Band	Name MHz	Uplink		Downlink		Bandwidth MHz	Duplex Spacing	Area
21	1,500 upper	1,447.9	1,462.9	1,495.9	1,510.9	15+15	48	Japan
11	1,500 lower	1,427.9	1,447.9	1,475.9	1,495.9	20+20	48	Japan
24 RD	1,600 L-band	1,626.5	1,660.5	1,525	1,559	34+34	−101.5	NAR
9	1,800	1,749.9	1,784.9	1,844.9	1,879.9	35+35	95	APAC
32	1,500 L-band	Downlink only		1,452	1,496	44		EMEA
1	2,100	1,920	1,980	2,110	2,140	60+60	190	All
2	1,900 PCS	1,850	1,910	1,930	1,990	60+60	80	NAR
10	AWS 1+	1,710	1,770	2,110	2,170	60+60	400	NAR
25	1,900+	1,850	1,915	1,930	1,995	65+65	80	NAR
3	1,800+	1,710	1,785	1,844.9	1,879.9	75+75	95	All

Table 8.17
FDD 2–4 GHz

Band	Name MHz	Uplink		Downlink		Bandwidth MHz	Duplex Spacing	Area
30	2,300 WCS	2,305	2,315	2,350	2,360	10+10	45	NAR
23	2,000 S-band	2,000	2,020	2,180	2,200	20+20	180	NAR
7	2,600	2,500	2,570	2,620	2,690	70+70	120	EMEA
22	3,500	3,410	3,490	3,510	3,590	80+80	100	EMEA

receiver front end. This is not intrinsically power efficient on a per-bit basis. A higher noise floor and any inherent phase distortion will reduce throughput. The additional bandwidth will disguise this as far as the user is concerned, although there will be a capacity cost to the operator that may or may not be compensated from the bandwidth multiplexing gain. More importantly, a higher noise floor will reduce range. Data rate has been traded against data reach. This will be directly noticeable to users. The network can compensate by implementing coordinated multipoint transmission (CoMP) schemes, but this reduces capacity, increases interference levels, and adds to backhaul cost.

Whatever way you look at the meter band, it is ultimately constrained by acoustic filter bandwidth (4% of center frequency) and antenna bandwidth (10% of center frequency). Either of these can be extended by using techniques like adaptive matching, but essentially you are fighting physics and there will be an associated performance cost and or an increase in performance variability.

Table 8.18
TDD < 1 GHz, 1–2 GHz, and 2–3 GHz

Band	Name MHz	Uplink		Downlink		Bandwidth MHz	Duplex Spacing	Area
TDD <1 GHz								
44	TD 700	703	803	703	803	100 MHz		
TDD 1–2 GHz								
34	TD 2,000	2,010	2,025	2,010	2,025	15		EMEA
33	TD 1,900	1,900	1,920	1,900	1,920	20		EMEA
37	TD PCS Center Gap	1,910	1,930	1,920	1,930	20		NAR
39	TD 1,900+	1,880	1,920	1,880	1,920	40		China
35	TD PCS Lower	1,850	1,910	1,850	1,910	60		NAR
36	TD PCS upper	1,930	1,990	1,930	1,990	60		NAR
TDD 2–3 GHz								
38	TD 2,600	2,570	2,620	2,570	2,620	50		EMEA
40	TD 2,300	2,300	2,400	2,300	2,400	100		China
41	TD 2,500	2,496	2,690	2,496	2,690	194		All
TDD 3–4 GHz								
42	TD 3,500	3,400	3,600	3,400	3,600	200		
43	TD 3,700	3,600	3,800	3,600	3,800	200		

An increase in hand capacitance effects (the effect of a user's hand on the outside of the phone) as bandwidth increases is an example of performance variability. Channel-to-channel variations across an extended passband due to filter ripple are another example. These effects can be mitigated, for example, hand capacitance effects can be countered by adaptive matching, passband ripple can be mitigated by a channel equalizer, but mitigation measures have their own cost both in terms of performance and in some cases digital noise from additional control lines.

While there is a clear trend towards wider passbands, for example, the extended 850 and PCS 1,900 and AWS bands in the U.S. band, there is an associated performance cost that has to be set against any realizable performance gain irrespective of the generation of radio air interface that is used.

Table 8.16 shows the channel bandwidth available at L-band with Band 3 as the best present option, the reason why it has been particularly well suited to LTE deployment.

Table 8.17 shows the FDD options for S-band up to 3 GHz. Band 22 at 3,500 MHz will add another 80+80 MHz of FDD bandwidth.

8.16 5G TDD in the Meter Band

Last but not least, there are the TDD options from 700 MHz through to Band 43 at 3,700 MHz shown in Table 8.18. These would superficially at least seem better suited to 4G or 5G deployment offering contiguous bandwidth options of 100 MHz or more.

TDD has a number of benefits. The uplink and downlink are reciprocal, which makes adaptive antenna design and associated channel sound easier, but in reality frequency-domain complexity translates more or less directly into the time domain. Theoretically, in a user device a time-domain split between transmit and receive could and should deliver similar isolation to a frequency duplex split. In practice it is hard to avoid residual noise leaking from the transmit path into the receive chain with an associated loss of sensitivity.

At network level, TDD works adequately well in dense networks at higher frequencies/shorter wavelengths, but wider area networks at lower frequencies/longer wavelengths ideally need to be clocked together with cosited base stations to manage and minimize internetwork intersymbol interference. This is the reason why no TDD networks have been successfully deployed at 700 MHz but internetwork ISI remains an issue for all wide-area 4G or 5G deployments. This can be resolved to an extent by increasing the time domain guard band, but this has an impact on network capacity. The asynchronous 5G physical layer proposals discussed in Chapter 2 might help to provide at least a partial solution to this problem. Simply put, core and edge timing can be improved. However, this also has a cost consequence.

8.17 The Coexistence Story Continued

There are substantial coexistence studies that remain ongoing within the ITU [21] within the meter band. These include sharing and compatibility studies between LTE/IMT and digital terrestrial broadcasting between 470 and 694/698 MHz, SAB/SAP (Services Ancillary to Broadcasting and Programming) from 694 to 790 MHz, sharing with radio astronomy at 608–614 MHz, 1,330–1,400 MHz, 1,400–1,427 MHz, 1,610.6–1,613.8 MHz, 1,660–1,670 MHz and 2,690–2,700 MHz, compatibility of L-band mobile broadband at 1,375–1,400 MHz and 1,427–1,452 MHz, compatibility with Earth exploration satellites at 1,400–1,427 MHz, compatibility with aeronautical telemetry systems at 1,429–1,535 MHz, sharing studies with meteorological satellites at 1,695–1,710 MHz and sharing with space-to-space links and Earth exploration satellites at 2,025–2,110 and 2,200–2,290 MHz.

If the 600-MHz incentive auction in the United States works as intended and releases new spectrum for mobile broadband below 700 MHz, then there would be a theoretical opportunity to deploy a rural 5G physical layer as an ultrasparse network, for example, reusing at least some of the existing terrestrial TV infrastructure. This might also provide the basis for low-cost 5G for developing markets. At present, this appears to be unlikely and most industry sources are looking at LTE in what is sometimes termed "third digital dividend spectrum." At WRC 2015 the decision was taken to delay discussion of any change of use of the UHF band in Region 1 (Europe and Africa) including below 694 MHz to WRC 2023.

8.18 4G Sub-1-GHz RAN Optimization: Antenna Tilt Techniques

The coexistence narrative and cost narrative come together in present LTE RAN optimization techniques with base station antenna tilt as a specific example.

Remote antenna tilt was first introduced 20 years ago [22] to allow for the coverage from specific base station sites to be changed in response to changing traffic loading. It has always been possible to change an antenna pattern by visiting a site and changing the mechanical position of a single or multiple antennas, which could include manually adjusting the tilt of the antenna or the spacing between multiple antennas. Changing the spacing between two or more antennas changes the phase relationship between the antennas and provides the ability to create nulls in directions from which interference is being received or caused and provides gain in directions where additional coverage is needed.

This can now be done remotely either by using mechanical actuation on the mast head and/or by changing the electrical phasing between the antennas (electrical rather than mechanical beam shaping). A recent (contemporary) iteration of this technique is shown in Figure 8.6.

Conventional electrical tilting for macro antennas is usually achieved by using multiple phase shifters at the radiating elements of the array to impose a progressive time or phase delay along the array length.

In this example, the same variable phase slop across the array is achieved by a single phase shifter with the phase shift providing independent tilt for multiple bands below 1 GHz, although the principle can also be applied to mid-band and high-band combinations.

From LTE Release 9 onwards (for TDD) and Release 10 (for FDD), there has also been an opportunity to exploit dual beamforming, which allows the LTE downlink shared traffic channel to use 2×2 spatial multiplexing MIMO and adaptive beamforming. For example, the same signal can be delivered down two feeders with each signal carrying the same information but with different phase weights.

Different physical resource blocks supported on different groups of LTE OFDM subtones can then be delivered with different azimuthally beamformed patterns. This means that each user receives their respective signals via individual

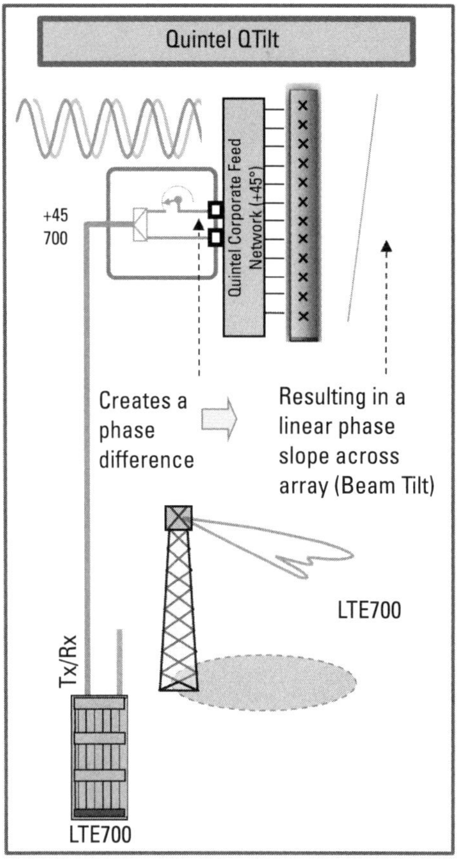

Figure 8.6 Beam tilt as an optimization technique. (Courtesy of Quintel.)

beamformed patterns. The potential to scale this technique to the centimeter and millimeter bands is obvious and shows that spatial multiplexing and adaptive beamforming can be complementary.

8.19 Summary

Deploying 5G into the meter band has a number of theoretical advantages including propagation gain relative to the shorter-wavelength centimeter and millimeter bands. In practice this gain is at least partially offset by antenna inefficiency due to wavelength versus space constraints in form factor-limited devices and base station antenna arrays. Figure 8.7 illustrates this relationship, although arguably implementation loss at longer wavelengths/lower frequencies can be higher. Larger cell radii are also dependent on edge of cell interference management and sufficient frequency reuse to meet capacity expectations.

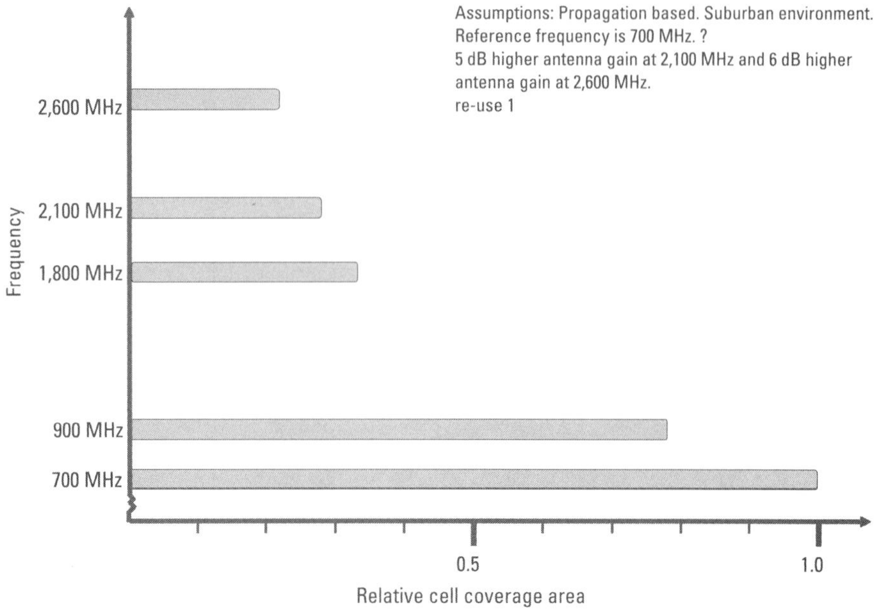

Figure 8.7 Propagation gain versus antenna gain [23]. (Courtesy of Ericsson.)

Meter-band channel bandwidth is determined by acoustic filter bandwidth and antenna bandwidth, which scales as a ratio as wavelength decreases. Band 8, 900-MHz GSM/LTE as a reference point, has a duplex upper and lower passband of 35 by 35 MHz (3.9% of center frequency). The upper and lower duplex separated by a duplex gap of 10 MHz means that the overall passband is 80 MHz (<10% of center frequency). It would be unusual to find a phone with a dedicated antenna for the 900-MHz band but this is theoretically the most RF efficient option.

Scaling these numbers to 1,800 MHz gives us a 70 by 70 MHz passband. Band 3 is close with a 75 by 75 MHz duplex (4.3% of center frequency) sitting in an overall passband of 170 MHz (<10% of center frequency). The same ratio at 3.6 GHz yields 140 by 140 MHz in a 340-MHz passband. A 10% bandwidth channel at 90 GHz yields a 900-MHz passband.

At present, the deployment of LTE Advanced is dependent on the realization of aggregated channels to achieve channel bandwidths up to 100 MHz. This is relatively easily achievable in LTE base stations and access points as they are generally configured to support multiple passbands in low-band, mid-band, and high-band spectrum allocations. It is significantly harder to achieve in user devices without adding cost and complexity and without introducing performance loss.

These constraints are wavelength-related and can be resolved by using beamwidth as an additional mechanism for achieving throughput and capacity gain. This is the underlying physical layer rationale for deploying 5G into the centimeter and millimeter bands, which brings us to our next two chapters.

References

[1] http://www.etsi.org/news-events/news/861-2014-new-industry-specification-group-on-millimetre-wave-transmission-at-etsi.

[2] https://5g-ppp.eu/fantastic-5g/ H2020-ICT-2014-2 Duration 24 months- start date July 2015.

[3] https://5g-ppp.eu/our-members/.

[4] http://company.nokia.com/en/about-us/our-company/our-story.

[5] https://on.google.com/hub/.

[6] http://www.etsi.org/technologies-clusters/technologies/radio/ultra-wide-band.

[7] http://www.gpsalliance.org/.

[8] Salkintzis, A., "Radio Resource Management in Cellular Digital Packet Data Networks," *IEEE Personal Communications*, December 1999, pp. 28–36.

[9] http://standards.ieee.org/findstds/standard/802.22-2011.html.

[10] ICT KTN and Cambridge Wireless RF Front End Technology Challenges Paper, December 2012.

[11] http://www.weightless.org/about/weightlessp.

[12] http://www.neul.com/neul/.

[13] http://www.gsma.com/connectedliving/gsma-iot-device-connection-efficiency-guidelines/.

[14] http://gsmr-info.com/gsm-r_history.html.

[15] http://www.p25.com/.

[16] https://www.apcointl.org/.

[17] https://www.hq.nasa.gov/alsj/alsj-NASA-SP-87.html.

[18] https://itunews.itu.int/en/4618-LTE-450MHz-technology-for-broadband-services-in-rural-and-remote-areas-BR-Case-study-of-Brazil.note.aspx.

[19] http://450alliance.org/.

[20] https://www.lora-alliance.org/.

[21] ITU R JTG Group approved sharing and compatibility studies.

[22] http://www.commscope.com/Solutions/Wireless-Solutions/Outdoor-Sites/Capacity-Enhancement-Solutions/RET-Antenna-Systems-for-Network-Optimization/.

[23] http://www.ericsson.com/res/docs/2013/ericsson-apt700-creating-a-truly-global-band.pdf.

9

The Centimeter Band: 3 to 30 GHz

9.1 Fixed Access Broadband at 3.5 GHz

In 1993, a license was granted to the U.K. Company Ionica to operate a fixed wireless access network (also known as radio in the local loop) at 3.5 GHz using a 17+17 MHz duplex band at 3,425 to 3,442 MHz (uplink) and 3,476 to 3,493 MHz downlink coupled to a 10-GHz backhaul allocated in 1995. The network was intended to provide last-mile connectivity. The company went into administration in 1998 defeated by unexpectedly high computer premises equipment (CPE) hardware costs, installation, and support costs. The spectrum reverted back to the Radio Communications Agency for reassignment.

This was U.K. Ministry of Defence spectrum to which the government had negotiated rights of access cosharing with military radio location. The lower parts of the band were also subject to out-of-band radar emissions. Ionica is a specifically U.K. example, but the fixed wireless band was generally available throughout Europe and a set of European standards were developed to cover time division multiple access (TDMA), frequency division multiple access (FDMA), direct sequence, and frequency-hopping code division multiple access (CDMA) and point-to-multipoint antennas.

In the United Kingdom, the 3.5-GHz band was intended for voice and narrowband data. The 10-GHz band was intended for services above 144 Kbps. In parallel there was consultation to consider the 28-GHz band (27.5–29.5) and 40-GHz band (40.5–43.5 GHz) for broadband digital delivery [1]. In 2008, Ofcom auctioned licenses at 10, 28, 32, and 40 GHz for a comparatively modest £1.5 million with T-Mobile acquiring two 80-MHz licenses at 10 GHz, two 252-MHz licenses in the 32-GHz band, and two 250-MHz licenses in the 32-GHz band. BT bought two licenses in the 32-GHz band. Other licenses included Arqiva, Digiweb, Faultbasic, MLL, Red M, Transfinite, and UK Broadband. The licenses were technology and service-neutral and tradeable and to date have been used for point-to-point and point-to-multipoint backhaul. In 2015, Ofcom issued a consultation on the suggested repurposing of the 10-GHz and 40-GHz bands for 5G.

With regard to fixed wireless access at 3.5 GHz, the failure of Ionica is at least partly explained by being regulated and licensed as a U.K.-specific deployment with limited regional scale potential and minimal global scale potential. A second wave of deployment from 2005 onwards was based on the WiMAX 802.16d standard at 3.5 and 2.3, 2.5, and 2.6 GHz. The 2.5-GHz deployments were in countries with fixed microwave assignments with the United States allocating 195 MHz at 2,495 to 2,690 MHz, the spectrum now being redeployed by Sprint as part of their Tri-Band LTE offer. A number of WiMAX networks were also deployed or partly deployed at 3.5 GHz, Ireland being one not particularly happy example.

We include this historical context on the basis of the third-time-lucky principle of radio network rollout based on the assumption that the industry finally works out a technical and commercial model in which fixed point-to-point and point to multipoint backhaul complements and coexists with fixed wireless access and mobile broadband service delivery.

This could include LTE FDD Band 22 and LTE TDD Bands 42 and 43 between 3.4 and 3.8 GHz (Table 9.1). At present, the backhaul for these yet-to-be deployed networks is assumed to be using existing allocations in the centimeter and millimeter bands. However, there is no specfic technical or commercial reason why backhaul could not be in band and indeed no specfic technical or commercial reason to prevent the service offer combining mobile and fixed access connectivity.

As of the time of this writing, Nokia and Google were exploring the technical and commercial practicalities of deploying LTE-U into the 3.5-GHz

Table 9.1
3–4 GHz LTE Allocations

FDD 3–4 GHz							
	Mob TX		Mob RX				
Band 22	3,410	3,490	3,510	3,590	80+80	100	EMEA
TDD 3–4 GHz							
Band 42	3,400	3,600	3,400	3,600	200		
Band 43	3,600	3,800	3,600	3,800	200		

band. In parallel, the FCC is proposing a Citizens Broadband Radio Service in the band within a 150-MHz passband.

At WRC 2015, CEPT proposed the 3.6–3.7-GHz band for Region 1 International Mobile Telecommunications (IMT) deployment but the proposal failed to progress. The United States, Columbia, Costa Rica, and Canada agreed that the band could be used, but what is effectively a limited regionally specific IMT deployment will make scale economy hard to achieve.

As with the meter band, it is hard to see how 5G could add value to these bands, but 5G could scale the concept of in-band backhaul integrated with mobile wide area and fixed area service provision into the relatively large amount of spectrum now allocated to backhaul across the rest of the centimeter and millimeter bands. This seems like a good point to look at point-to-point backhaul systems in more detail.

9.2 The Point of Point-to-Point Backhaul: Scheduler Efficiency and 4G and 5G Network Economics

This potential repurposing of 4G backhaul hardware and software into 5G mobile broadband base stations and user devices is a fairly obvious opportunity but would have to complement rather than compromise existing and future point-to-point and point-to-multipoint backhaul performance.

Latency and delay variability between base stations are becoming progressively more important within 4G networks to support coordinated multipoint transmission and intercell interference coordination.

Backhaul has always been and will continue to be a critical part of mobile broadband networks and a major cost component, accounting for up to 30% of network capital and operational (Capex and Opex) cost. Self-evidently, backhaul bandwidth needs to keep pace with the growth in mobile broadband traffic and network density and network functionality, but this is only part of the picture. The 4G mobile broadband network economics are increasingly determined by scheduler efficiency and the same dependency will almost certainly translate across into 5G networks.

Scheduler efficiency is determined by a combination of access control and dynamic allocation of bandwidth in the time, frequency, power, and spatial domain. Dynamic bandwidth allocation in 4G LTE networks has to be managed at the subframe level every millisecond. This is more onerous than legacy scheduling, which is either implemented every frame (10 ms) or every other frame (20-ms semipersistent scheduling for voice).

This means that in LTE Advanced and 5G networks, microseconds of additional backhaul delay, particularly delay variability on the control plane (signaling bandwidth), will have a significant impact on mobile broadband network efficiency and network economics. Put more positively, any reduction in backhaul delay or delay variability should translate into more capacity and a better user experience. Scheduling can be optimized for throughput (maximum efficiency) or cell edge performance (optimum quality of service).

The trick is to achieve an optimum compromise between these two not entirely complementary objectives, like getting an optimum match on a Smith chart. It is backhaul performance that determines how effectively this optimum compromise can be achieved and maintained.

Backhaul bandwidth therefore needs to be designed to ensure that end-to-end delay and delay variability is kept to a minimum. This requires careful implementation of channel coding and error correction and higher layer send again protocols. It also implies that backhaul bandwidth has to be provisioned to avoid buffering under all load conditions including traffic peaks. This raises the question as to whether backhaul should be implemented as a relay or repeater.

Relay hops demodulate and decode user and signaling traffic. This has the advantage of reducing end-to-end bit error rates, a function of the modulation and coding gain. The cost is the delay involved in demodulating and decoding, then modulating and coding the backhaul, and then performing the reverse process at the other end.

An alternative is to bring backhaul in band and treat backhaul as a repeater system rather than a series of relay hops. The coding gain disappears but then so does the modulation and coding latency.

Bringing backhaul in band would also mean that mobile broadband scale economics could be applied to the presently fragmented backhaul product sector. Making the link a repeater link rather than a relay would reduce cost further. Either there can be endless disputes as to whether or how existing point-to-point fixed wireless backhaul bandwidth should or could be repurposed for mobile wide-area broadband or there could be a sensible debate as to whether a cosharing and integration model could deliver technical and commercial gain to all parties.

Backhaul band allocations are every bit as complicated as LTE band allocations (Table 9.2). The meter band (300 MHz to 3 GHz) includes 1.4-GHz licensed links using Yagi antennas to provide terrestrial DAB and utility point-to-point and smart grid connectivity. The 2.4-GHz ISM band is also used, cosharing with Wi-Fi. The centimeter band includes unlicensed point-to-point backhaul in the 5-GHz Wi-Fi band and licensed bands at 11, 13, 15, 18, 23, 26, and 28 GHz. The millimeter band includes unlicensed point-to-point backhaul in the 60-GHz Wi-Fi band and licensed or lightly licensed bands at 32 GHz, 38 GHz, 40 GHz, 42 GHz, 45 GHz, 52 GHz, 55 GHz, 65 GHz, 71–76 GHz, 81–86 GHz, and 92–95 GHz. The licensed and lightly licensed

Table 9.2
Point-to-Point and Point-to-Multipoint Backhaul Frequencies, Channel Bandwidths, Passbands, and Throughput

Meter Band	Centimeter Band	Millimeter Band
Center frequencies	Center frequencies	Center frequencies
1.4 GHz and 2.4 GHz	11,13,15,18, 23, 26 and 28 GHz	32 GHz, 38 GHz, 40 GHz, 42 GHz, 45 GHz, 52 GHz, 55 GHz, 65 GHz, 71–76 GHz, 81–86 GHz, 92–95 GHz
Channel bandwidth	Channel bandwidth	Channel bandwidth
3.5-MHz channels in a 24-MHz passband	28 or 56 MHz channels	28 or 56 or 112-MHz channels Up to 5-GHz passbands
Throughput	Throughput	Throughput
9 Mbps	400 Mbps	1,000 Mbps

bands are allocated on a country-by-country basis with little regional harmonization either in terms of band plan or physical layer implementation, which is generally vendor-specific.

9.3 The Lack of Harmonized Standards in Point-to-Point Backhaul

The lack of harmonized spectrum in present backhaul radio links, including a lack of harmonization by region or country, adds to the cost of radio link hardware. Costs are increased further by a lack of harmonized technical standards.

In the meter band, a 1.4-GHz link would typically use 32-level QAM, producing a peak throughput of 9 Mbps per 3.5-MHz channel within a 24-MHz passband.

In the centimeter band, licensed link equipment at 28 GHz typically uses 512 QAM to deliver 400-Mbps peak throughput through a 56-MHz channel with 38 dBi of gain from a dish antenna.

In the millimeter band, licensed link equipment at 38 GHz typically uses a 28-MHz channel with 1,024 QAM to give a gross bit rate of 250 Mbps per channel with channel aggregation (for example, to 56 MHz) to support throughput of 500 Mbps. Higher antenna gain (>50 dBi) also helps the link budget.

A 42-GHz or 70–80-GHz band might typically be implemented with 112-MHz or 250-MHz spacing. A single 112-MHz channel at 42 GHz should support a gross bit rate of 1 Gbps using 1,024 QAM modulation. A 70-GHz or 80-GHz link could have 4 250 MHz channels aggregated together: 1 GHz of bandwidth supporting 1 Gbps of data throughput.

9.4 Backhaul Data Rates, Latency, and 4G and 5G Network Efficiency

Small-cell backhaul today has a typical peak bandwidth requirement of 100 to 200 Mbps. The general assumption is that this will increase to around 1 Gbps for LTE-A/Wi-Fi backhaul. This order of magnitude increase might seem intimidating, but should be supportable on the basis of the combination of additional band allocations and higher-order 2,048 QAM and 4,096 QAM modulation schemes.

Meeting latency budgets may be trickier. The 4G mobile broadband networks are increasingly dependent on link adaption mechanisms to achieve scheduler efficiency. From an operator perspective, scheduler efficiency is a proxy for cell spectral efficiency. From a user perspective, the scheduler also has to deliver acceptable cell edge performance. Achieving these conflicting objectives is dependent on close control of the feedback mechanisms that determine admission control.

In high-speed downlink packet access (HSDPA), the scheduler allocates time-domain and code-domain resources, but always occupies the full (5-MHz) channel bandwidth. In LTE, admission control is also performed in the frequency domain implemented at the resource block level. A resource block is 12 × 15 kHz subcarriers giving a resource block bandwidth of 180 kHz. In the time domain, the basic frame length within LTE is 10 ms. Voice is supported with semipersistent scheduling every 20 ms, but from Release 10 onwards, data is scheduled at each 1-ms subframe with additional scheduling possible at each of the two slots (0.5 ms) within the subframe. The dynamic allocation of resource blocks is usually described as frequency-domain scheduling but is in practice a combination of frequency-domain and time-domain bandwidth allocation.

Interference management in the time domain is implemented at the subframe level. In the frequency domain, resource block subcarriers can be allocated on the basis of channel quality (channel-dependent scheduling) or can be chosen to minimize interference to other users or adjacent radio systems; reduction of out-of-band emissions is an example.

Interference management is implemented in the code domain with orthogonal cover codes to support different layers of spatial multiplexing and to discriminate between different terminals shared on the same resource within a cell or in neighboring cells. This is integral to spatial domain interference management based on antenna beam-forming combined with various options of transmit diversity and spatial multiplexing. Interference management in the power domain remains an open debate, at least for 5G networks.

Close control of transmit power from user devices has been a fundamental part of cellular voice network design for 30 years. The principle is that mobile devices should never use more power than needed to overcome path loss. This has been an effective way to manage interference and has generally helped increase user battery life.

Schedulers now have the option of operating devices at a fixed power output and changing the amount of available resource block bandwidth to accommodate variable traffic rates.

Alternatively, physically and spectrally efficient user devices can be run at different power levels with interference cancelled out by using successive interference cancellation at the eNode B. This is usually described as nonorthogonal multiple access (NOMA).

Admission control decisions are based on all of the above including frame-by-frame changes in coding and modulation. The decisions are made by the eNodeB or, more correctly, by groups of eNodeB's cosharing the required information and include handover, load management, interference management, and network and mobility optimization. This is done over the X2 interface.

The legacy rule of thumb on end-to-end signaling delay is that 10-ms round trip delay is acceptable and 5 ms desirable. This might be acceptable for the Release 8 intercell interference coordination, which provides semistatic coordination of resources every few seconds. It is not adequate for LTE Advanced scheduling/beam-forming, which potentially requires the dynamic coordination of frequency, time, power, and beam-forming resources at subframe level.

The signaling bandwidth is relatively trivial (less than 1 Mbps), but delay or delay variability on the X2 interface translates directly into reduced scheduler efficiency. This brings us back to the issue of future backhaul requirements. The delay budget on a well-designed point-to-point radio is of the order of 250 μs, which could be regarded as trivial but fails to take into account buffering delay, including the delay of taking traffic and signaling off the mobile broadband network and modulating and demodulating and encoding/decoding over the backhaul link.

The backhaul is therefore functioning as a relay. This has the benefit of improving the signal to noise ratio but has the crucial disadvantage of introducing additional delay and delay variability. It may therefore be more appropriate for backhaul systems to function as repeaters rather than relays. Repeaters are now (more or less) standardized within the LTE specifications. Additionally, there may be merit in considering bringing backhaul in band to allow wide-area mobile networks to coshare point-to-point bandwidth in the centimeter and millimeter bands.

This would have scale economy and functional benefits. With the exception of the 23-GHz and 38-GHz bands, all other licensed and lightly licensed point-to-point bands are subscale, making it hard to justify the development of lower-cost, higher-performance component and packaging technologies such as surface mount GaAs PHEMT MMICS. Adding mobile broadband volume to these bands would help resolve component cost and performance issues.

Bringing backhaul in band would also imply a rationalization of the present combination of higher-layer time division multiplex (TDM), asynchronous transfer mode (ATM), Internet Protocol (IP), Ethernet, IP over Ethernet, and related IEEE1588 PTP transport and timing protocols. However, it is hard to see how else the latency requirements and technical and commercial requirements of advanced 4G and 5G networks would require significant attention to core and edge timing accuracy and stability.

9.5 Dynamic Beamforming: Adaptive Point-to-Point Backhaul

However, point-to-point backhaul is, as the description suggests, a point-to-point connection. Repurposing and/or cross-subsidizing the same hardware and software approach to mobile broadband will require the addition of adaptive array antennas to support dynamic beamforming to support connectivity between a fixed point (a 5G base station) and a moving point (a person, car, boat, train, or plane). For some applications, including military 5G, there will be a need to support mobile-to-mobile connectivity. For all mobility conditions, Doppler shift (an increase or decrease in frequency) will need to be accommodated.

Smart adaptive antennas require extremely close control of phase and linearity and a lot of maths. Smart adaptive antenna arrays work in Wi-Fi, but the RF power is relatively low (10 mW for user devices) and users are either slowly moving or stationary and usually within a few meters of the base station. Smart adaptive antenna arrays in wide-area mobile broadband networks have to work over a wider dynamic range (>70 dB) and respond to a radio channel that is being influenced by a wide range of local and wide-area propagation and reflection mechanisms.

The beamforming capability of the array is directly a function of wavelength versus the overall aperture of the array. It is physically more practical to implement a complex array at shorter wavelengths both for small-form factor user devices and dimensionally compact base stations and for access points and relays or repeaters.

Designing adaptive antenna arrays is a complex task, but many engineering teams have already done it including design teams from the satellite industry, near-space and deep-space communications industry, and radar industry. Even radio astronomy can be raided for useful tips on correlating multiple inputs from multiple radio sources via multiple antennas each supported by individual receive paths.

Each of these industries works across our three bands of interest for 5G. Smart adaptive antenna arrays deliver link budget gain (sensitivity gain).

Smart adaptive antenna arrays deliver selectivity, which, in turn, determines coexistence cost.

9.6 Fifth Generation Inmarsat Spot-Beam Antennas

Most contemporary and planned satellite systems have spot-beam antennas, which support the ability to provide wide-area coverage to discrete areas on Earth or focused beam coverage for Earth-to-space or space-to-space links.

In Chapter 6 we included TRDS [the National Aeronautics and Space Administration (NASA) Tracking and Data Relay Satellite System] as an example of geosynchronous satellites operating as relays and repeaters. TRDS satellites have S-band, Ku-band, and Ka-band transceivers. The S-band antenna array has 32 receive antenna elements and 15 transmit antennas.

In December 2013, Inmarsat launched its first Global Express Inmarsat 5 (fifth generation) satellite into geostationary orbit, with a second successful launch in February 2015. The satellites built by Boeing weigh over 6,000 kg and have a solar panel wingspan that is larger than a 737 passenger jet. Each satellite produces 15 kW of power and has an expected lifespan of 15 years. The third and fourth satellites in the constellation were scheduled to launch by 2016. The $1.6 billion investment is substantial in the context of the satellite industry, although it is relatively modest when compared to terrestrial mobile broadband investment costs, but the important point to make is that the antenna technology used in these platforms substantially increases the value of the services that they offer based on a dual payload consisting of 89 spot beams per satellite for global service coupled with steerable beams that can focus power and bandwidth on an on demand basis. The satellites are deployed at 17.7–20.2 GHz and 27.5–30 GHz.

It could be argued that satellites are different. Geostationary satellites as an example are 35,000 km away from their users or supported objects. Even if objects are moving relatively quickly, for example, an aircraft, the relative rate of change as seen by the satellite will be slow, and anyway the job of a spot beam antenna is to focus coverage geographically rather than provide a focused beam on each individual user. The purpose of spot beams generally is to increase capacity through frequency reuse and maintain an acceptable link budget.

Tracking fast-moving individual users or fast-moving individual objects in a dense terrestrial network will be significantly more challenging. It becomes easier when users are relatively distant from a cell site. The assumption is that users in a dense network, for example, in a dense urban network, will be slow-moving. Fast-moving users will be more likely in a sparser network, for example, rural or suburban/semiurban.

Accepting that caveat, the principles of spot beam antennas deployed on satellites in the centimeter band, particularly the combination of wide beam and narrow beam coverage, are of direct technology relevance to 5G system design. Satellite systems also have direct commercial relevance as they provide backhaul bandwidth to the mobile broadband industry. There are therefore already existing mutual interest business models that are based on mobile broadband and satellite industry cooperation. It would be costly and self-defeating to compromise these arrangements as a result of disputes over primary or shared access to centimeter-band spectrum.

9.7 Satellite Spectrum in the Centimeter Band

As a summary, satellites are deployed in the following parts of the band

9.7.1 C-Band

The frequencies 3.4 to 4.2 GHz are used for fixed satellite service (FSS) and TV broadcast satellite downlinks in some countries. The band overlaps LTE FDD Band 22 (3,410–3,490 and 3,510–3,590 MHz) and LTE TDD Bands 42 (3,400–3,600 MHz) and 43 (3,600–3,800 MHz). The FSS and TV uplinks are at 5.9 to 6.4 GHz.

9.7.2 X-Band

The frequencies 8 to 9 GHz are used for space research, deep-space operations, and environmental and military communication satellites. Satellites/spacecraft often have S-band and X-band transceivers.

9.7.3 Ku-Band

The frequencies 10.7 to 11.7 GHz support fixed satellite services, 11.7 to 12.2 GHz supports domestic TV Broadcast Satellite Service (BSS) downlinks including DVB-S, 14.5 to 14.8 GHz is the uplink feed for the Ku downlink, and 17.3 to 18.1 GHz is an alternative BSS uplink.

9.7.4 Ka-Band

Arabsat, Avanti, EchoStar, Eshaisat, Eutelsat, Gascom, Hispasat, Inmarsat, Intelsat, Nilesat, Nigcomsat, O3b, RSCC, SES, Telenore, Telesat, Thaicom, Turksat, Viasat, and Yahsat and the Brazilian, Australian, and French governments operate or plan to operate satellite systems at 24.65–25.25 GHz/17.3–17.8 GHz/21.4–22 GHz and 27.0–30.00/17.7–20.2 GHz Ka-band frequencies.

Many of these systems are regarded as being critical to national interests. Many of these systems also provide backhaul services to terrestrial mobile broadband networks in addition to emergency and safety support, media distribution, automatic identification systems, aeronautical broadband, and telemetry and telecommand services.

The frequency 23–27 GHz is increasing in popularity as fixed-link, broadcast, environmental, and space operations satellites move from lower bands to gain more bandwidth. Water vapor and rain absorption limit the usefulness of this band in the tropics. Automotive radar is also in this band.

9.8 Earth-to-Space and Space-to-Earth Links

The European Space Agency provides a comprehensive listing of communications frequencies specific to spacecraft [3]. Wavelength/frequency ranges can be summarized in terms of the antenna footprint on the spacecraft. The choice of wavelength/frequency is largely determined by the atmospheric weather conditions between the spacecraft and terrestrial radio transceiver (Table 9.3).

9.9 C-Band and X-Band Coexistence with Weather and Aviation Radar

In Chapter 8 we highlighted the need to manage coexistence with weather radar at the S-band. Weather radar is also implemented in the centimeter bands at C-band and X-band (Table 9.4).

The choice of band for weather radar is determined by the size of the raindrop to be measured, the heavier the rain, the bigger and the raindrop. Raindrops should correctly be described as hydrometeors. The 5-cm wavelength

Table 9.3
Earth to-Space and Space-to-Earth Satellite Spectrum Allocations and Antenna Configurations

Frequency Range	Wavelength Centimeters	Link Direction	Spacecraft Antenna	Used When the Weather Is
1–2 GHz	30–15	Earth-to-space uplinks	Wide-beam, low-gain antenna	Clear or rainy
1–4 GHz	30–8	Space-to-Earth downlinks		Rainy
1–6.5 GHz	30–5	Space-to-Earth downlinks	Wide beam low gain antenna	Clear
3–5.9 GHz	30–5	Space-to-Earth downlinks	High gain antenna	Rainy
6–16 GHz	5–2	Earth-to-space uplinks	High gain antenna	Rainy
11.5–35.5 GHz	3–0.84	Earth-to-space uplinks	High gain antenna	Clear
26–40 GHz	1.1–0.74	Space-to-Earth downlinks	High-gain antenna	Clear

Table 9.4
Weather Radar Bands in the Meter and Centimeter Bands

Band	Wavelength (cm)	Frequency (MHz)	Range	Application
S-band	11.7–10.33 (10 cm)	2,700–2,900	300 km	Heavy rain
C-band	5.7–5.24 (5 cm)	5,250–5,725	200 km	Light rain
X-band	3.22–3.155 (3 cm)	9,300–9,500	50 km	Drizzle

C-band weather radar is good for rain detection up to 200 km. The 3-cm wavelength X-band radar is more sensitive than C-band or S-band and therefore better at detecting light rain or small raindrops but is limited to a range of 50 km.

New meteorological micro rain radar (MRR) applications are being developed at 24 GHz. These systems are optimized for measuring hydrometeor drop size distribution. Cloud composition radar measurement systems known as cloud radar are being developed at 35 GHz [4].

9.10 5G Options in the Centimeter Band

So what are the practical options for 5G between 3 GHz and 30 GHz? The CEPT candidate bands for WRC 2015 identified the already allocated LTE bands between 3.4 and 3.8 GHz and suggested 3.8–4.2 GHz, 5.725–5.85 GHz, and 5.925–6.425 GHz for further study. The 5.35–5.47 GHz band was specifically excluded.

Whether this constitutes 5G spectrum is open to debate. The main impetus for lobbying for more bandwidth in the 5-GHz band has been the introduction of IEEE 802.11ac implemented into wider-band 160-MHz channels, which cannot be accommodated at 2.4 GHz (Figure 9.1).

The problem in Europe is that the 5-GHz spectrum is shared with other primary services including radar. The present 5-GHz spectrum allocation in Europe only supports two 160-MHz channels and four 80-MHz channels. Wi-Fi uses a dynamic frequency sensing (DFS) algorithm and Transmit Power Control (TPC), also known as a polite protocol or listen-before-transmit. DFS

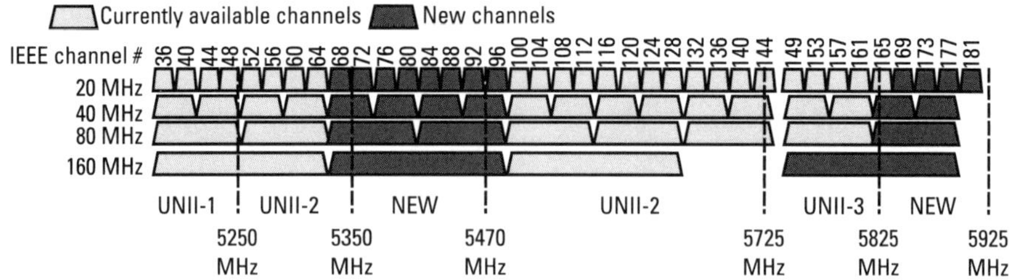

Figure 9.1 Potential European 5-GHz band extensions.

detects channels already in use by other Wi-Fi users or channels in use by other primary users. The algorithm is set so that there is a 30-minute time elapse before another attempt is made to access the channel. In a 5-GHz Wi-Fi one Watt EIRP hot spot, for example, at an airport or in a busy town center with multiple Wi-Fi access points, this can result in a loss of capacity and throughput (Figure 9.2). Allowed power is also determined by whether the access point is indoors or outside.

Other primary users include weather radar. If radar signals are detected from weather radar, then Wi-Fi devices have to vacate the channel. In practice this would mean that 802.11ac would be used more frequently on 20-MHz or 40-MHz channels, which would mean minimal differentiation with 802.11n.

For the 3 years prior to WRC 2015, ITU-R coordinated sharing studies between Wi-Fi and Earth exploration radar, bistatic aeronautical radar (bistatic radar has a transmitter and receiver separated by a distance comparable to the target distance; a monostatic radar has collocated transmit and receive paths), frequency-hopping radar, and terrestrial ground-based radar. Not a lot of progress was made and the process will now be repeated for WRC2019. The United States is marginally more proactive, with the FCC and NTIA required to study the use of unlicensed U-NIII devices in the 5.35–5.57-GHz and 5.85–5.925-GHz bands. The prospect of implementing LTE-U (LTE Unlicensed) or LTE A (LTE Assisted) presently being standardized in 3GPP Release 13 or 5G into these bands seems at the moment remote.

At the lower end of the band, there is an ITU-R sharing and compatibility study between LTE and radio astronomy at 4.8–4.99 GHz and 4.99–5 GHz [5].

In the United States, 50 MHz of spectrum between 4.949 GHz and 4.990 GHz was allocated by the FCC [6] in 2002 for fixed and mobile services with the band designated to be used to support public safety with an RF power output of up to 3W. Example applications were wireless LANS for incident scene management, mesh networks, temporary fixed communication, and fixed point-to-point links.

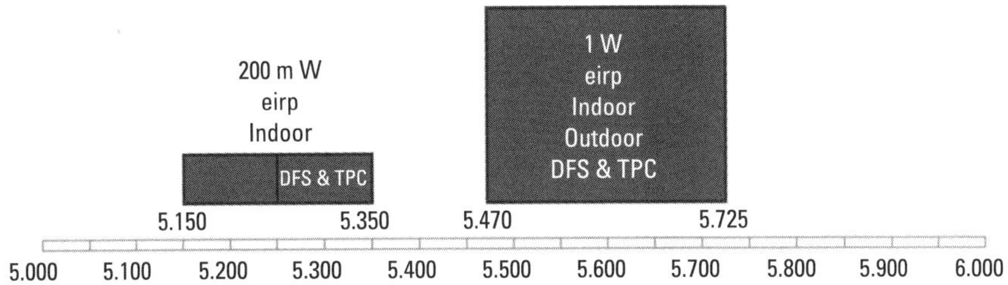

Figure 9.2 Outdoor hot spot versus indoor 5-GHz coverage.

At the other end of the band, the automotive connectivity standard 802.11p is being developed for implementation between 5.874 GHz and 5.925 GHz based on 20-dBm (100-mW) power class devices.

Theoretically at least, there is scope for 5G to integrate either with existing 5-GHz Wi-Fi spectrum and/or the mid-band extension from 5.35 to 5.47 GHz and/or the 50-MHz public safety band at 4.949 GHz and 4.990 GHz below the existing Wi-Fi band and/or 802.11p above the band, but substantial political and technical work would be needed to make all or any of these happen.

Other 5G relevant ITU R co sharing studies for this part of the centimeter band are IMT and fixed service sharing at 3.4 to 4.2 GHz [7], cosharing with geostationary satellite fixed satellite services at 3.4–4.2-GHz and 4.5–4.8-GHz bands [8] and cosharing with fixed satellite services at 5.85–6.425 GHz [9].

The Ofcom consultation process [10] on spectrum use above 6 GHz produced the responses listed in Table 9.5 divided into spectrum considered good for 5G by mobile operators and/or vendors (supported bands) and bands where incumbents required ongoing protection and objected to any proposed change of use (nonsupported bands).

Table 9.5 highlights the range of industries and industry interests that the mobile broadband industry will need to work with to bring 5G into the centimeter band.

This is just one regulator in one country asking for country-specific responses and does not begin to capture differences from region to region and country to country. Even when countries have overlapping mobile service (MS), fixed service (FS), and fixed service satellite (FSS) spectrum, there are usually different supported passbands. Table 9.6 shows the differences between the United States, the European Union, and South Korea.

In terms of coexistence, the positive part of this is that fixed point-to-point links can be implemented in the same spectrum as satellite and radio astronomy. The challenge for mobile broadband will be to prove that coexistence with 5G mobile broadband is equally feasible. This implies the need to demonstrate that there is no technology cost projected on existing incumbents (imposed technology costs) or, if there are costs, that these are more than out balanced by other benefits.

Those other benefits could include commercial benefits delivered through an incentive auction or shared-use compensation, for example, the DOD (U.S. Department of Defense) compensation process in the recent Advanced Wireless Service (AWS 3) auction. It would be useful if the technology cost could be shown to be negligible at least to the point where compensation expectations could be managed to be affordable.

Imposed technology costs are primarily coexistence costs so the starting point is to ensure that the 5G physical layer is coexistence-friendly. This

Table 9.5
Supported and Nonsupported Bands Between 6 GHz and 30 GHz

Band	6–10 GHz		10–13 GHz		14–16 GHz		17–20 GHz		21–24 GHz		25–30 GHz*	
Supported	Vodafone	5.925–8.5	Ericsson	10.125–10.225, 10.475–10.575	Alcatel Ericsson	15	Alcatel	18	Alcatel Intel	21.2–23.6, 24.25–24.45	Alcatel, Cost IC 1004 Huawei Samsung, Intel	25.25–29.25
Not supported												
BBC		7.11–7.25, 7.3–7.425		10.7–12.75			19.7–20.2				29.5–30	
ESOA and UKSA and SES		5.925–8.5		10.7–11.7	14.4–14.8		17.8–19.7		UKSA	21.2–23.6	ESOA, Inmarsat, Iridium, SES, UKSA	25.25–29.5
MOD		7.25–7.75, 7.9–8.4		10.9–11.7, 12.25–12.75		13.75–14.5		20.2–21.2				30 GHz
Jodrell Bank		5.925–8.5		10.5–11.7		14.4–15.35			21.2–23.6			
Met Office				10.68–10.7		15.35–15.4			23.6–24			

*Noting the need to share with microwave links.
Source: [11].

Table 9.6
MSS, FSS, and FSS Spectrum in Korea, the United States, and the European Union

United States	27.5 GHz	29.5 GHz
European Union	26.5 GHz	29.5 GHz
Korea	27 GHz	29.5 GHz

might involve some loss of spectral efficiency. As we pointed out in Chapter 4, improved out-of-band performance, for example, lower leakage ratios, will generally impose performance cost on in-band performance. However, it is also worth reflecting why point-to-point links can coexist with other systems including satellite and radio astronomy and the answer is that the links are point-to-point. The closer 5G can get to present point-to-point directivity in elevation and azimuth, the easier it should be to minimize system to system interference on a practical and statistical basis.

However, what is effectively spatial separation is a minimization and mitigation measure and does not remove the need for frequency, phase, and time-domain selectivity and stability, the process of separating wanted from unwanted signal energy in radio systems that are spectrally and geographically proximate.

9.11 Centimeter-Band Transceivers: Back to the Superheterodyne to Deliver 5G RF Selectivity[1]

In Chapter 8 we studied acoustic filters as the workhorse of the meter band, unbeatable as a mechanism for defining duplex separation and a steep rolloff across the duplex gap and across the guard band to adjacent channels or bands.

Acoustic filters become awkwardly small over 4 GHz to the point at which they become mechanically fragile and vulnerable to heat damage and detuning. Their power-handling capacity also reduces.

This is not to say that acoustic filters will disappear from the centimeter-band user and Internet of Things (IoT) devices, but they are more likely to be used for intermediate frequency filtering both on the transmit and receive chains. The same performance constraints apply to active devices including power amplifiers on the transmit path and low noise amplifiers on the receive path. This assumes that there is an intermediate frequency (IF), which means that the front-end architecture will need to be a superheterodyne rather than a direct conversion transceiver.

1. With thanks to *Analog Devices aand Agilent and Microwave and RF Magazine* [12].

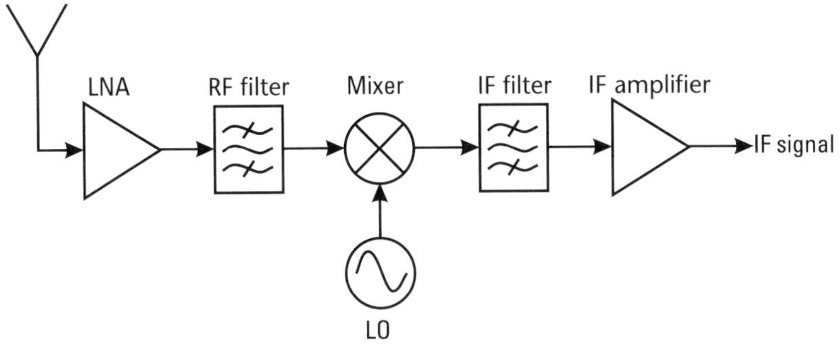

Figure 9.3 Superheterodyne front-end receiver at 9 GHz [13].

Direct conversion receivers were introduced in the meter band with the express purpose of reducing filter counter count (by eliminating IF filters). Amplifier technology had improved sufficiently over 30 years to deliver low noise and high gain at low cost up to at least 3 GHz.

For the time being at least, it would seem likely that the superheterodyne will enjoy a renaissance, which is excellent news for RF designers and mixer component vendors.

For example, you give your RF design team the job of producing a 200-MHz channel bandwidth, low-cost receiver front end working at 9 GHz. They should be able to source acceptable low noise amplifier, RF and IF filters, and a local oscillator module bringing the 9 GHz down to an IF of 2 GHz to provide the option to reuse Band 1 meter-band components. In this design a gallium arsenide (GaAs) high electron mobility (HEMT) amplifier is chosen, fabricated with standard printed circuit board techniques and low cost PCB laminates. The noise figure should be better than 3.5 dB and the input return loss should be better than 19 dB. You would be looking for an overall conversion gain in the receive front end of 30 to 35 dB and an input 1-dB compression point better than −30 dBm with an error vector magnitude (EVM) at least sufficient for present LTE systems (Table 9.7).

The IF filtering is done with high-selectivity passive filters used with an image rejection filter implemented as a microstrip digital filter with the bandpass filter constructed as an array of quasi TEM mode transmission line resonators [14].

The local oscillator (LO) module has a frequency synthesizer integrated with a low noise digital phase frequency detector, a precision charge pump, a passive third-order loop, and a voltage controlled oscillator (VCO) chip. The phase locked loop (PLL) for the receive front end is designed for an output center frequency of 7 GHz. Within the PLL loop bandwidth, the PLL phase detector is the dominant noise source. Outside the loop bandwidth the VCO noise is usually dominant. The phase noise is better than −70 dBc/Hz offset 1

Table 9.7
Link Budget for a Centimeter-Band Receiver Front End at 9 GHz

Component	Noise Figure (dB)	Gain (dB)	IIP 3 (dBm)
LNA	1	22	–8
RF filter	3	–3	>80
Mixer	9	–9	21
IF filter	3	–3	>80
IF amplifier	2.7	24	12
Total	2.2	32.6	>–9

kHz from the carrier, better than –80 dBc 10 kHz from the carrier and better than –90 dBc/Hz offset 100 kHz from the carrier.

Having done all this, you can hand the project over to the DSP engineering team to digitize the 200-MHz channel. The DSP team has to achieve a clean analog-to-digital/digital-to-analog conversion and keep the device power budget close or equivalent to or ideally lower than those of existing LTE user devices. Processing instantaneous bandwidths of 200 MHz or greater will be challenging for the analog-to-digital conversion, but at least the superheterodyne will have reduced some of the front-end dynamic range.

9.12 Summary

There are a number of potential coexistence issues or perceived coexistence issues that remain to be resolved in the centimeter band and significant resistance from existing incumbents to any significant change in access rights anywhere in the band.

There is a clear opportunity to reuse fixed link spectrum and fixed link hardware for 5G wide-area deployment but substantial spatial processing will be needed both to minimize coexistence issues and to achieve a sufficient link budget for wide-area coverage. Samsung as an example has a Matchbox-sized 64-antenna element array designed to work at 28 GHz [15] with custom-built signal processing to allow the signal phase at each antenna to be dynamically changed to generate a 10° beamwidth antenna pattern. The prototype base station was able to send data up to 1 Gbps to two line-of-sight receivers moving at 8 km per hour up to 2 km away. Nonline of sight reduced the range to 200m to 300m. Power levels from the base station were similar to an LTE eNode B.

The antenna array implementation is combined with a number of possible alternative physical layer proposals other than orthogonal frequency division multiplexing (OFDM) including nonorthogonal and asynchronous options that can be used in narrow segments of spectrum. An alternative approach might be to accept the cyclic prefix overhead implicit in OFDM and decide on physical layer options on the basis of assuming that contiguous spectrum can be allocated to 5G, albeit on a shared-use basis with fixed service and fixed service satellite provision. It could also be argued that there will be a need to be able to support nonline of sight beyond 200m in order to allow centimeter-band 5G to scale outside dense urban environments already served by Wi-Fi at 2.4 GHz, 5 GHz, and 60 GHz.

Crucially, centimeter-band user devices and centimeter-band IoT devices will need to be delivered at a cost and power budget that adds value over LTE Advanced options presently being designed and developed. It may be that this is only achievable by bringing centimeter-band and millimeter-band technologies together in an integrated physical layer.

References

[1] http://www.ofcom.org.uk/static/archive/ra/topics/pfwa/3-4ghz/consult/3_4ghz/3_4ghz.htm.

[2] https://www.fcc.gov/rulemaking/12-354.

[3] http://www.esa.int/Our_Activities/Telecommunications_Integrated_Applications/Satellite_frequency_bands.

[4] www.erodocdb.dk/Docs/doc98/official/pdf/ECCRep166.pdf.

[5] Draft new report ITU R RA (RAS-IMT).

[6] 47 C.F.R. Part 90, Sections 90.1201 to 90.1217, http://www2.fcc.gov/pshs/public-safety-spectrum/4-9GHz-Public-Safety-Band.html.

[7] ITU _R F IMT-FS 3 -400-4 200 MHz sharing.

[8] ITU _R F 4400-4990 MHz 990 MHz cosharing.

[9] ITU _R F FSS IMT C band uplink.

[10] http://stakeholders.ofcom.org.uk/consultations/above-6ghz/.

[11] Ofcom.

[12] 9 GHz Case study with thanks to Analog Devices and Agilent and Microwave and RF Magazine.

[13] http://mwrf.com/mixed-signal-semiconductors/low-cost-front-end-receives-9-ghz.

[14] http://www.rfwireless-world.com/Terminology/difference-between-TEM-wave-and-quasi-TEM-wave.html.

[15] http://global.samsungtomorrow.com/samsung-announces-worlds-first-5g-mmwave-mobile-technology/.

10

The Millimeter Band: 30 GHz–300 GHz

In April 2015, Ofcom, the U.K. regulator, published a study on possible 5G spectrum bands above 6 GHz [1] (see Table 10.1). The frequency bands 10.125–10.225/10.475–10.575 GHz, 31.8–33.4 GHz, 40.5–48.9 GHz, and 66–71 GHz were proposed to be considered for study at the World Radio Congress 2015 under an agenda item on 5G mobile broadband for WRC2019.

The study observed that it would be hard to find contiguous allocations of 1 GHz or more below 30 GHz with the 10-GHz band yielding at best 2 × 100 MHz.

Ofcom had auctioned the 10-, 28-, 32-, and 40-GHz bands in 2008. These bands are now widely used for mobile broadband backhaul. Not surprisingly, there were a range of views from the mobile broadband industry and satellite industry on preferred and nonpreferred options. These are summarized in Table 10.2 (see also Table 9.5). The responses also included inputs from COST IC1004 [2] and ISG mWT [3].

The study uncovered differences between stakeholders in terms of supported and nonsupported candidate bands. These are summarized in Table 10.3. The highlighted bands are the bands already widely used for mobile broadband backhaul. (See also Chapter 9.)

The point to make is that this is just one regulator in one country. The lack of consensus in one country becomes much more of a problem when scaled regionally and globally. The FCC for example proposed the 28-GHz, 37-GHz, 39-GHz, and 64–71-GHz bands as preferred 5G candidate bands for study at WRC 2015. The lack of regional and country-specific consensus made it much harder to achieve material progress on 5G spectrum at the congress and it is hard to see things being very different for WRC 2019.

Table 10.1
Proposed Candidate Bands for 5G per Ofcom Study, April 2015

Centimeter Band (10–1 cm)		Millimeter Band (10–1 mm)			
10.125–10.225 GHz	10.475–10.575 GHz	31.8–33.4 GHz	40.5–43.5 GHz	45.5–48.9 GHz	66–71 GHz

Source: [1].

The same divergence applies to mobile broadband backhaul. Table 10.3 shows U.K. market preferences expressed by a cross-section of stakeholders.

10.1 Repurposing Point-to-Point RF Hardware for 5G Radios: The Need for Global Scale

In Chapter 9 we discussed how fixed point-to-point hardware scaled in terms of wavelength and peak data rate with licensed link equipment at 28 GHz typically delivering 400-Mbps peak throughput through a 56-MHz channel with 38 dBi of gain from a dish antenna. A 38-GHz link with a 56-MHz aggregated channel supports 500 Mbps with 50 dBi of antenna gain. The 42-GHz, 70-GHz, or 80-GHz channels use 112-MHz or 250-MHz channel spacing with high-level modulation to deliver 1 Gbps. The 70 or 80-GHz links can also achieve the headline 1-Gbps data rate by aggregating 4 × 250-MHz channels together. The additional bandwidth means lower-order modulation can be used.

These systems are implemented in licensed or lightly licensed bands. The licensed bands are typically below 50 GHz, and the lightly licensed bands are typically above 50 GHz. Fixed point-to-point systems are also available in the unlicensed 60-GHz band (covered later in this chapter). We argued that it would be an obvious opportunity to implement 5G mobile broadband in the licensed and lightly licensed bands on the basis that RF and baseband hardware was already available with a higher allowable EIRP compared to unlicensed 60-GHz Wi-Fi.

This is only going to be economic if global scale can be achieved both in terms of band allocation and technology.

However, harmonizing spectrum on a global basis would be challenging. Table 10.5 gives an example of country-to-country differences at 38 GHz, although in all cases there is probably enough overlap to support global roaming.

10.2 European Research Programs

At a European level, there are studies on 5G physical layer candidates for millimeter-band network deployment including Millimetre-Wave Based Mobile Radio Access Network for Fifth Generation Integrated Communications (mm-

Table 10.2
Preferred and Nonpreferred Options for 5G: Summary of Industry Responses

	Centimeter Band	Millimeter Band
Wavelength	10–1 cm	10–1 mm
Frequency	3–30 GHz	30–300 GHz
Infrastructure Vendors		
Alcatel Lucent	Priority Bands: 27–29.5 GHz shared with microwave links Lower priority: 5.925–8.5 GHz, 15 and 18 GHz, 21.2–23.6 GHz, 25.35–27 GHz	Priority Bands: 36–37.5 GHz, 39.5–40.5 GHz, 42.5–52.6 GHz excluding 50.2–50.4 GHz, 55.78–66 GHz Lower priority: 36–40.5 GHz Bands for further study: 31.8–33.4 GHz, 40.4–43.5 GHz
Ericsson	10 GHz, 15 GHz	Above 30 GHz
	Prioritize bands not allocated to passive (receive only) services on a primary basis. Bands allocated to broadcasting services on a primary basis should be investigated to determine if they should be considered.	
Huawei	Priority Bands: 27.5–28.35 GHz, 29.1–29.25 GHz	Priority Bands: 37–38.6 GHz, 64–71 GHz, 71–76 GHz, 81-86 GHz Bands for further study: 31.8–33.4 GHz
Silicon vendors		
Interdigital Europe		Priority Bands: 55–71 GHz
Samsung Electronics UK	Priority Bands: 25–30 GHz focusing on 28 GHz	Priority Bands: 30–43.5 GHz Bands for further study: 40.4–42.5 GHz
Intel	Bands for further study: 24.25–24.45 GHz, 25.95–25.25 GHz	Bands for further study: 31–31.3 GHz, 42–42.5 GHz
Mobile Broadband Operators		
Vodafone	Priority Bands: 5.925–8.5 GHz	Priority Bands: 43.5–47 GHz, 51.4–52.6 GHz, 72–77 GHz, 81–86 GHz Bands for further study: 77–81 GHz
EE		Bands for further study: 31.8–33.4 GHz As per METIS submissions: https://www.metis2020.com/
Confidential response	Priority Bands: 6–30 GHz for mobile	Priority Bands: Above 50 GHz for backhaul
Confidential response		Bands for further study: 37–39 GHz, 43.5–47 GHz, 57–64 GHz, 70–80 GHz
ESOA		Priority Bands: 37–39 GHz, 43.5–47 GHz, 57–64 GHz, 70–80 GHz
Research Groups		
COST IC 1004	Priority Bands: 25.25–29.5 GHz	Priority Bands: 36–40.5 GHz, 55.78–76 GHz, 81–86 GHz, 92–100 GHz
ISG mWT		Priority Bands: 31.8–33.4 GHz

MAGIC) [4] with a study brief from 6 GHz to 100 GHz [5]. The physical layer work focus is on developing novel adaptive and cooperative beamforming and beam-tracking techniques. This is similar to the Samsung 32 antenna ar-

Table 10.3
Supported and Nonsupported Bands and Bands Used for Mobile Broadband Backhaul

Band GHz	31.8–33.4	36–40	40.5–43.5	43.5–47	47–52	52–55	55–70	70–80	80–92
Supported	Alcatel, Huawei, Intel	Alcatel, Huawei, COST	Alcatel, Samsung, Intel	Alcatel, Vodafone		Vodafone 51.4–52.6	Alcatel, Huawei	Huawei, Vodafone, 72–77, 77–81	Huawei, Vodafone, 81–86 GHz
Mobile broadband backhaul	32 GHz	38 GHz	40 GHz	42 GHz, 45 GHz	52 GHz	55 GHz	65 GHz	71–76 GHz	81–86 GHz, 92.95 GHz
Not supported	Met Office	Jodrell Bank, UKSA	Jodrell Bank, UKSA	Jodrell Bank, UKSA, MOD	Radio Society of Great Britain (RSGB), Met Office	Met Office, 48.94, 49.4, 50.2–50.4, 52.6–54.25	Jodrell Bank, UKSA	RSGB, 75.875, 76	Met Office, 86–92, Jodrell Bank, 81–86, UKSA, 81–86
Other users	Deep space and intersatellite	Cloud radar at 35 GHz	Fixed service satellite, Broadcast, Mobile satellite, 40.5–41 GHz	Fixed, fixed satellite, radio astronomy	Mobile satellite, 47.2–48.9, Amateur satellite, 47–47.2	See above	See above	Automotive radar, 77–81 GHz	See above

Table 10.4
Point-to-Point Backhaul: Channel Bandwidths and Peak Data Rate, Vendor Examples

28 GHz	38 GHz	42 GHz	70 GHz	80 GHz
56-MHz channels	56-MHz channels	112 or 250 MHz or 4 × 250 MHz		
38-dBi gain	50-dBi gain	>50-dBi gain		
400 Mbps	500 Mbps	Up to 1 Gbps		

Table 10.5
Mobile, Fixed Service, and Fixed Service Satellite Spectrum Between 30 and 40 GHz: Five Country Comparisons

United States	CEPT	Korea	Russia	Japan
38.6–40 GHz	37–39.5 GHz	38–39.5 GHz	36–37, 39.5–40 GHz	38.06–39.48 GHz

ray system demonstrated in the centimeter band at 28 GHz. Ideally, the 10° beamwidth of the Samsung test system would reduce over time to the typical beamwidth of a point-to-point dish (1° or 2°).

10.3 Smart Adaptive Arrays and Millimeter-Band Coexistence Control

If this was achieved, the coexistence conditions for millimeter mobile broadband in any of these bands would be similar to the existing coexistence conditions of fixed point-to-point links and other users. This would help to minimize the regulatory challenges of 5G system deployment.

10.4 Beamforming for the Millimeter Band

The one thing that would not work particularly well would be MIMO, as the one thing that beamforming helps to minimize is multipath, which is how MIMO achieves higher throughput. There is also a need to manage nonline of sight (NLOS) connections. Table 10.6 shows how the line of sight and the NLOS link budget changes with wavelength and the additional attenuation caused by light rain and heavy rain and the impact of oxygen resonance attenuation at 60 GHz.

This explains why many of the 5G millimeter physical layer study items include flexible routing from device to device (mesh networks) to overcome the NLOS propagation loss and to mitigate rain fade.

Table 10.6
Path Loss Increase Due to Wavelength, Line of Sight/NLOS, Rain Fading, and Oxygen Resonance

Frequency GHz	Wavelength	Path Loss (dB)		Rain Attenuation		Oxygen Absorption
		Line of sight	NLOS	5 mm/h	25 mm/h	At 200m
28	1.07 cm	1.9–1.9	4.5–4.6	0.18 dB	0.9 dB	0.04 dB
38	7.89 mm	1.9–2.0	2.7–3.8	0.26 dB	1.4 dB	0.03 dB
60	5 mm	2.23	4.19	0.44 dB	2 dB	3.2 dB
73	4.1 mm	2	2.45–2.69	0.6 dB	2.4 dB	0.09 dB

Source: [6].

10.5 60-GHz Wi-Fi: The Same but Different

If you want to minimize cell-to-cell interference in an ultradense network and have the primary objective of delivering high data rates over short distances, then the oxygen resonance peak at 60 GHz is a help not a hindrance. This is the basis of the 802.11ad extension of the Wi-Fi standard also known by its trade name of WiGig [7]. The standard divides the 57-GHz (5.25-mm wavelength) to 64-GHz band (4.68-mm wavelength) into four 2.16-GHz channels/subbands with an orthogonal frequency division multiplexing (OFDM) physical layer supporting peak data rates up to 7 Gbps.

It is a Wi-Fi TDD physical layer using the same frequency for the uplink and downlink, so the propagation is the same in both directions, which helps channel sounding for MIMO and beamforming. The standard can be used for fixed point-to-point links but is not presently able to support mobile users.

This highlights the point that there is not much point in trying to make a 5G physical layer do what a Wi-Fi physical layer does well: provide high data rate connectivity over small distances to users who are either stationary or move around slowly and not very far. It also highlights the need to differentiate 5G from Wi-Fi including what will effectively become a Wi-Fi tri-band, tri-wavelength 2.4 GHz/5 GHz/60 GHz service offer (see Table 10.7).

Table 10.7
Tri-Band Wi-Fi Bandwidth and Throughput

2.4 GHz	5 GHz	60 GHz
12.49 cm	6 cm	5 mm
80-MHz passband	2 × 160-MHz channels	4 × 2.16 GHz
150 Mbps × 1 radio × 2 streams 300 Mbps, multiple input, multiple output (MIMO) with two spatial streams	150 Mbps 2 radios × 3 streams 900 Mbps, MIMO with three spatial streams 1.755 Gbps, three spatial streams on 80-MHz-wide channels; reduces to 600 Mbps for outdoor access points	7 Gbps

Source: [8].

An example of a 60-GHz 802.11ad modem for outdoor use deployed as an access point on a lamppost is pictured in Figure 10.1.

A 60-GHz active phased array antenna is used to deliver electronic beam steering with 2 × 12 element antenna arrays for transmit and receive operation. Several channel bandwidths are provided up to the full 802.11ad channel width of 1,760 MHz with typical ranges of >200m at 1 Gbps. Samsung has a similar 60-GHz platform capable of 4.6 Gbps.

10.6 Differentiating the 5G Physical Layer from Wi-Fi

The obvious points of differentiation are that the 5G physical layer will need to support larger cell radii and mobile users and or mobile objects. This means that user devices and Internet of Things (IoT) devices have to handle power outputs substantially higher than the 10-mW power output of Wi-Fi devices. The 5G base stations have to handle higher powers than the 250 mW typically supported in Wi-Fi access points.

Wi-Fi physical layers support MIMO or beamforming optimized for small cells. The 5G devices and base stations need to support beamforming optimized for wider area cells with mobile users and or moving objects. These mobile users and mobile IoT devices can either be close in to the cell center (fast-changing beamforming algorithms) or further out from the cell center (more slowly changing beamforming algorithms).

These are different design and specification start points and they have a profound impact on hardware and software cost and complexity.

Figure 10.1 Outdoor lamppost-mounted 60-GHz Wi-Fi hot spot access point [9]. (Courtesy of Mark Barrett, Blu Wireless.)

10.7 RF Hardware Requirements of the Millimeter Band

In Chapter 9 RF hardware and architectures were discussed, including the need to realize gain efficiently at centimeter and millimeter wavelengths and to manage noise and linearity particularly in active components, RF amplifiers, low noise amplifiers, and mixers. We heralded the reappearance of the superheterodyne as the present most obvious way forward for at least initial iterations of 5G user and IoT devices in the centimeter band.

These constraints scale directly to the millimeter band. The good news is that RF devices get smaller; the bad news is that, assuming that they also need to handle similar power levels, they will get hotter, which means they get noisier and less stable. This implies a need for materials and packaging innovation.

10.8 RF Power Amplifier Transistor Options for the Millimeter Band

In an ideal world, all RF power amplifiers would use complementary metal-oxide semiconductor (CMOS) on the basis of its low production cost, choice of fabrication sources (minimized time to market and supply chain risk), the ability to integrate functions in a single chip with on chip calibration and self-test, and wide choice of design and simulation tools. The disadvantage is that CMOS has a low breakdown voltage, low device gain, and relatively poor linearity, all properties that become more important at shorter wavelengths and device geometry.

The 5G power amplifier requirements are likely to range from very low-power, highly linear devices directly coupled to individual elements in an adaptive antenna array through to high-power single devices for repeaters and relays.

There are many materials and combinations of materials other than silicon that can be used to create a more efficient power amplifier or low noise amplifier (LNA). They cost more so the choice is directly driven by the optimum compromise point for every application.

New material mixes are generally used initially in military and space and satellite applications. Gallium arsenide as an example was originally used in military radio systems in the 1970s and only became widely adopted in the mobile broadband industry in the late 1990s partly to meet increased demand for higher-frequency (>2 GHz) amplifiers but also to meet the requirement for wider bandwidth operation and increased linearity and to manage problems of heat dissipation and energy consumption in small-form factor devices.

Gallium nitride is a relatively new option first demonstrated in the 1990s but now is becoming more readily available initially for higher-power RF applications including phased array radars and electronic warfare and weapons systems. Gallium does not exist freely in nature but is a byproduct of the production process of zinc and aluminum. The gallium nitride compound for RF applications is formed by gallium and nitrogen atoms combined in a lattice structure. The combination is achieved in a high-temperature (1,100°C) metal organic chemical vapor deposition process or a molecular beam epitaxy process. The end result of either option is a layer of gallium nitride on a silicon carbide substrate.

Semiconductor materials are compared in terms of the amount of energy required to free an electron from its orbit around the nucleus and allow it to move freely through the solid. This is known as the bandgap and is measured in electron volts (eV). Gallium nitride has a high bandgap value of 3.4 eV compared to 1.4 eV for gallium arsenide and 1.1 eV for silicon.

This means that gallium nitride has a higher power density (measured in watts per millimeter). This is matched to the good thermal conductivity and low RF loss of the carbide substrate. The devices also have high electron mobility and high saturation velocity, the two parameters that determine how fast the electrons move in the solid.

Gallium nitride is piezoelectric so shares many of the properties of a surface acoustic wave (SAW) or film bulk acoustic resonator (FBAR) acoustic filter.

The higher power density means that a smaller device can meet a given power requirement but also means that a circuit designer can design an amplifier with a wider bandwidth with lower combining losses. The high electron mobility and saturation velocity mean that the device is more efficient at higher frequencies. Gallium nitride is also used for field effect transistors for amplifying weak RF signals.

Gallium nitride can deliver a power density of 20 watts per millimeter at high frequencies, but presently anything above watts per millimeter becomes problematic in terms of heat dissipation.

Running any transistor at elevated temperatures will raise the noise floor of the device and introduce mechanical reliability issues. Large amounts of heat, even if moved efficiently away from the device, cause stability problems in adjacent circuit functions, for example, frequency drift in resonant devices including filters. Thermal design and stress analysis become at least as important as electrical design and can be a particular challenge when designing and building gallium nitride devices. Improving device performance is never a particularly easy process.

10.9 Filter Options for the Millimeter Band

In Chapter 8 we covered acoustic filters as the work horse of the RF front end in the meter band but pointed out that the devices could not scale to the centimeter band due to the mechanical structure becoming too small and fragile to handle mobile broadband power requirements.

Filters for centimeter and millimeter bands are typically realized using low temperature cofired ceramic (LTCC). The low temperature means less than 1,000°C; the cofiring refers to the process of combining the ceramic with aluminum or copper. The material is sometimes referred to as glass ceramic as it is mainly composed of glass and alumina.

The manufacturing challenge is to produce mechanical functions such as cavity resonators within closely managed tolerances and to maintain surface flatness. Both of these become harder to manage as wavelength reduces/operational frequency increases. LTCC substrates are widely used in the automotive industry for controller modules. For RF applications a low dielectric constant is required combined with low resistance.

LTCC devices are already widely used in point-to-point backhaul hardware as bandpass and duplex filters including duplex filters for the 60-GHz band separating the receive path (59 to 61.5 GHz) and transmit path (62.5 to 64 GHz), a relatively stringent isolation requirement. The electrical performance of these devices is easily compromised by parasitic effects, interconnection, and radiation losses.

The devices can be made to be compact by vertically stacking the cavity resonators with microstrip feed lines vertically coupled through rectangular slots etched on the input and output resonators. If carefully designed, the device can combine the air cavity with integrated dual-polarized, cross-shaped patch antennas. The microstrip lines function as the feed structure to excite the cavity resonators via coupling slots that couple energy magnetically from the microstrip lines into the cavity. The cavity lengths of these millimeter-band components are of the order of 2 mm, the width around 1.3 mm, and the height around 0.100 mm. This level of precision mechanical manufacturing can be hard to scale for mass-market low-cost consumer products, but that is the whole point of technology innovation and has been a determining factor in the wireless industry for at least 100 years. Moving to higher frequencies/shorter wavelengths introduces new design and manufacturing problems but opens up profit opportunities for the companies that solve those problems cost-efficiently.

10.10 Printed Circuit Boards and Substrates for the Millimeter Band

While it is possible to design and build individual millimeter wavelength components and realize good performance on a test bench, it is more challenging to integrate them with other functions in the RF front end. A millimeter wavelength antenna and duplexer needs to talk to adjacent devices and it has to do this via the printed circuit board.

In 1982 when Motorola started production of first generation cellular phones, one of the challenges was to source low-cost flame-retardant woven fiberglass (FR4) printed circuit board material with sufficient quality and consistency to realize stable RF designs at 850 MHz. Thirty years later, we take good-quality FR4 for granted as the default material used for standard printed circuit board layouts at cellular RF frequencies in the meter band. This is not to say that circuit board laminate materials would not benefit from improved performance particularly if cost is the same or less.

The limitations of FR4 are the batch-to-batch consistency of the dielectric constant, impedance stability over frequency, signal loss, and thermal conductivity when supporting active devices providing high linearity. Minimal temperature expansion is also important. More highly specified FR4 will have lower loss but higher cost.

Materials like graphene, covered later in this chapter, might help due to their extreme ability to transfer heat. Whatever we do, it is safe to say that the workhorse of the industry, the printed circuit board, needs some fundamental innovation.

One starting point is to look at what will be needed to realize a low-cost, power-efficient, smart, Matchbox-sized smart antenna array that can scale across centimeter and millimeter wavelengths and support $<10°$ beamforming in elevation and azimuth. It will need to be small enough and sufficiently low cost to be used in a user device, IoT device, base station, repeater, or relay.

Note we are assuming that 5G smart phones will be smart enough not to attempt to transmit RF power at centimeter and millimeter wavelengths through a human head or hand. The 4G smart phones know whether they are horizontal or vertical. The 5G smart phones will need to have three-dimensional RF spatial awareness including angle of arrival, required angle of departure, angular energy, and signal polarization. We are also assuming the device will need to produce hundreds of milliwatts of radiated RF power in a user device or IoT device form factor or watts of power in a base station, relay, or repeater.

This suggests a need for the Elon Musk approach to RF component design. Faced with the challenge of improving electric car batteries, Mr. Musk decided his Tesla cars should be powered by hundreds of small batteries all designed to be produced at low cost. System power efficiency is achieved by individually controlling each battery to maximize power output and to allow for fast (under a half hour) recharging. As with RF devices, the big challenge is heat management and stability.

Applying the same principle to a 32 array or 64 or 256 antenna array implies that each array has its own RF power amplifier on the transmit path and LNA on the receive path. Each amplifier can be individually phase controlled to provide the required beamforming. Each individual amplifier produces a few milliwatts or microwatts or picowatts of power, but that power is focused across the whole antenna array with exquisite efficiency. As our friends from the deep-space radio space industry would say, it is all about aperture, the magic of millimeter-wavelength RF.

At the moment, much of this magic disappears into the substrate. At this point, we need to briefly digress into transmission line theory [with thanks to Tarun Amla of the Isola Group and Paul Cooper of Qorvo].

10.11 Microstrip Line Theory

The microstrip line is transmission-line geometry with a single conductor trace on one side of a dielectric substrate and a single ground plane on the opposite side. In a microstrip line, the electromagnetic (EM) fields exist partly in the air above the dielectric substrate and partly within the substrate itself. The effective dielectric constant of the line is therefore expected to be greater than the dielectric constant of air (1) and less than that of the dielectric substrate.

There are three types of losses that occur in microstrip lines: conductor (or ohmic) losses, dielectric losses, and radiation losses. An idealized microstrip line, being open to a semi-infinite air space, acts similarly to an antenna and tends to radiate energy. Substrate materials with low dielectric constants (5 or less) are used when cost reduction is the priority.

Similar materials are also used at millimeter-wave frequencies to avoid excessively tight mechanical tolerances. However, a lower dielectric constant translates into a lower concentration of energy in the substrate region and, hence, higher radiation loss. Radiation loss depends on the dielectric constant, the substrate thickness, and circuit geometry.

The use of high-dielectric-constant substrate materials reduces radiation losses because most of the EM field is concentrated in the dielectric between the conductive strip and the ground plane. The benefit in having a higher dielectric constant is that the package size decreases by approximately the square root of the dielectric constant. This is an advantage at lower frequencies but may be a problem at higher frequencies due to manufacturing tolerances. In most conventional microstrip designs with high substrate dielectric constant, conductor losses in the strip conductor and the ground plane dominate over dielectric and radiation losses.

Parameters related to the metallic material forming the strip, ground plane, and enclosing walls, for example, conductivity, surface roughness, and skin effects, determine the conductor losses. When designing antennas at millimeter wavelengths, the same effects need to be taken into consideration, although the design aim is to maximize rather than minimize radiation loss.

10.12 Practical Printed Circuit Board Design Issues in the Millimeter Band

The physical size of a high-frequency transmission line is dependent on the dialectic constant of the printed circuit board (PCB) material. The constant is the ratio of the permittivity of the material compared to free space. A lower ratio means the material concentrates electric flux more efficiently. For 50-ohm impedance, the width of the transmission line reduces as the dielectric constant reduces. The resulting circuit dimensions are more compact but can be hard to fabricate and variations in the dielectric constant across the circuit board can introduce phase distortion. The constant needs to stay constant over temperature and time.

Reinforcement materials used in the circuit board, such as glass weave, can disturb signal propagation velocity. These impairments are familiar to engineers requiring phased matched channels in radar systems and are now becoming, or should be becoming, more familiar to 5G design teams.

Millimeter-wave circuits require low dissipation. Any roughness of a copper surface will produce high conductor losses at higher frequencies. Thinner laminates are generally needed to minimize unwanted resonances, but this, in turn, can cause fabrication issues including tolerance and yield.

Printed circuit boards started being used in consumer radio receivers after World War II. The thermal plastic polymer, polytetrafluoroethylene (PTFE), had been discovered in April 1938 at the Dupont Research Labs. Combined with woven glass, these lithographically printed circuit boards were robust enough to be used in ordnance applications including anti-aircraft proximity fuses.

In the 1950s, a thermosetting industrial fiberglass composite laminate of filament glass cloth with an epoxy resin known as FR4 was introduced and is still in widespread use today. It is strong and moisture-resistant and has excellent electrical properties, at least up to microwave frequencies.

The basic principle of a printed circuit board is to produce a sandwich of thin, electrically conductive layers and insulating layers of polymer, glass, ceramic, or polymer filled with glass or ceramic. Over time, the number of layers has increased together with the need to support RF and digital technologies and the mechanical and thermal properties of the dielectrics have become more important as has the requirement to handle more power.

Millimeter wavelengths (10 mm to 1 mm) and frequencies (30 to 300 GHz) are particularly challenging for PCB materials. The standardized methods and techniques for measuring dielectric constant become increasingly inaccurate and unreliable. The designer is faced with conflicting performance requirements including dielectric loss, thermal stability, thermal management, layer count, twist and warp resistance, stability with humidity and temperature, thermal cycling tolerance, power-handling capability, and passive intermodulation.

The skin depth of copper at millimeter frequencies is extremely small. This means that surface roughness translates directly into attenuation loss producing conductor losses, which can exceed dielectric loss. This can be reduced by using smooth high purity copper, although this increases cost.

Products shipped at consumer price points or at lower frequencies will generally use a mix of FR4 and RF optimized substrates but scaling this approach to millimeter frequencies particularly for devices combining high-speed digital and RF functions is problematic.

It is generally assumed that PTFE is stable with temperature; however, the crystalline structure of the material and the manufacturing process (sintering) produce a range of crystallinity that will produce batch-to-batch variation. The material can also suffer permanent deformation, also known as creep, at close to room temperature. The impact of this is relatively trivial at lower frequencies but significant at frequencies above 70 GHz and has been a particular design concern for automotive applications with a subzero to 85°C temperature gradient.

The lower expansion coefficient of copper also introduces a risk of fatigue failure from high plastic strains. Hybrid mixes of FR4 and ceramic-filled substrates will have a tendency to delaminate due to temperature cycling and higher layer board combinations are likely to oxidize. These problems can be mitigated by using different filler materials but these are often highly abrasive and result in high drill wear for the interlayer through holes. The drilling costs can be higher than the material cost. Alternatives such as plasma drilling can be used but are equally expensive.

All of the above has led to work being done on finding combinations of materials that deliver a more optimum compromise between electrical and thermal mechanical properties and cost, essentially thermoset materials that behave like standard FR4 but with properties that do not degrade with temperature. The materials need to have low conductor loss both to minimize power consumption and to limit heat rise on the board. Conductor surface roughness not only increases parasitic capacitance but also results in a phase constant that will change in frequency, affecting phase and group velocity.

These effects can be mitigated by using low-profile copper which also reduces insertion loss and heat rise, but this in turn requires a thermoset that provides high peel strength even when smooth, low-profile copper is used. Lower loss also increases power-handling capability. The thermoset needs to have good predictable dimensional stability, low drilling cost, and similar flow and fill behavior to FR4. The desired and required outcome will be low-cost multilayer printed circuit boards optimized for mixed RF/digital signal processing at millimeter frequencies across extended temperature gradients.

10.13 Graphene and 2-D Materials for 5G

While we are discussing materials and manufacturing, we may as well discuss single atom layer 2-D materials including graphene. Graphene is a monolayer, hexagonal arrangement of carbon atoms. When assembled as a multilayer structure, it could provide the building blocks for ultrafast transistors and fast, efficient data and energy storage including high-density, lightweight, fast-recharge batteries. Graphene has the ability to conduct heat more efficiently than copper and when stacked, doped, chemically reduced, or electrically/magnetically biased can be an efficient (low-resistance) conductor of electricity. Graphene also has excellent mechanical and gas barrier properties. In the telecoms industry it could provide the basis for more efficient routers including highly optimized low-cost optical devices. It could also improve the RF performance of smart phones and 4G and 5G user devices by improving EMI shielding and thermal management.

Graphene is also being explored as a possible substrate for terahertz antennas due to its ability to support the propagation of surface quasi particles known as Plasmon polaritrons producing surface confined waves at the low end of the terahertz spectrum, essentially a terahertz SAW filter [10]. A European project team is researching similar applications at microwave frequencies [11].

Graphene is in practice not one material but a family of materials, with each material being a product of the manufacturing process used. This, in turn, determines the properties and performance of the material. Graphene can be produced by physical exfoliation. Applying adhesive tape to a piece of graphite and pulling it off will isolate multilayer graphene on the tape. By applying a fresh adhesive tape and repeating the process, it will eventually produce few layer and monolayer graphene. The isolated sample can be then be deposited on to a substrate such as a quartz or silicon wafer; however, the result is neither scalable nor consistent.

Alternatively graphite can be broken into flakes, for example, by ultrasound and shaken and stabilized in a liquid suspension to produce a range of graphite inks of varying flake size. These can be used to dip or spray substrates. These are relatively low-cost to produce but do not have the performance or consistency of graphene produced from more direct production processes. The process is also subject to changes in the cost of graphite.

Similar constraints apply to reduction techniques producing graphene oxide from graphite oxide. Graphite oxide is a compound of carbon, hydrogen, and oxygen molecules produced by treating graphite with a strong oxidizer or combination of oxidizers, for example, sulfuric acid, sodium nitrate, potassium permanganate, or phosphoric acid. Graphene oxide is a by-product of this reduction and oxidation process. Problematically, the oxidation process compromises the quality of the graphene. Research is ongoing to find a process that is fast but effective and efficient, probably some combination of electrochemical process, preferably avoiding toxic waste.

More direct processes are essentially various forms of chemical vapor deposition that involve disassociating carbon atoms from a suitable gas, for example, methane, acetylene, or carbon dioxide using heat in a furnace to transfer the atoms directly on to a substrate. This disconnects the process from the direct material cost of mined graphite.

The challenge is to ensure that the carbon atoms do not cluster together (forming soot). Creating the right carbon structure requires high levels of heat of the order of more than a 1,000°C. Typically, a catalyst will be used for the reduction process but this introduces additional compounds into the combustion chamber and can result in unwanted reactions, for example, the carbon atoms dissolving into nickel.

Getting the graphene on to a suitable substrate is also complex. Copper is one substrate option. A mix of copper and mechanically and chemically weak copper oxide allows the graphene to be recovered and the copper to be reused. Other options include the use of polymers to facilitate the transfer process, for example, polymethyl methacrylate (PMMA). The ideal end result is a uniform layer of graphene, but this can be frustrated by the convection and turbulence of the carrier gas. The fluid dynamics of the gas can mean that the reactants are depleted before the gas reaches the further end of the substrate.

Another option is synthesized graphene powder. This is sprayed into a furnace and converted into graphene platelets 1 to 2 microns in size with a thickness of typically less than 5 nm. The powder can be added to a suspension or added to other composite materials to improve electrical and heat conductivity. The advantage with this process is that it can be scaled to several tons per year and should produce polymer composites that have less defects and cracks than polymers using exfoliated or reduced graphene, which may contain graphite lumps.

An alternative is to heat silicon carbide (SiC) to a high temperature (>1,100°C) at low pressure. The output is epitaxial graphene with dimensions that are dependent on the size of the silicon carbide substrate. This determines the thickness, mobility, and carrier density of the graphene. Epitaxial graphene has been used by IBM to build a microwave GFET (graphene field effect) mixer [12].

Graphene has also been discussed as a potential transistor material. One of the potentially useful benefits of graphene transistors is that they overcome the short channel effects that occur as Si is scaled down; the short channel effect is a condition where the channel length is the same order of magnitude as the width of the depletion layer resulting in a change of behavior when compared to the source and drain in a conventionally dimensioned metal-oxide-semiconductor field-effect transistor (MOSFET). This should result in higher f_t devices. Graphene transistors have their own set of problems which include lack of saturation (cannot set bias to get maximum f_t) and lack of power gain (very low f_{max}) as well as unusual ambipolar transfer characteristics.

Silicon will have similar issues as and when it hits the 10-nm node around 2020, but in practice graphene is more likely to be useful in passive applications including antennas, shielding, interconnects, metamaterials, absorbers, and thermal management. Thousands of engineers will be solving the silicon scaling problem. Graphene engineers have other priorities.

Although integration levels have increased over time, it is also true to say the number of discrete active and passive devices has either stayed the same or increased, at least in higher end-user phones. Mixes of materials that provide improved conductivity and improved isolation, for example, keeping digital noise out of front-end receive paths, would be particularly useful. CVD graphene, for example, potentially provides significantly higher isolation than gold film and monolayer graphene could potentially shield as much as 97.8% of the unwanted electromagnetic energy (electromagnetic interference).

Although graphene has enjoyed the most attention (and a Nobel Prize), it is not the only 2-D material. Reducing silicon to an atom thick, popularly described silicene produces a honeycomb structure not dissimilar to graphene and like graphene allows electrons to move as if they were massless, which means that they move very quickly. It could also be used as a transistor in its natural form. It has only recently been synthesized and is presently harder to manufacture than graphene and is unstable under ambient atmospheric conditions [13].

Reducing germanium, the original transistor material from the 1940s to a single layer of atoms produces germanene [14]. This conducts electrons five times faster than germanium and ten times faster than silicon and might be more compatible with existing scale production processes than present graphene manufacturing techniques. As with silicene, germanene has stability problems that presently limit its usability.

Molybdenum disulfide [15] is a 2-D structure similar to graphite, but can be restructured as a single atomic layer sandwiched between two sulfur atoms producing a natural form that could function as a higher-efficiency transistor.

It is not impossible to consider some combination of all of these materials to produce composites that could act as optimized conductors, semiconductors, or insulators. The behavior of 2-D materials in general remains relatively unexplored and sometimes unexplainable with observed behavior not always consistent with existing quantum theory. The behavior of combinations of these materials is presently even more arcane.

Graphene may have a role to play in producing thermally and electrically optimized printed circuit boards, although present research is focusing more intently on energy conversion and storage where the combination of high electrical conductivity, physical flexibility, and high surface-to-weight ratio opens up particular opportunities in electric charge storage in batteries and supercapacitors and as catalysts in solar and fuel electrodes.

By comparison, thermoset PCB material innovation may seem prosaic but is likely to have a more fundamental short-term impact on the cost and performance economics of 5G E-band network and user devices.

The cellular market in 1982 was crucially dependent on the availability of stable low-cost, high-quality FR4 capable of working efficiently at 800 and 900 MHz, a key enabler for base stations and user equipment. Thirty-five years on the same material, mechanical and manufacturing constraints apply and need to be factored into 5G mobile broad spectrum planning, technology planning and the economic modeling of wide-area high data rate delivery cost.

10.14 Automotive Radar: Solving the RF Hardware Problem at E-Band and Related 5G Translation Opportunities

Fortuitously the 5G industry does not need to start with a blank sheet of paper on this problem as it is already being addressed by the automotive industry for E-band 77-GHz automotive radar. Over a million people die in road accidents every year, so there is a massive human and financial motivation to improve automotive safety using a combination of infrared, optical, and RF processing [16].

Automotive radars use a form of chirp radio in which the transmitted signal moves in frequency continuously across the channel. The frequency difference between the sent and received signal is captured as a beat frequency, which is then used to establish the time delay and hence range and movement of the detected object.

Chirp radars are cheaper than pulsed radars and use less power. Chirp-waveform, 77-GHz automotive radars consume about 2.5W and are capable of differentiating objects and their size, speed and direction over a distance of 150m to 200m at speeds of up to 160 kmh.

Once the preserve of high-end cars, automotive radars are now available as optional extras or as standard equipment on mid-market vehicles for collision avoidance (lower insurance premiums) and are being integrated with laser (LIDAR), ultrasound, and imaging systems to move motoring towards a highly managed (safer) more automated experience.

Radar systems have a number of advantages over optical systems; the units can be hidden behind plastic bumpers and can detect nonreflective objects (dirty cars) and work in adverse (foggy) weather. They can point forward, sideways, and backwards and can detect speed, distance, direction, and elevation (differentiating a bus from a sports car and a baby carriage). More complex systems are presently being designed with up to 11 separate radar transceivers (Figure 10.2).

The 24-GHz and 26-GHz systems have met with deployment issues due to the use of this band for weather sensing (24 GHz is the mechanically resonant frequency of water vapor). The 122-GHz and 244-GHz systems are both interesting but challenging in terms of RF component performance.

The 77-GHz system is attractive in that it has plenty of bandwidth (to provide good radar resolution) and is benefiting from substantial component and system level investment [17, 18]. This investment is yielding RF component innovation, spatial processing algorithmic innovation, and new approaches to smart antenna design, all potentially useful for 5G user and network devices.

Short-range radars are good at range accuracy; mid-range and long-range radars can look further away. Short-range and mid-range radars with ranges of a few tens of meters are used for stop-and-go applications in urban areas. Long-range radars (hundreds of meters) are used in cruise control systems and provide enough accuracy and resolution for relatively high speeds (up to typically 120 km per hour).

Short-range and long-range radars are capable of measuring relative velocity with high accuracy. Long-range radar can detect objects 200m away. The angular resolution is the same for both systems but the long-range systems will have a higher tangential error. Short-range radar provides a wide view typically greater than 30° with good spatial resolution (<10 cm). Long-range radars are usually narrowband, and short-range radars are usually wideband.

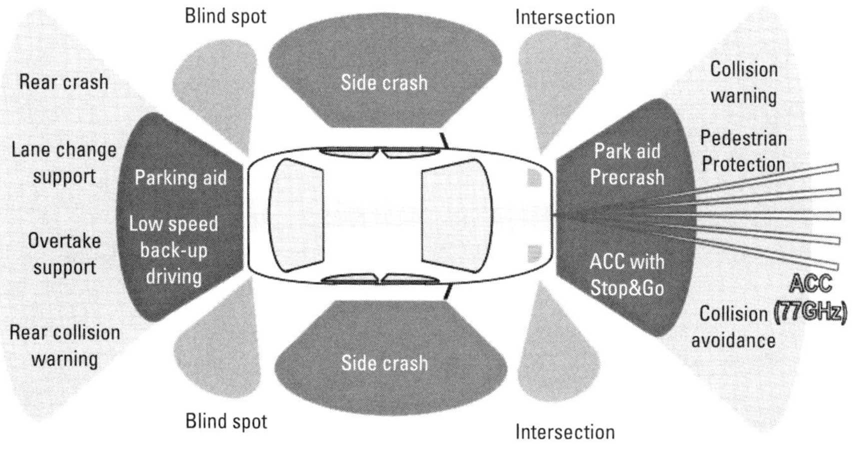

Figure 10.2 Millimeter-band automotive radar.

10.15 Automotive Bands and Subbands

The 24-GHz band consists of two bands, one centered on 24.125 GHz with a bandwidth of 200 MHz and the other centered on 24 GHz with a bandwidth of 5 GHz. Both of these bands can be used for short-range and mid-range radars. The 77-GHz band consists of two subbands, 76–77 GHz for narrowband long-range radar and 77–81 GHz for short-range wideband radar.

As wavelength reduces, the size of the active elements in the antenna array reduces and therefore angular resolution improves. The higher carrier frequency also means that the Doppler frequency increases proportionally relative to the velocity of the target. Millimeter-wave automotive radar therefore supports higher-speed resolution. Range resolution depends on the modulated signal bandwidth; the wider the bandwidth, the better the range resolution.

10.16 Impact of Automotive Radar on Millimeter Component Availability

The automotive radar industry has had a significant pull-through effect on silicon germanium with integrated Si GE transistors now available with cutoff frequencies (ft, ft,max) of over 250 GHz.

To meet RF front-end phase noise requirements, fundamental oscillators are often designed at lower frequencies, for example, 20 to 40 GHz, and then upconverted using frequency multipliers. Dielectric resonator oscillators are also an option to realize high-quality tank circuits.

10.17 Automotive Radar Emission Regulations

Power outputs/spectral densities in the millimeter band for pulsed and frequency-modulated continuous-wave radar (FMCW) are specified by ETSI for Europe (Table 10.8) and the FCC for the United States.

ETSI allows −15 dBm/MHz to −3 dBm/MHz mean spectral density at 79 GHz and 46.2 to −55-dBm peak EIRP for short-range radar at 79 GHz. For the FCC, the state of the vehicle determines the restrictions on transmitted output power. For a stationary vehicle, the spectral density in any direction must not exceed 0.2 μW/cm^2 in any direction. For a moving vehicle, the allowed spectral density is 60 μW/cm^2 looking forward and 30 μW/cm^2 for side-looking and rear-looking directions. The maximum field strength determined by the FCC is 500 μV/m at a 3-m distance equivalent to an EIRP power spectral density not exceeding −51.3 dBm/MHz.

Table 10.8
ETSI Fixed Antenna Structure EIRP and Out-of-Band Emissions for 77-GHz Automotive Radar

Band	76–77 GHz	
EIRP (FMCW)	50 dBm (mean)	55 dBm (max)
EIRP (Pulsed)	23.5 dBm (mean)	55 dBm (max)
3-dB beamwidth (typical)	5°	
Out-of-band emission	73.5–76 GHz	0 (dBm/Hz)
	77–79.5 GHz	0 (dBm/Hz)

Regulatory agencies have been encouraging migration to the millimeter band by restricting emissions in the 24-GHz band so 24-GHz systems are likely to be phased out over time, at least in Europe.

10.18 Automotive Radar and Millimeter Mobile Broadband: Commercial Cooperation Opportunities

There would therefore seem to be several technology and commercial touchpoints between automotive radar and 5G in the millimeter band. The signal processing and frequency-domain and time-domain spatial signal analysis used in automotive radar would seem to be directly translatable to 5G physical layer development. The range requirements are similar with a typical intersite distance (ISD) of 150m in 5G equivalent to the 150m to 200m in automotive radar with similar resolution and rate requirements. Coexistence issues have been addressed, so there is no real reason for not working together on common technology interests.

There is RF component and transceiver architecture commonality. Research and development budgets in the automotive industry are similar in scale to the telecommunications industry; VW on its own spends over $12 billion per year on research and development. An increasing percentage of this budget is being spent on radar and sensing systems, which will need to integrate and coexist with future radio systems. This implies a parallel need to consider the coexistence of radio networks and cars equipped with multiple radar transceivers. It therefore makes sense for the automotive radar supply chain and mobile broadband community to work together.

The potential common interest is determined by the spectral adjacency of the 77 GHz automotive radar band to the fixed link bands at 72–77 GHz and 81–86 GHz. These bands offer channel bandwidth availability (5+5 GHz of spectrum), power (up to 3W for user devices), and link budget (military communication systems in the United States use 2° beamwidth spot beam antennas to provide 60-km clear weather range).

From an automotive perspective, it also makes potential sense to have car connectivity in the millimeter band. The alternative is the top end of the 5-GHz band but the 72–77-GHz and 81–86-GHz bands offer significantly more bandwidth than 802.11p (5,850–5,925 MHz) with less onerous adjacent channel requirements and could have the advantage of being a standardized 5G compatible physical layer. Economically the spectral positioning of automotive radar (77–81 GHz) midway between the 72–77-GHz and 81–86-GHz bands opens up technology scale and integration opportunities.

The robust but cost-effective test regimes used by the automotive industry could also be beneficially applied to help reduce 5G RF test costs; testing the linearity of the sweep waveform, for example, is a critical parameter. Conversely, military investment in E-band radio systems is yielding materials innovation (including improved gallium nitride-based devices) and network topology innovation (adaptive routing techniques), which can translate across into automotive radar and connected car applications.

It would be particularly intriguing to explore the potential of an integrated network that coupled automotive radar and a spectrally adjacent 10-Gbps 5G mobile broadband network if automotive is extended to include anything that moves.

On a first pass it might seem problematic to have 4-GHz bandwidth automotive radars centered on 79 GHz coexisting geographically with a 5 + 5 GHz bandwidth 5G radio network between 72 and 77 GHz and 81–86 GHz. Interference management is essentially spatial awareness, which, in turn, can be used to beam form to discriminate between wanted signal energy and unwanted signal energy.

There is no reason why cars and other moving objects including delivery vehicles, buses, trains, boats, and planes cannot function as mobile repeaters and relays. If every Ford in the world had an integrated 5G repeater and or relay painted black, you would have by default a perfectly adequate global 5G network with virtually nonexistent estate management costs. Tractors could be included as well.

In the context of the contemporary supply chain and network delivery economics of 4G LTE, there is an already well-established coupling of mobile broadband with emerging connected car applications. Network operators are developing market offers linked to car manufacturers using automotive connectivity, support, and safety platforms as a service and value differentiator. Tesla electric cars have their software upgraded over a 4G radio interface.

In the context of the network and supply chain economics of 5G, it is reasonable to assume that automotive integration will become progressively more important with the integration of automotive radar being a key part of the integration process.

Many existing safety features such as collision avoidance are standalone with limited reliance on network connectivity. In the longer term, this limits functionality and user value. Using cars as observant machines sharing data in real time with other road users would be a significant step forward, but implies a need for wide-area, high-data-rate, low latency/stable latency networks.

The vehicles themselves can be part of that network. The relay and repeater standards evolving in 3GPP Release 14-16 should help to take that process forward, providing standards support for the Internet of moving objects.

There is a school of thought that 5G is not a new physical layer but rather an abstraction of existing and evolved personal, local and wide-area radio technologies. While this may be partially true, there will almost certainly be a need to differentiate 5G in terms of social and economic value and this is probably only achieved by a step function increase in data rate and data reach based on large cell connectivity.

The radar community is adept at developing innovative waveforms and digital signal processing and techniques linked to front-end transceiver technologies that can deliver significant broadband performance at any frequency from VHF to E-band.

The automotive industry and automotive supply chain is adept at repurposing those technologies and techniques into robust low-cost, high-value products. The collaborative opportunities would therefore seem to be apparent.

10.19 Military Radio in E-Band

There is a Defense Advanced Research Projects Agency (DARPA) project to deploy millimeter-band communications systems from subspace platforms providing instantaneous battlefield communication over areas up to 1,000 km2. These systems are integrated with terrestrial LTE in the 700-MHz band using standard LTE 700 base station and user device hardware. The millimeter radio has a 2° beamwidth antenna with a claimed clear weather range of 60 km.

The component support work flow includes the development of low Coefficient of Thermal Expansion (CTE) polymers with good electrical characteristics for E-band transceiver printed circuit boards and digital CMOS amplifiers with a claimed power added efficiency of 25%.

10.20 Summary of Millimeter-Band RF Hardware Enablers

RF hardware is already widely available for the millimeter band including fixed point-to-point hardware. Figure 10.3 shows an antenna for V-band and E-band mobile backhaul applications from Huber and Suhner.

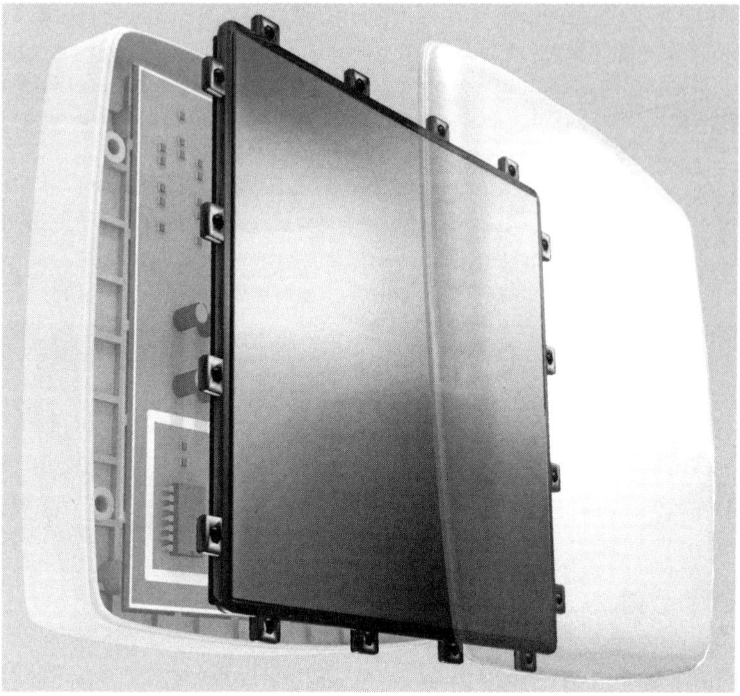

Figure 10.3 Huber and Suhner millimeter-band antenna.

Table 10.9 gives the typical gain realizable from these products at V-band and E-band.

Implementing 5G into these fixed point-to-point bands would transform the scale economy of these existing backhaul products and would pull through component and packaging innovation.

The challenge will be to deliver efficient gain with sufficient linearity and phase stability with low front-end noise budgets and good dynamic range at low cost. Materials innovation is important in components such as RF amplifiers particularly for higher-power applications but requires parallel innovation in circuit board material and construction techniques. The automotive industry is

Table 10.9
Gain from V-Band and E-Band Antennas

Antennas	Gain	Frequency Range
V-band antenna without housing	38 dBi	57–66 GHz
E-band antenna without housing	38 dBi	71–76 and 81–86 GHz
E-band antenna without housing	38 dBi	71–76 and 81–86 GHz
E-band antenna without housing	43 dBi	71–76 and 81–86 GHz

Source: [20].

successfully addressing these cost and performance issues in a harsh application environment with stringent safety and quality requirements.

High-count antenna arrays can be designed with individual power ampifiers per active element. Each individual amplifier has a low power output but the EIRP from the whole array when beamformed should adequately support wide-area 5G connectivity in the millimeter band.

The 71–76 GHz and 81–86 GHz bands are of particular technical interest for 5G but would also appear to offer potential commercial translation opportunities with 77-GHz automotive radar. Automotive radar signal processing should also have relevance to 5G antenna array design.

U.S. military investment in these bands suggests additional military-to-consumer and consumer-to-military technology and commercial translation opportunities.

10.21 Millimeter-Band, Wide-Area Radio

A number of indoor and outdoor 5G trials and studies are underway in the 71–76 MHz (4.2 mm–3.94 mm wavelength) band [21]. This includes a joint study between Nokia and NTT DoCoMo on a potential combination of meter-band, centimeter-band, and millimeter-band radios (Table 10.10).

The throughput rate available from LTE Advanced is assumed as tens of gigabits per second per square kilometer bewteen 2020 and 2025 from spectrum in the meter band plus the 3.4–3.8-GHz band. From 2025 this would be combined with a 5G centimeter-band, 500-MHz carrier bandwidth physical layer providing a combined throughput of several hundred gigabits per second

Table 10.10
System Configuration for LTE Advanced and Centimeter-Wave and Millimeter-Wave 5G

Parameter	LTE Advanced	Centimeter Wave	Millimeter Wave
Frequency Band	<6 GHz Includes 3.4 to 3.8 GHz, + 4.4–4.9 GHz	6–30 GHz 28 GHz	30–300 GHz 39 and 73 GHz
Wavelength	<5 cm	5 cm–1 cm	10 mm–1 mm
Carrier bandwidth	100 and 200 MHz	500 MHz	2 GHz
Modulation	64 QAM	256 QAM	64 QAM
MIMO combination	8 × 8	8 × 8	2 × 2
SU MIMO rank	8	8	2
MU MIMO rank	2	2	2
Antenna configuration	10 × 1 AAS 8 antenna ports MIMO Macro	Omnidirectional, 4 antenna ports	4 × 4 AAS 4 sectors, 2 antenna ports

Source: [23].

per square kilometer. Spectrum options being studied include the 4.4–4.9-GHz band.

From 2030 this would be combined with a 5G millimeter-band, 2-GHz carrier bandwidth physical layer providing several terabits per square kilometer. The millimeter-wave radio would provide backhaul in small cells with a maximum of two hops within a mesh network to meet an assumed radio latency requirement of <1 ms. Very large antenna arrays configured as an adaptive antenna system (AAS) will be used to compensate for the higher path loss. The intersite distance (ISD) for dense networks is assumed to be between 75m and 100m, although complemented by a wider area solution and in building solution.

A need for low-cost devices is also identified [22] plus a need for low-energy IoT devices though Nokia point out that LTE Release 13/14 supports a 10-year battery life for machine-type communication.

The peak data rate of a 5G system is assumed as 10 Gbps with a required cell edge rate for 95% of users of at least 100 Mbps. The spectrum requirements are based on 30 GB of personal data per day, a subscriber density of 100,000 users per square kilometer and a busy hour loading equivalent to 10% of the daily traffic load. Centimeter-wave radios would use eight-stream MIMO, multiuser MIMO, and 256 QAM modulation. Millimeter-band radios would use lower-rank (two to four stream), single-user MIMO, multiuser MIMO, and beamforming.

The study work to date also suggests that cell edge data rates (data reach) can be significantly improved by having simultaneous dual-band (centimeter-band and millimeter-band) connectivity.

The comparison between load balancing with a single data connection and multi (dual) connectivity is shown in Table 10.11.

The 73-GHz radios would deliver 1 Gbps of backhaul capacity over the two-hop mesh. This would limit the number of access points requiring wired backhaul to about 20%, mainly macro sites.

Table 10.11
Average Throughput and Cell Edge Throughput Comparisons

		Load Balancing		Multiconnectivity	
Centimeter Wave, 10 GHz, 3 cm	**Millimeter Wave, 73 GHz, 4.1 mm**	**Average Throughput**	**Fifth Percentile Throughput**	**Average Throughput**	**Fifth Percentile Throughput**
Intersite distance					
100m	100m	1.4 Gbps	87 Mbps	1.5 Gbps	286 Mbps
75m	75m	2.1 Gbps	210 Mbps	2.6 Gbps	784 Mbps
50m	50m	3.1 Gbps	420 Mbps	4.1 Gbps	1,300 Mbps

Source: [23].

As an example of an ultradense network deployment, Nokia analyzed a real urban environment in Tokyo based on an LTE Advanced macro layer at an intersite distance of 240m with 100 MHz of aggregated channel bandwidth at 2 GHz and below, a small cell centimeter band physical layer at 10 GHz with 500 MHz of bandwidth colocated with a small-cell, millimeter-band physical layer at 73 GHz with 2 GHz of bandwidth. The small cells are deployed at an intersite distance of 75m (Figure 10.4).

10.22 Summary: The Millimeter Band

The millimeter band has a number of advantages for 5G. The short wavelengths enable compact Matchbox-sized phased array antennas to be built that provide 2° beamwidth coverage in elevation and azimuth delivering a robust link budget, high immunity to interference, and minimal coexistence issues.

Development costs including RF component and subsystem development can be cross-amortized across fixed point-to-point radio, satellite, military radio, and automotive radar applications. The 3-D spatial processing algorithms used in automotive radar can be repuposed for 5G beam processing.

This does not preclude deployment into the centimeter and meter bands but suggests that there will be a clear differentitiation in terms of passband bandwidth and channel bandwidth.

Millimeter-band 5G will be implemented in 5-GHz passbands supporting 1–2-GHz channel spacing with headline data rates of 2 Gbps but with edge of cell rates of 100 to 200 Mbps.

This could be achieved with relatively simple modulation and lightweight channel coding. This would improve RF power efficiency and reduce delay and delay variability through the channel decoder. The minimization of multipath

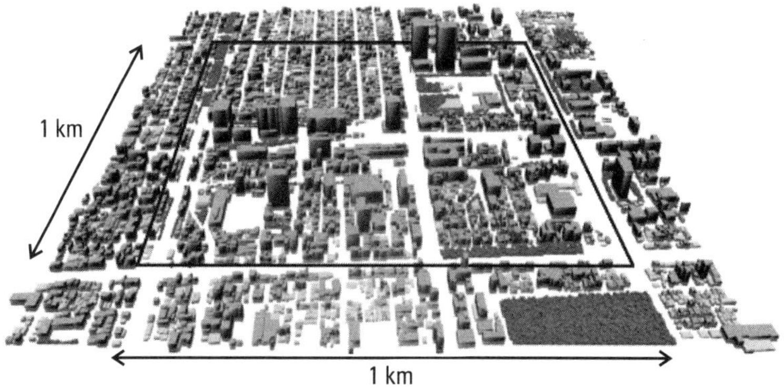

Figure 10.4 5G in Tokyo modeled relative to 4G LTE. (Courtesy of Nokia.)

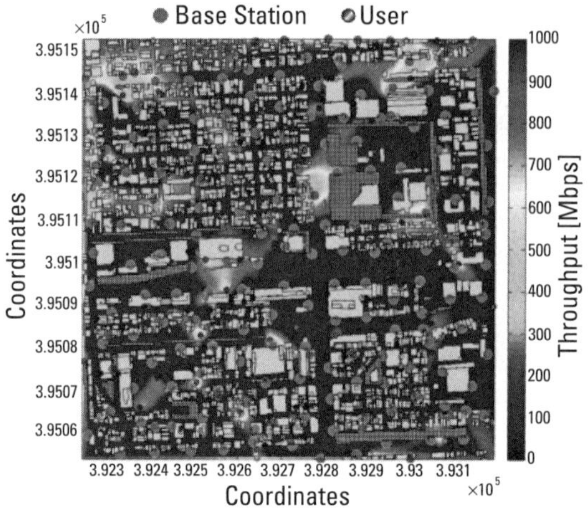

Figure 10.5 Coverage and throughput maps of 5G Tokyo model. (Courtesy of Nokia.)

achieved by beamforming would make OFDM unnecessary saving clock cycles and reducing power drain.

Centimeter-band 5G might need to be implemented in narrower pass-bands due to coexistence issues and it is unlikely that channel spacing above 500 MHz can be supported.

Meter-band 5G could be implemented in 100-MHz or 200-MHz bandwidth channels, but it is not immediately obvious what the benefits would be of introducing yet another physical layer to coexist in an already overcrowded part of the radio spectrum.

Centimeter-band and millimeter-band 5G integrated with meter band 4G as a tri-band service offer would seem to be an optimum technical and commercial compromise. Figures 10.6–10.8 show a visual conceptualization of a 5G network based on macro cell backhaul (Figure 10.6), urban deployment (Figure 10.7), and wireless to the home (Figure 10.8).

There remain a number of unanswered questions on how 5G performance objectives will be achieved cost-effectively. Reducing performance parameters such as latency and packet loss normally have an associated cost or impose a performance compromise in another area. For example, if throughput gain and capacity gain are achieved through network densification, it becomes harder to meet end-to-end latency targets. This is particularly the case if flexible routing is used to realize backhaul cost savings. Scheduling and multiplexing increase throughput efficiency but introduces end-to-end delay and delay variability. Increased reliability, however it is measured, will generally introduce additional cost due to the need to provision additional bandwidth and spare hardware and software resources.

Figure 10.6 Macro cell backhaul for 5G. (Courtesy of Huber and Suhner.)

Figure 10.7 5G urban deployment. (Courtesy of Huber and Suhner.

Commercial issues often frustrate access to street furniture that could be potentially repurposed for 5G. Ownership and access rights are often town- or zone-specific.

In terms of technology challenges, at the physical layer, wider channel bandwidths increase multiplexing gain, particularly for high rate bursty traffic but are not well suited to narrow bandwidth communication including voice. These trade-offs often become apparent in the analog-to-digital and

Figure 10.8 5G wireless to the home. (Courtesy of Huber and Suhner.)

digital-to-analog signal conversion process and in physical layer digital signal processing. This brings us to our next chapter.

References

[1] http://stakeholders.ofcom.org.uk/consultations/above-6ghz/.

[2] IC1004 Cooperative Radio Communications for Green Smart Environments.

[3] ETSI Industry Specification Group on millimeter Wave Transmission (ISG mWT).

[4] https://5g-ppp.eu/mmmagic/.

[5] Horizon 2020 –ICT 2014-2.

[6] ITU, https://www.itu.int/rec/R-REC-P.838/en.

[7] http://www.wi-fi.org/discover-wi-fi/wigig-certified.

[8] Cisco.

[9] http://www.bristolisopen.com/.

[10] http://www.newelectronics.co.uk/electronics-technology/making-massive-mimos-for-high-speed-short-range-comms/111039/.

[11] http://project-nanorf.com/.

[12] http://phys.org/news/2011-06-ibm-graphene-based-circuit.html.

[13] http://arxiv-web3.library.cornell.edu/pdf/1310.6820v1.pdf.

[14] http://pubs.acs.org/doi/pdf/10.1021/nn4009406.

[15] http://www.nature.com/nnano/journal/v6/n3/pdf/nnano.2010.279.pdf.

[16] http://www.sabertek.com/automotive-radar.html.

[17] http://www.bosch-mobility-solutions.com/en/de/_technik/component/CO_PC_DA_Adaptive-Cruise-Control_CO_PC_Driver-Assistance_2434.html?compId=2496.

[18] http://www.freescale.com/webapp/sps/site/application.jsp?code=APLADDRAS.

[19] http://www.millitech.com/www.darpa.mil/program/mobile-hotspots.

[20] Huber and Suhner.

[21] http://company.nokia.com/en/news/press-releases/2015/03/02/ntt-docomo-nokia-networks-showcase-high-speed-data-transmission-in-5g.

[22] http://networks.nokia.com/file/28771/5g-white-paper.

[23] Nokia.

11

5G DSP

11.1 The Impact of Wider Channel Bandwidths on Digital Signal Processing

In Chapters 9 and 10, we reviewed the device and RF component and packaging challenges of moving to the centimeter and millimeter bands. In this chapter we look at the analog-to-digital/digital-to-analog conversion process and the digital signal processing and baseband processing requirements associated with these shorter-wavelength, wider-bandwidth channels.

Channel bandwidths of up to 100 MHz are supported in LTE Advanced, but wider options being considered include 200-MHz and 500-MHz channel bandwidths in the centimeter band and 1 GHz and 2 GHz in the millimeter band. In Chapter 9 we included a case study of a superheterodyne receiver at 9 GHz with an intermediate frequency (IF) channel bandwidth of 200 MHz. Present point-to-point systems in the centimeter and millimeter bands are typically 56 MHz or 112 MHz.

Vendor views of how rapidly the industry will move to wider bandwidth channels vary, but there is a general presumption that power-efficient, wide dynamic range, low-cost devices capable of processing 100-MHz and 200-MHz channel bandwidths will be available by 2020 increasing to 500 MHz by 2025 and 1 to 2 GHz by 2030.

WiGig Wi-Fi radio systems with channel bandwidths of 2 GHz at 60 GHz exist today and provide a reference point for wide-area radio design, but the additional power and dynamic range requirements of wide-area radio need to be accommodated. Wide-area radio implies a higher path loss, more multipath, more delay spread (delay consequent on the bandwidth of the channel and the channel path), and Doppler spread (the frequency offset introduced by the relative movement of the user device to the base station).

The economics of wide-area radio also require various forms of diversity to be exploited to improve throughput per megahertz of allocated or auctioned spectrum. This includes spatial diversity and polarization diversity and or adaptive beamforming in 5G systems. These factors, when combined together, increase digital signal processing and memory overhead.

11.2 15 Years Ago: The Challenge of 5-MHz Channel Bandwidths

In 2000, the industry was grappling with the challenge of implementing 3G base stations and 3G user devices. Although an increase in headline data rates was important, it was not the only design requirement and there was an explicit expectation that voice channel costs would be reduced relative to legacy GSM [1].

The existence of two competing standards (CDMA2000 and W-CDMA) meant that it was important to have at least some software configurability. It is generally the case that some of the physical layer processing tasks, for example, the turbo decoder and fast Fourier transform (FFT), at least initially, cannot be realized in a standard digital signal processor (DSP) and need to be implemented in an application-specific integrated circuit (ASIC), typically as a nonprogrammable hardware accelerator.

This applies both for processors used in base stations and processors in user devices. The power constraints and heat rise limits and noise issues in a user device are key performance parameters, but the DSP only has to extract the signal of interest of a single user or session, whereas the DSP in a base station or access point will be handling tens or hundreds of simultaneous sessions across the whole passband rather than just one channel.

For example, the chip rate processing in a 2001 3G base station including the spreading codes, despreading, acquisition, and path delay estimation was too computationally intensive for a DSP. Similarly, the symbol rate processes including the forward error correction and convolutional and turbo decoding were more efficiently processed in flexible semiprogrammable coprocessors.

A 64-user 3G base station typically had a chip rate processing overhead of 30 BOPS (billion operations per second) and required a clock speed of 1.1 GHz for symbol rate processing. By comparison, 2G modem processors in 2001 were running at a clock speed of 40 to 50 MHz. The channel bandwidth for W-CDMA was (and is) 5 MHz in a typical passband of 35 MHz.

This distance between the relatively light processing load and flexibility of the 2G GSM legacy physical layer and the relatively heavy processing load and relative inflexibility of a new (3G W-CDMA) physical layer had a fundamental impact on the user experience, which translated directly into slower market adoption.

On a positive note we showed in Chapter 5 that market adoption of 4G has been faster than 3G. This can be at least partially explained by the simple fact that the FFT used in the OFDM physical layer is relatively easy to realize in a DSP.

11.3 4G DSP Today

Moving on 15 years from the early years of 3G, the story has not fundamentally changed, although the numbers have become larger. Table 11.1 illustrates bit throughput increases across the 17 categories of Release 12 LTE user devices per 1-ms downlink transmission time interval (TTI). The number of layers corresponds to the number of ports available for MIMO processing.

Table 11.1 only shows some of the options. In Release 12, the user equipment downlink and uplink categories are decoupled. In total there are now 17 downlink and 14 uplink categories including a new category 00 for low cost Internet of things (IOT).

Category 15 and 16 were introduced in June 2015 to cover 750/800 Mbps and 1,000/1,050 Mbps. The maximum downlink data rate is achieved in Cat-14 at almost 4-Gbps peak data rate. It is achieved with 8 MIMO layers and 256 QAM modulation with five aggregated 20-MHz carriers (100-MHz channel bandwidth).

The highest combined uplink and downlink data rate is supported in Category 8 devices. The relative uplink and downlink rates per category are shown in Figure 11.1. As with Category 14, this requires 100 MHz of bandwidth. Given that most networks are presently deployed with 5-, 10-, or 20-MHz channel bandwidths, it may be that Category 7 devices are more immediately achievable.

Irrespective of what hardware actually arrives in the market, the standards have moved in 15 years from user devices specified to deliver up to 384 Kbps for a mobile device in a 5-MHz shared channel to 3 Mbps from a 100-MHz channel, a tenfold increase in peak data rate and twentyfold increase in channel bandwidth, although actual device availability traditionally lags several years behind the standard.

There is also an issue of compatibility overhead. This is the additional processing overhead introduced by the need or perceived need to provide backwards compatibility with legacy wide-area mobile systems including 3G and

Table 11.1
LTE Device Categories*

DL Category	Max. Number of DL SCH Transport Block Bits Received Within a TTI	Max. Number of Bits of a DL SCH Transport Block Received Within a TTI	Total Number of Soft Channel Bits	Max. Number of Supported Layers for Spatial Multiplexing in DL	Support for 256 QAM in Downlink
0	1,000	1,000	25,344	1	No
1	10,296	10,296	250,368	2	
2	51,024	51,024	1,237,248	2	
3	102,048	75,376	1,237,428	2	
4	150,752	75,376	1,827,072	4	
5	299,552	149,776	3,667,200	2 or 4	
6	301,504	75,376 (2 layers), 149,776 (4 layers)	3,654,144	As above	No
7	301,504	As above	As above	As above	No
8	2,998,560	299,856	35,982,720	8	No
9	452,256	As above	5,481,216	As above	No
10	452,256	As above	As above	As above	No
11	603,008	75,376 (2 layers 64 QAM), 97,896 (2 layers, 256 QAM), 149,776 (4 layers, 64 QAM), 195,816 (4 layers, 256 QAM)	7,308,288	As above	Optional
12	603,008	As above	As above	As above	Optional
13	391,632	97,896 (2 layers), 195,816 (4 layers)	3,654,144	As above	Mandatory
14	3,916,560	391656	47,431,680	8	Mandatory
15	798,800	75,376 (2 layers 64 QAM), 97,896 (2 layers, 256 QAM), 149,776 (4 layers, 64 QAM), 195,816 (4 layers, 256 QAM)	9,744,384	2 or 4	Optional
16	1,051,360	75,376 (2 layers 64 QAM), 97,896 (2 layers, 256 QAM), 149,776 (4 layers, 64 QAM)	12,789,504	2 or 4	Optional

**Source:* [2].

4G LTE (and 2G GSM in the present systems), forwards compatibility, the need to allow for future iterations of the standard and sideways compatibility, the need to provide compatibiity and or interoperability with parallel radio systems including local area Wi-Fi and satellite and or subspace radio systems.

These factors determine the clock cycle overhead, memory overhead, and processor architecture including the amount of parallel processing required.

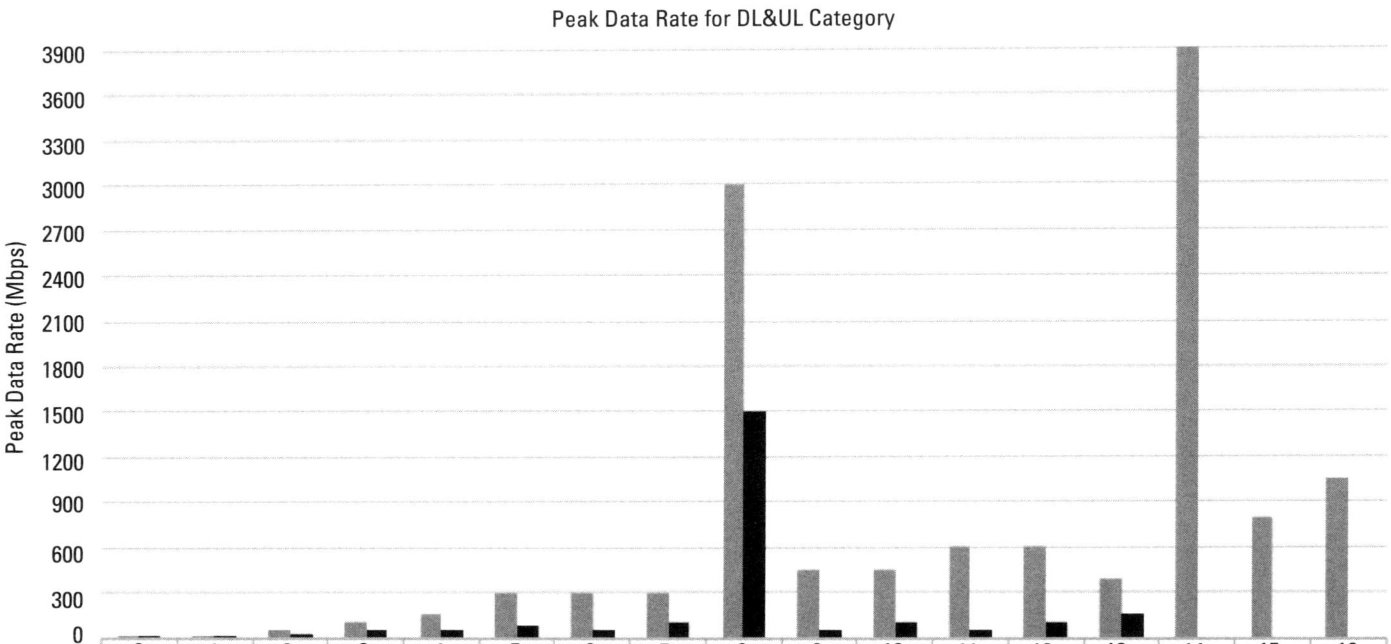

	0	1	2	3	4	5	6	7	8	9	10	11	12	13	14	15	16
UL	1	10	51	102	150	299	301	301	2998	452	452	603	603	391	3916	798	1051
DL	1	5	25	51	51	75	51	102	1497	51	102	51	102	150			

Figure 11.1 Peak data rates for downlink and uplink categories.

11.4 The Next 15 Years

On a 15-year forward view, it is not unreasonable to have a 5G wide-area physical layer expectation of peak data rates of the order of 1 or 2 Gb from a channel of between 1 and 2 GHz within a passband of 3 to 5 GHz in the centimeter or millimeter band.

These peak data rates could be realized from a mix of radio systems, for example, parallel traffic streams delivered over wide-area, local-area Wi-Fi, and satellite radio. Alternatively, 5G could be considered and provisioned as a clean slate standalone air interface, although the same overall bandwidth constraints will apply.

From Shannon-Hartley theorem, for a band-limited channel with Gaussian noise, the maximum rate to transmit information over a link is constrained by the bandwidth and the signal to noise ratio of the link.

$$C = B \cdot \log_2\left(1 + \frac{S}{N}\right)$$

where C is the capacity of the channel in bits per second, B is the bandwidth of the channel in hertz, S is the signal power, and N is the noise power.

It suggests that channel bandwidth limits the peak data rate. The peak rate can be increased linearly by increasing the channel bandwidth if the signal-to-noise ratio (SNR) is fixed; alternatively, if the bandwidth is fixed, higher modulation schemes can be introduced to increase spectrum efficiency. For instance, if SNR is fixed at 10 dB, with $B = 5$ GHz in the centimeter or millimeter band, theoretically one can get $C = 17.3$ Gbps, while in the narrowband case when $B = 200$ kHz, one can only get $C = 691.9$ Kbps.

The above equation effectively targets a single input single output (SISO) channel where both transmitter and receiver have one antenna. With an orthogonal multiple input multiple output (MIMO) system, the capacity could be N times the capacity of the SISO channel [where N is min (TX, RX) antenna].

The question is whether this can be achieved cost-efficiently and power-efficiently across a broad range of user and IoT devices and wide-area and local-area base station and access point hardware and the extent to which digital signal processing and analog-to-digital conversion constraints determine this rate of transition.

The higher the bandwidth, the higher the sampling rate and hence the higher the digital front-end processor complexity. As suggested by Nyquist, the minimum required sampling rate is a function of the bandwidth of the signal as well as its position in the frequency spectrum. Using LTE as an example, the sampling rates for 1.4-MHz to 20-MHz system bandwidth are 1.92 Mbps and 30.72 Mbps, respectively. This introduces extra cost in front-end filtering, decimation, and the initial synchronization process.

The performance of a signal processor can be measured in MIPS (millions of instructions per second) or MOPS (millions of operations per second), but real-life performance in practice is determined by the efficiency of the instruction set, which, in turn, is determined by the mix of tasks it is expected to perform.

11.5 The Widening Gap Between Moore's Law and Algorithmic Complexity

Instruction set efficiency generally has to be traded against task flexibility and task complexity. On past experience the net efficiency gain yields a performance gain of around 1.5 times every 2 years. It is reasonable to expect this to continue for as long as Moore's Law applies, which is probably at least 10 years, but past experience also suggests that algorithmic complexity is increasing twofold every 2 years [3]. The disparity between processor performance and algorithmic complexity is therefore increasing over time [4].

11.6 Gene's Law of Power Consumption

Gene's Law [4] is the equivalent of Moore's Law but applied to power consumption. As with Moore's law, the present assumption is that the power needed per computational MIP is halved every 18 months, but the sustainability of that decrease beyond 2025 is presently in some doubt.

11.7 Latency Requirements

There is also a trade-off between the amount of channel coding, the complexity of the channel coding, latency, and silicon area (hardware cost). Latency is a function of the number of iterations per codeword and the clock frequency, and more iterations per code word should result in better physical layer performance, although with some throughput delay and additional die cost.

11.8 CEVA/ARM Device Software Optimized (DSO) Platform as a Contemporary Reference Point

In Figure 11.2, the example from CEVA/ARM shows a block diagram of the baseband processor for a Category 7 LTE advanced modem. The Layer 1 func-

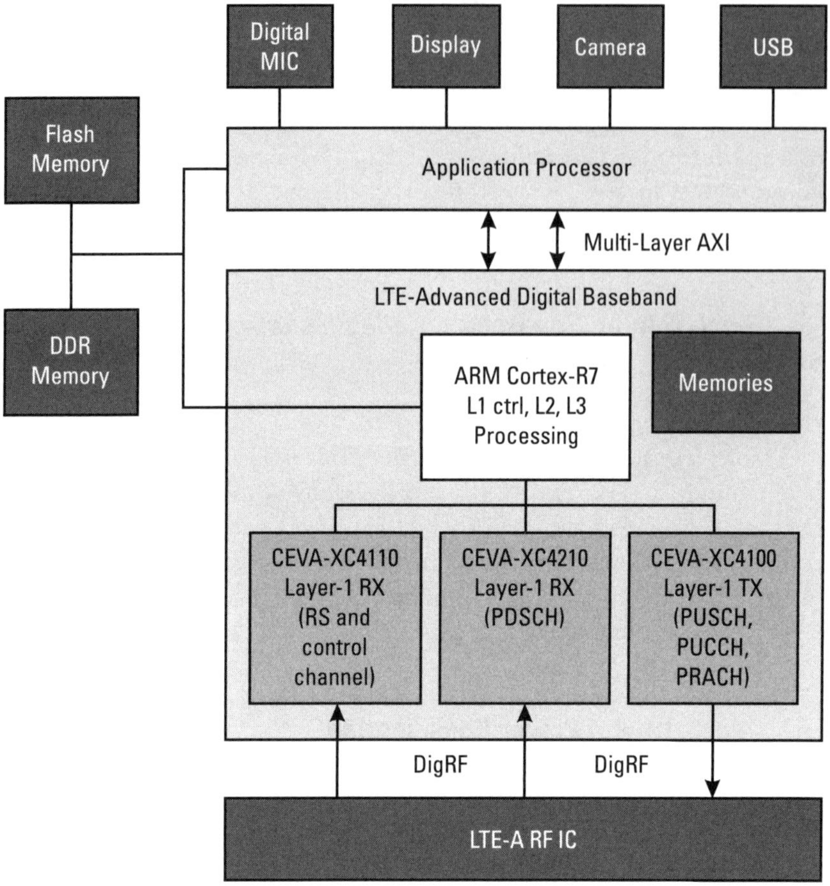

Figure 11.2 LTE Advanced modem for CAT 7. (Courtesy of CEVA/ARM.)

tions include forward error correction, interleaving and bit stream manipulation, constellation modulation, MIMO encoding, OFDM signal modulation, and RF IC signal conditioning.

The layer 1 RX and TX physical layer processing is realized in the three CEVA processors (XC4110/XC4210/XC4110) with associated control and management functions implemented in an ARM CPU (central processing unit).

Upper layer (Layer 2 and 3) processing is performed in the ARM Cortex R7 processor and includes the Medium Access Control functions, Packet Data Convergence Protocol, radio link control, and radio resource management. The Cortex R7 processor interfaces to the applications processor, which in this example is running the Android OS (operating system).

The processor is clocking at speeds greater than 1.5 GHz and is implemented on a 28-nm process.

Much of the system design is focused on ensuring coherency between the functional areas in the modem and minimizing delay and delay variability between functions. This includes low-latency random access memory (RAM) used for critical software and the data required for Interrupt Service routines, routines that have to be executed immediately rather than via an Advanced eXtensible Interface (AXI) bus potentially blocked by large data transactions.

Table 11.2 shows the range of memory types required in the modem including the H = ARQ^2 buffers used for recombining data stored as soft bits (bits stored in log likelihood ratios of a 1 or a 0 rather than in binary bits) and whether the memory is on chip or off chip.

Figure 11.3 shows the software mapping in the device. The layer 1 functions look after the forward error correction, interleaving, and hybrid ARQ, which manages the retransmission of data that has not been correctly received. The combination of higher data rates and low latency means that these buffers need to be large and can become expensive.

The Cortex core manages the layer 1 scheduling running on an LTE subframe of 0.5 ms. Note that support for voice over LTE (VoLTE) and HD (high definition) voice and the need to manage voice protocols including semipersistent scheduling introduces additional software requirements. There is a particular need to ensure that voice services are power-efficient.

Figure 11.4 shows the TX and RX functional block diagram showing interconnects and interfaces.

Table 11.2

Memory in an LTE Advanced Modem

Memory	Location	Size	Comment
I Q receive buffer	On chip	450 KB for Cat 7	Buffers RX IQ
H-ARQ (Hybrid Automatic Repeat Request) Cache	On chip	344 KB	LLR soft bits for combining
Layer 2 cache	On chip	128 KB ARM	Typical size but can be optimized through code profiling
TCM	On chip	Core 1 (RS processing), 192 KB, Core 2 (PDSCH processing), 320 KB)	
Layer 1 I and D Cache	On chip	32 KB for DSP I cache per DSP	
DDR Memory	Off chip	128 MB ARM	1 Gb LP-DDR-2
Flash Memory	Off Chip	128 MB ARM	I Gb NAND
H-ARQ buffer	Off chip	3.67 MB	Cat 7 requirements located in off-chip DDR

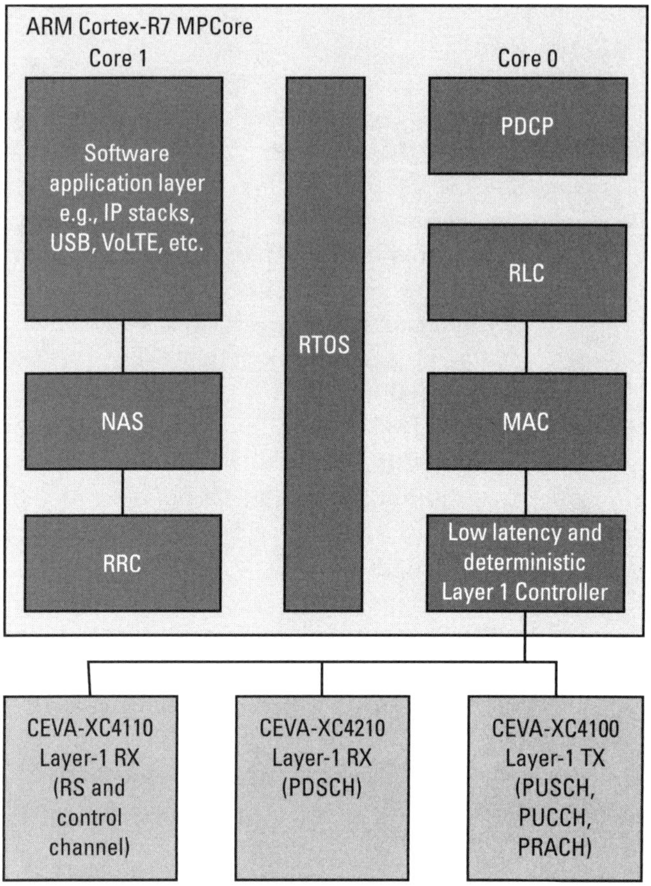

Figure 11.3 LTE Advanced modem software mapping.

11.9 Wi-Fi at 60 GHz as a Precursor to Wide-Area 5G

An 802.11ad WiGig baseband has a digital sample rate of 2.64 GHz, significantly higher than present LTE requirements. Given that advanced silicon on chip clock frequencies are in the range of 500 to 1,000 MHz, then it is obvious that parallel processing is a necessity rather than an option.

Present solutions (Blu Wireless [5], for example) combine fixed function DSP blocks with highly optimized parallel vector digital signal processors to provide a pool of DSP processors with fixed function blocks arranged in clusters with each cluster having its own heterogeneous controller. The controller has high-level software that dispatches software threads in order as a virtual pipeline in a series of interlocking threaded subtasks with units switched off between tasks to save power. The real-time data flow defines the execution timing and order.

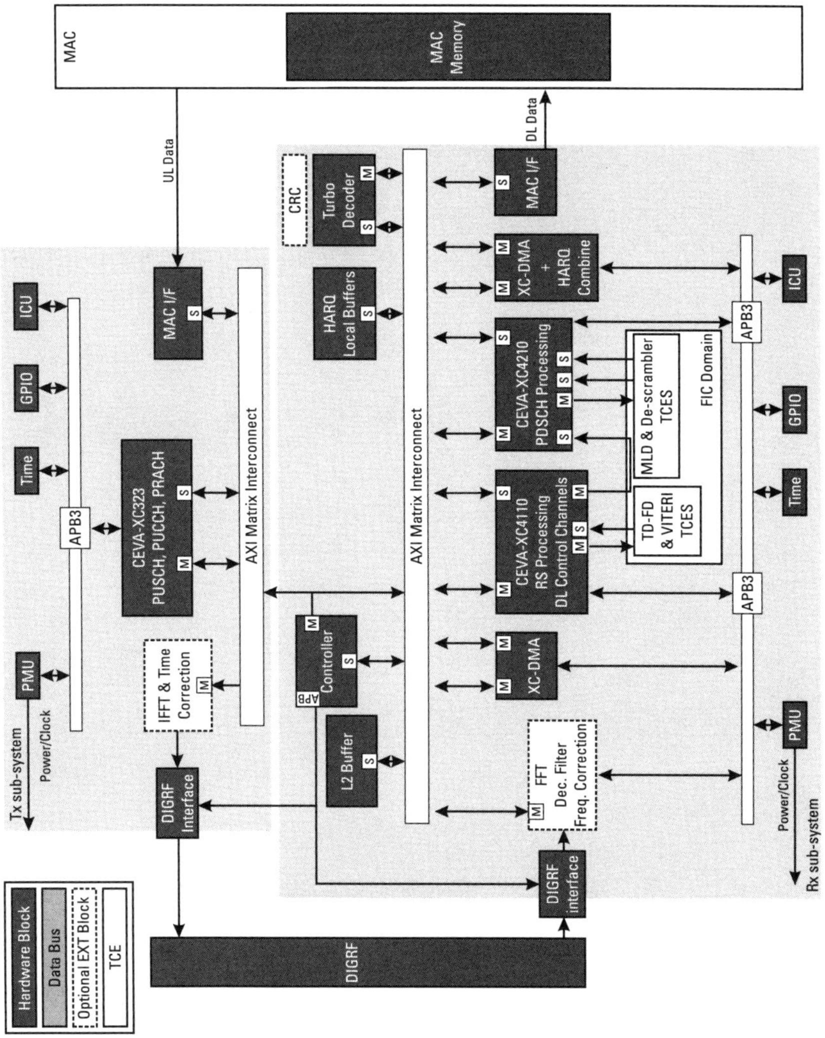

Figure 11.4 TX and RX subsystem.

Each of the data path functions is kept as busy as possible without over use of intermediate register files or local memory. The vector data path is replicated typically up to 128 times across the clusters with each cluster having a unique VLIW instruction dispatcher. Table 11.3 shows the present band plan for WiGig based on 2-GHz bandwidth channels at four center frequencies though the channels could be subdivided to increase range. The spectral mask allows for single-carrier or OFDM modulation. The architecture allows the same silicon resources to be reused for each of the separate TX and RX DSP pipelines including SC, SC FDE, OFDM, the control physical layer, and all modulation schemes.

Table 11.4 shows the modulation rates, code rates, data rates, and sampling rates.

The 8-bit IQ ADC and DAC is working at a 2.64-GHz sample rate. The SC FDE modem operates at over 2 Teraops. This particularly product is targeted at both the backhaul and access market.

Table 11.3

WiGig Channels and Channel Boundaries

GHz	**57**	**58**	**59**	**60**	**61**	**62**	**63**	**64**	**65**	**66**
USA	———	———	———	———	———	———	——→			
Canada	———	———	———	———	———	———	——→			
Korea	———	———	———	———	———	———	——→			
EU	———	———	———	———	———	———	———	———	———	——→
China						———	———	——→		
Japan						———	———	———	———	——→
Boundaries	57.24–59.4 GHz			59.4–61.56 GHz		61.56–63.72 GHz		63.72–65.88 GHz		

Table 11.4

Modulation Rates, Code Rates, Data Rates, and Sampling Rates

Modems	**Modulation**	**FEC Code Rate**	**Data Rate**	**Sample Rate**
Single carrier	BPSK QPSK 16 QAM	1/2, 5/8, 3/4, 13/16	385 Mbps–4.62 Gbps	1,760 MHz
OFDM	Spread QPSK QPSK 16 QAM 64 QAM	1/2, 5/8, 3/4, 13/16	693 Mbps–6.76 Gbps	2,640 MHz
Control	DBSK	1/2, 32 chip spreading	27.5 Mbps	1,760 MHz

11.10 5G DSP Summary

The DSP has been a workhorse of the industry for almost 20 years. Each new generation of technology has required some functions to be performed in hardware coprocessors but over the lifetime of each standard, programmable DSPs have managed to handle most Layer 1 channel coding and decoding functions and higher-layer protocols.

From standard to standard, Layer 1 and higher-layer protocols have become more sensitive to delay and delay variability. This has meant that particular attention has had to be paid to bus and memory architectures and bus and memory performance and memory location (on or off chip).

However, if algorithmic complexity increases at a faster rate than processor capability, there can be no absolute guarantee that DSP performance wil be able to keep pace with physical-layer performance expectations over the expected 15-year lifespan of 5G. One solution is to implement optimized DSP architectures that combine general-purpose functions with optimized fixed functions. Present WiGig devices illustrate the present rate of progress.

11.11 5G Analog-to-Digital Function as a Constraint for Wide-Area 5G Systems

Realizing a WiGig physical layer is not the same as implementing a wide-area radio system. The typical range expectation of WiGig at 60 GHz is a few meters up to a maximum of just over 300m for a 1-Gbps channel in clear weather with 40 dBi of gain. By comparison, a wide-area network may need to support a range between a few meters and several kilometers.

Figure 11.5 shows where the analog-to-digital and digital-to-analog function is positioned within a generic transceiver block diagram.

The motivation of moving to wider bandwidth channels is partly to support higher user data rates and partly to deliver more multiplexing gain (more users on the same channel).

In LTE, the multiplexing gain is achieved by supporting devices on a varying number of resource blocks at varying power levels. The combination of wider channels and more users per channel increases the amount of dynamic range required in the ADC. Because LTE is a wide-area network, the difference in received power is significantly higher than a Wi-Fi network. As a result, an ADC capable of handling a 20-MHz LTE channel needs at least 60 dB of dynamic range.

The dynamic range needed in the ADC determines the bit width of the ADC. The bit rate of the ADC is a function of the bit width and sampling frequency, which must be at least twice the signal bandwidth. The bit rate of the ADC and associated signal processing together determine the amount of power

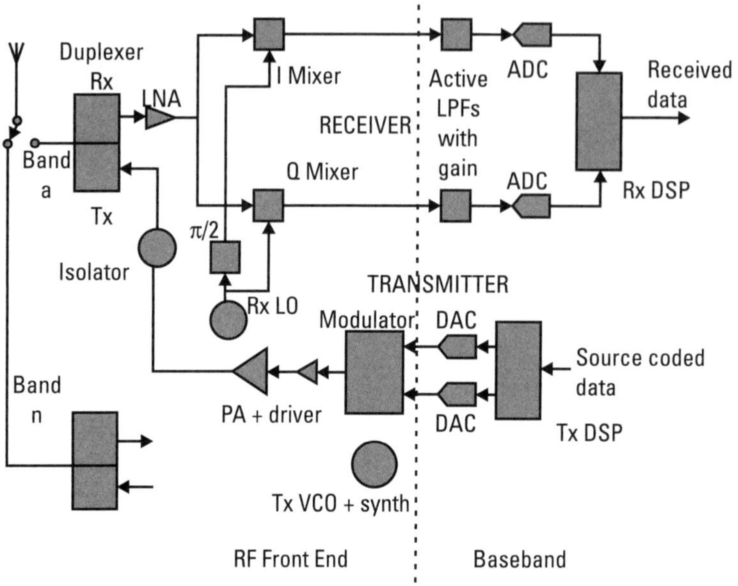

Figure 11.5 Generic transceiver block diagram.

consumed. The bit width has to allow for additional resolution to accommodate RFIC imperfections including direct conversion (DC) offsets and adjacent channel interference.

11.12 Pipeline ADCs

The majority of ADCs to date are Successive Approximation Register (SAR) ADCs. The input analog voltage is tracked and held and then compared with prior samples using a binary search algorithm. The power dissipation scales with the sample rate.

For LTE the two alternative options are either the sigma delta ADC or pipeline ADC.

A sigma delta ADC produces a high-resolution signal and low-resolution signal and uses error feedback to compare the two signals. A pipeline ADC, as the name implies, produces a high-resolution description of an analog signal from a series of lower-resolution stages, with the first stage working on the most recent sample and the following stages working in analog remainder voltages left over from previous examples.

All ADCs generate quantization noise and are sensitive to clock jitter. Noise and jitter become particularly important when demodulating 16 QAM, 64 QAM, or 256 QAM signals.

Power efficiency is therefore a composite of conversion efficiency and conversion effectiveness expressed as a signal-to-noise ratio, which, in turn, determines error vector magnitude, which determines throughput efficiency.

The benefits of a well-designed, well-behaved front end can therefore be compromised by a poorly implemented ADC. Conversely, a well-designed ADC can compensate for a poorly implemented (noisy and nonlinear) front end, although the additional resolution required may result in unnecessarily high-power consumption.

Specifying an ADC to handle the dynamic range required in a 20-MHz channel will mean that the ADC will have substantial headroom when processing narrower band channels. This might allow for a relaxation of analog filtering, which would reduce component count, component cost, and insertion loss. Alternatively, dynamically reducing the bit width of the ADC for narrower band channels reduces ADC power drain.

These performance trade-offs and costs scale to higher channel bandwidths and higher modulation options where additional noise can rapidly increase channel error rates.

A pipeline ADC is an open loop architecture with a latency of between 4 and 6 clock cycles. They are normally implemented in complementary metal-oxide semiconductor (CMOS) using switched capacitor discrete time circuitry. A relatively complex analog antialiasing filter is needed, which consumes power and silicon area. The pipeline ADC is generally the most efficient option for bandwidth input signals of 10 MHz to 100 MHz.

Note that the receive path is usually split into two components I and Q, which require individual ADCs generally known as IQ ADCs (see Figure 11.5).

In wireless applications, up to half of the effective number of bits can be needed to handle unwanted signals, illustrating the typical trade-off between front-end RF analog filtering and ADC specification.

Just as a reminder, ADCs perform an amplitude quantization of an analog input signal into binary output words of finite length, a nonlinear process. The nonlinearity shows up as wideband noise in the binary output also known as quantization noise. Quantization noise can be reduced by oversampling and dithering, but both options have an associated cost in terms of power consumption.

In the RF domain there is an advantage in moving to higher frequencies because additional gain can be achieved from compact short wavelength antennas. However, the bandwidth ratio also increases as frequency increases. For example, in previous chapters we have discussed the analog filtering requirements of a 35-MHz passband at 900 MHz. This is a bandwidth ratio of 3.8% (35 MHz is 3.8% of 900 MHz as a center frequency). Unsurprisingly, the same ratio at 9 GHz yields a channel bandwidth of 350 MHz and 3.5 GHz at 90 GHz. This is not to say that RF filtering at 9 GHz or 90 GHz is easy. Designers have to deal with parasitics and matching and loss and noise, but at least the resonance ratios remain similar.

In an analog-to-digital converter, we have the fundamental constraint that usable system bandwidth is dependent on the analog-to-digital converter's sample rate and system bandwidth cannot be greater than half the converter's sample rate, which suggests a sample rate of 4 GHz to digitize a 2-GHz channel at 90 GHz.

High-performance, 16-bit analog-to-digital converters were available in 2015 with a sample rate of 200 M/sample/s giving a usable bandwidth of 100 MHz and a signal-to-noise ratio of 79 dB. However, with all high-performance analog-to-digital converters, a nonlinear charge is produced in the sampling process, which is reflected into the input network each time the sampling switches close and there is always a risk that this will be resampled. Avoiding this requires a very carefully matched (50-ohm) network.

Most analog-to-digital converters are differential to provide good common-mode rejection. Any loss of symmetry, for example, due to board layout and interconnects, will show up as second-order harmonic distortion. Differences in ground current on adjacent ground planes will add to this distortion.

The effects of direct sampling on the source of the analog signal can be minimized by using an amplifier (a filter with gain; see Figure 11.5) to absorb the charge from the sampling process. If the amplifier is located close to the converter, the reflections can be reflected multiple times before the sampling period ends reducing the impact of glitches on the converter's spurious free dynamic range. However, this requires amplifiers with a large gain bandwidth product.

It becomes apparent that the specified performance of an ADC on a specification sheet measured in laboratory conditions may not match a real-life device and there will generally be some implementation loss.

11.13 Impact of Wider Channel Bandwidths on Adaptive Tracking Including Envelope Tracking

The use of envelope tracking to improve power amplifier (PA) energy efficiency in existing LTE radio systems has already been covered in Chapter 8 but provides a further example of the processing complexity that is introduced as channel bandwidth increases.

Envelope tracking requires a high bandwidth envelope reference signal generated by the modem or RF IC in a dedicated high-speed digital-to-analog converter (DAC).

The reference signal is generated by a relatively simple processing chain, usually from the digital 1/Q signals but the sampling rate will typically need to be six times the RF channel bandwidth, for example, 120 MSPS for a 20-MHz LTE carrier.

Additionally, the relationship between the instantaneous RF amplitude at the PA input and the supply voltage at the supply terminal of the PA must be precisely controlled and the relative timing of the signals must be controlled with subsample precision.

The envelope processing function is relatively small, occupying 50,000–100,000 gates of digital logic and around 10 mW of power but the requirements for timing alignment accuracy are directly related to the channel bandwidth, typically ±0.5 ns for a 20-MHz LTE channel.

Prior to calculating the magnitude of the IQ signal, the signal must be up sampled and filtered to provide sufficient bandwidth for the magnitude calculation. LTE IQ signals are normally generated at 1.536 times the channel bandwidth, for example, 30.72 MSPS for a 20-MHz channel. The magnitude calculation of the unshaped envelope must be performed at a higher sample rate to avoid aliasing, which requires interpolation of at least four times to avoid unwanted distortion and ideally six times (184.32 MSPS) or eight times (245.76 MSPS).

The analog-to-digital converter and digital-to-analog converter must therefore be considered as one of the critical components that could slow the implementation of wider channel bandwidth 5G radio systems with the constraint extending to RF PA optimization techniques such as envelope tracking. The positive side to this is the market opportunity produced by the need to deliver innovative analog-to-digital architectures including optimized filtering and amplification.

It also strengthens the rationale for keeping 5G modulation options relatively simple in order to make the overall throughput of the radios more tolerant to phase noise and distortion and to reduce the need for correction systems such as envelope tracking that are inherently bandwidth-limited [6].

11.14 Differentiating Digital Signal Processing and Digital Signal Processors

It is useful to consider digital signal processing and digital signal processors as two related but separate topics. Digital signal processing has been fundamental to every generation of digital cellular technology from GSM onwards and has had arguably the single largest impact on spectral and power efficiency.

Initially, spectral and power efficiency gains were achieved by exploiting the redundancy implicit in voice traffic with progressive improvements in speech encoding and decoding. Digital voice encoding combined with channel coding and digital error control techniques delivered additional performance improvements.

As data became progressively dominant in the traffic mix, there was a need to implement wider bandwidth channels in order to realize higher per user peak data rates and to deliver multiplexing gain, particularly important with bursty data exchanges.

Third and fourth generation cellular systems therefore implemented channel coding and channel equalization techniques, which allowed the introduction of 5-MHz channels in 3G systems and potentially 100-MHz aggregated channels in 4G LTE.

There has always been a lag between standards support for channel functionality and practical device availability. This is because digital signal processors initially struggle to support the required processing tasks and generally consume too much power or have memory constraints that introduce delay and delay variability in task processing. Given that many processing tasks are required to be strictly deterministic, this can be problematic.

The usual solution is to use hardware accelerators, but this solution conflicts with an increasing need for software upgrades for new features both at the network side and in user devices. Devices also need to switch between physical layer standards including legacy standards. This places a premium on programmability. ASIC hardware is also dependent on standards being stable, but unless technology stops evolving, this is unlikely to happen in the foreseeable future. Standards also need to achieve global scale for ASICs to be commercially viable.

The commercial success of a new physical layer is therefore generally dependent on having enough clock cycles available in a DSP at a sufficiently low-cost and sufficiently low-power budget. This constraint also extends specifically to the analog-to-digital conversion process where there is an inherent trade-off between clock cycles and power consumption and channel quality.

The processing load in an analog-to-digital conversion can be lowered by reducing the bit width used to digitize the incoming waveform but this reduces dynamic range. Front-end filtering and AGC can be used to manage this but adds cost and complexity. Similarly, it is possible to reduce processing load by reducing sampling rates, but this introduces quantization error.

The combination of these practical constraints suggest that the analog-to-digital and digital signal processor and digital signal processing are a crucial part of the critical path determining the rate at which a new 5G physical layer can be introduced, particularly for applications where cost and power efficiency are dominant considerations.

References

[1] Gatherer and Auslander, (eds.), *Application of Digital Signal Processors in Mobile Communications*, New York: John Wiley, 2002.

[2] 3GPP TS 36.306 V12.6.0 (2015-09).

[3] Rabaey, J. M., "Silicon Platforms for Next Generation Wireless Systems," *Field Programmable Logic and Applications Conference*, 2000, pp. 277–285.

[4] Frantz, G., "Digital Signal Processor Trends," *IEEE Micro*, Vol. 20, No. 6, November/December 2000, pp. 52–59.

[5] http://www.bluwirelesstechnology.com/.

[6] Indiga Consulting.

12

Is 5G the End of the Story?

12.1 The Third Radio Access Network

In Chapter 2, we highlighted the various definitions proposed for 5G including functional parameters. We cited the outputs of METIS 2020 (Mobile and Wireless Communication Enablers for the Twenty-Twenty) Information Society as one example [1], specifying the requirements for a 5G physical layer that could deliver:

- 10 to 100 times higher typical user data rate in a dense urban environment with a typical user data rate ranging from 1 to 10 Gbps;
- 1,000 times more mobile data per area (per user) with a volume per area (per user) of 100 Gbps/km2 based on a typical individual usage per month of 500 Gb;
- 10 to 100 times more connected devices;
- 10 times longer battery life for low-power massive machine communications where machines such as sensors or pagers will have a battery life of a decade;
- Support of ultrafast application response times, for example, for the tactile Internet where the end-to-end latency will be less than 5 ms with high reliability;
- A similar cost and energy dissipation per area as today's cellular systems.

In Chapter 6, we argued that 5G needed to be deployable into developing economies but that this required a reduction in cost and energy dissipation rather than cost and energy parity.

In 4G, some cost saving and capacity gain has been achievable through a closer integration of LTE and Wi-Fi. Additionally, Wi-Fi is being explored by Google as a basis for delivering Internet access from subspace balloons in the stratosphere each side of the equator. We are at least closer than we were to an integrated two-part radio access network (RAN) solution.

In Chapter 9, we suggested that a key part of the 5G story would be integrating 4G LTE and Wi-Fi with low-Earth, medium-Earth, and geostationary satellites. It is hard to see how the performance objectives proposed by METIS could be achieved without the addition of this third RAN as a complementary delivery platform. These systems already fulfill a backhaul function for terrestrial radio, suggesting that a more comprehensive technical integration should be relatively straightforward.

12.2 The 3-D RAN

The addition of the third RAN also means that coverage effectively becomes three dimensional (3-D), a 3-D space based on the integration of terrestrial, subspace, and space-based communications systems.

The 3-D RAN can also be conceptualized in terms of 3-D mobility based on the integration of mobile, fixed and nomadic communication; the delivery economics of Third World are going to be dependent on a well-executed 3-D radio topology.

12.3 To Shorter Wavelengths: Terahertz Radio

Similarly, high data rates and data density are going to be dependent on materials and manufacturing innovation. This includes materials that can support applications in the submillimeter band.

In the millimeter band, for example, in 77-GHz automotive radar, the antenna is connected directly to the receiver amplifier chip through an internal printed circuit substrate avoiding the need to use a waveguide. The substrate materials are typically ceramic-based or quartz and Teflon, but all of these exhibit high anttenuation at shorter submillimeter wavelengths.

Fujitsu have fabricated a demonstration transceiver at 300 GHz using a polymide heat-resistant synthetic polymer material with signals from the antenna transmitted to the receiver amplifier chip through a connecting circuit on the substrate. To minimize loss and maintain stability, the top and bottom faces of the printed circuit substrate are grounded and connected using through hole metalized vias. Together with the connected circuit, this forms a grounded coplanar waveguide structure that acts as a transmission pathway for the short-wavelength signal propagation.

Signal interference from the printed circuit substrate is minimized by spacing the vias apart by one-tenth of the signal wavelength, less than a few tens of micrometers. Although the polymide material has a signal loss that is 10% higher than quartz, it has a material processing accuracy that is four times better than quartz. This allows the vias to be placed closer together, which halves the overall signal loss compared to a quartz substrate.

The module has an overall volume of 0.75 cm^3 and therefore could be included in a smart phone. Download speeds of 20 Gbps have been demonstrated over a meter with a longer-term potential of reaching 100 Gbps (see Figure 12.1) [2, with thanks to Yasuhiro Nakasha, research manager at Fujitsu's Devices & Materials Lab].

12.4 RF to Optical Transforms

Most of us use an infrared transmitter at least once a day when we watch a TV program, but in the future we are just as likely to be using optical wireless communication. One hundred years ago, the light bulb was the enabling technology that brought us the radio valve. The equivalent enabling technology today is the light-emitting diode (LED), which conveniently can be used as a detector as well as a modulated light source.

Free-space optical wireless systems work in the visible light spectrum from 400 THz (780 nm) to 800 THz (375 nm).They are used for multigigabit

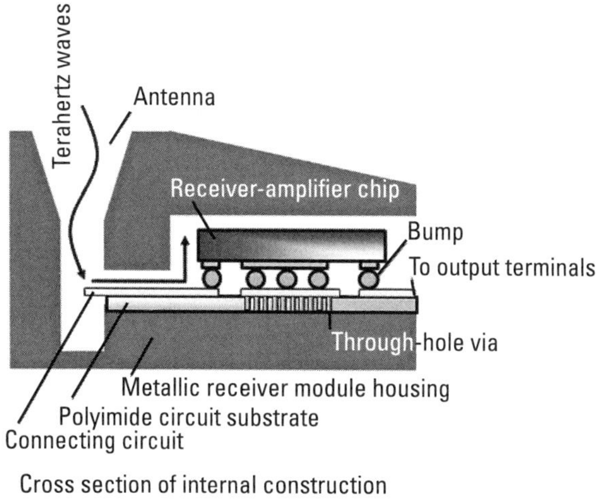

Figure 12.1 Fujitsu 300-GHz (submillimeter wavelength) demonstration transceiver.

point-to-point link connectivity [3] but are also increasingly promoted as an option for indoor and outdoor communication.

These systems use commercially available red, green, and blue LEDs as emitters to send data and as photodiode detectors to receive data. Next generation LED devices are claimed to be able to support data rates of several gigabits over 10m [4].[4] As LEDs penetrate the home and business lighting market, it becomes more credible to consider light sources as a new low-cost but localized communications option.

There are other potential touch points between 5G and the optical domain. In Chapter 4, we talked briefly about analog and digital interference cancellation as a mechanism for realizing full in-band duplex (simultaneous and continuous send and receive on the same radio channel).

The problem with present RF analog and baseband digital cancellation techniques is that they are bandwidth-limited and become less effective as bandwidth increases. An alternative option would be to introduce an RF to optical domain transform into the 5G processing chain. Translating RF signals into the optical domain is a well-established principle used in RF over fiber in distributed antenna systems, allowing multiple RF signals to be combined into a low-loss and low-cost transport layer. Examples include RF over fiber in cable TV systems where the RF, typically between 54 and 870 MHz, is converted to modulated light using 1,310-nm or 1,550-nm laser optics.

To realize optical domain cancellation, the receive signal is routed through an unbalanced drive and efficiently modulated on to an optical carrier [5]. The transmit signal is routed through a balanced drive and inefficiently modulated on to the optical carrier. The two components can then be separated in the optical domain and remixed back to RF. Achievable isolation is of the order of 40 dB over four decades of bandwidth into a 50-ohm load.

The photonic components include a continuous-wave (CW) laser, an optical modulator, and a photodetector. These devices would previously have had noise figures of the order of 20 dB and high insertion loss, but photonic links are now available that have lower noise and positive link gain. The suggested achievable noise figure operating between 700 MHz and 6 GHz is of the order of 4 to 6 dB with 17 to 21 dB of gain. Mismatch is managed by connecting a variable impedance to an additional antenna balance port, which replicates antenna impedance versus frequency.

This analog cancellation process on its own is unlikely to provide sufficient TX/RX isolation due to limitations of the optical modulator and differences between the antenna impedances and the balance. However, provided that there is sufficient suppression of the transmit signal in the analog front end to preserve RX linearity, then digital cancellation should be able to do the rest of the job. The transceiver knows what it is transmitting, so it should be able to produce an adequate estimate that can then be subtracted [6].

The power budget implications of this are not described but may be significant. The real benefit of an optical transform is extremely wide bandwidth. Even 100 GHz of RF bandwidth is a small fractional bandwidth when modulated on to a 200-THz optical carrier. As we do not actually need this amount of RF bandwidth for at least the foreseeable future, it is likely that electronic rather than optical solutions will remain generally better matched to mobile handset applications.

It is also unlikely that the global operator community would or could precipitately abandon traditional frequency division duplex (FDD) and the prospect of having some but not all devices transmitting and receiving in a transmit or receive passband might cause user to user interference issues that would be hard to address. That aside, improvements in traditional analog and digital interference cancellation techniques, combined with an optical transform, could deliver an interesting alternative to conventional RF front-end design.

12.5 The 5G Optical Smart Phone

Last but not least, every 4G smart phone includes an optical processing engine called a camera. Contemporary smart phones [7] have complementary metal-oxide semiconductor (CMOS) imaging sensors that capture 40-megapixel optical images.

CMOS imaging sensors are capable of measuring infrared and ultraviolet wavelength transmissions, so it seems a pity not to do something with this capability. Ultraviolet exposure measurement is also proposed as a standard feature in higher end devices and is already supported as an iPhone application [8].

12.6 Y-Fi

On Earth we are protected from ionizing radiation including x-ray and gamma-ray/Y-ray radiation by the magnetosphere. The atmosphere protects us at least partially from ultraviolet radiation. However, as with space observation, there are many opportunities to observe and measure the physical world around us.

12.7 Telesensing

Low-cost CMOS sensor chips are also becoming available that can detect the presence and levels of particular gases in the air that we breathe [9]. This is done by constructing a micro hot plate on the chip that can be heated at anything up to 1,000°C in 25 ms. This can be used as a source of infrared light. Alternatively, gases become highly reactive at specific temperatures, the heat equivalent of spectral lines.

Metal oxide sensing of carbon dioxide, for example, is reactive at 260°C, that of ethanol and other volatile organic compounds are reactive at 300°C. This opens up applications such as carbon monoxide sensing from smart phones, indoor and outdoor air quality measurements correlated to GPS positioning, and biosensing applications based on breath analysis. Breath analysis includes alcohol detection and acetone detection, an indicator of fat breakdown in the body, potential enablers for next generation wearable fitness devices. Whether this is specifically a 5G application is debatable, but heat, light, and radio are all interrelated and heat and light-based innovation could at a minimum be a source for future 5G-added value.

12.8 The Sensor Web and 5G Internet of Things Smart Sensing: Down-to-Earth Space Technology

Space observation is moving towards measuring the universe across the whole of the electromagnetic spectrum with measurements correlated from multiple measurement platforms including deep space exploration spacecraft, Earth-orbiting satellites, and Earth-based radio and optical telescopes.

In space these multiple platforms are being connected by multiple radio links deployed across multiple paths including low Earth orbit (LEO) to medium Earth orbit (MEO) to geostationary orbit (GSO) to Earth relays and repeaters using multiple radios operating across the meter, centimeter, and millimeter bands. This is sometimes described as the sensor web in space.

The sensor web on Earth includes space-facing optical and radio telescopes. It also potentially includes a new generation of 5G Internet of Things (IoT) devices equipped with chemical sensors that can be used in a broad range of body-worn health monitoring or remotely installed environmental monitoring.

There is considerable innovation taking place in space observation and space communication. Much of that innovation has relevance to 5G network and user device development. At the network level, the space sector industry has experience and expertise integrating observation and communications systems across L-band, S-band, C-band, X-band, Ku-band, and Ka-band. This is relevant to mobile broadband operators and vendors looking to develop radio systems that span meter-band, centimeter-band, and millimeter-band spectrum.

The extreme demands of deep radio communication have required innovations at the component and system levels that are relevant to 5G transceiver and system design. Parallel developments in earth and spaced based radar in

S-band, C-band, and X-band and micro rain radar at 24-GHz in K-band and cloud radar at 35 GHz are also potentially useful.

The space sector has also been commercially innovative, particularly in the development of mixed payload business models. These could be potentially relevant to 5G business modeling.

12.9 The Downside of Adversarial Auctions

The bandwidth requirements of space observation and communication technologies including Earth-based space observation and communication are increasing over time. Each new generation of radio telescope or radar or deep-space or near-space communications system requires more sensitivity and more bandwidth than its preceding generation. This suggests that there will be resistance to mobile broadband spectrum allocation in the centimeter and millimeter bands.

The 4G mobile broadband industry had many potentially beneficial complementary technical and commercial touch-points with the terrestrial broadcasting industry. The battle for spectrum in the 700-MHz and 800-MHz bands compounded by an adversarial auction process has made it hard to realize these benefits. There are likely to be similar tension points between the mobile broadband community and satellite and space sector due to the competing requirements for additional bandwidth in the centimeter and millimeter bands. This tension will make it harder to realize the potential benefits to be realized from integrating space and terrestrial communication and observation technologies.

12.10 Five Generations of Cellular: Is This the End of the Story?

What lessons can we learn from four generations of cellular technology and commercial innovation, and what does the story so far tell us about what happens next?

Each generation of cellular technology has been more complex than the previous generation but equally importantly the complexity gets shifted between the analog and digital domain and across the frequency domain, time domain, code domain and phase domain.

First generation cellular in the 1980s was complicated to implement in the frequency domain; an Advanced Mobile Phone System (AMPS) 25+25 850 MHz-band handset had to be capable of accessing 833 RF 30-kHz channels, a big challenge for frequency synthesizer design.

Second generation cellular simplified some of this frequency domain complexity both in the user device and at network level [200-kHz channels replacing 30-kHz or 25-kHz Total Access Communications System (TACS) channels]. The complexity moved into the digital domain. The big initial challenge was to get enough digital signal processing power to support the voice codec.

Third generation cellular theoretically simplified the RF domain further by shifting part of the sensitivity and selectivity task in to the code domain and increasing channel spacing from 200 kHz to 5 kHz. However, maintaining the phase argument in the modulated code stream required close control of phase noise and phase stability. This coincided with a relatively rapid increase in the number of frequency bands that needed to be supported, which introduced additional cost and performance loss.

Fourth generation cellular theoretically simplified RF domain processing by increasing channel spacing from 5 MHz to 10 or 20 MHz or potentially aggregated bandwidths of 100 MHz. This is achieved by the use of orthogonal frequency division multiplexing (OFDM), but this requires close control of RF subcarriers at 15-kHz or 7.5-kHz spacing reintroducing RF domain complexity.

The 5G wide-area mobile broadband physical layer potentially replaces much or at least some of the complexity of these prior generations by moving the heavy lifting to the spatial domain. If the 2° beamwidth smart antenna arrays used in E-band military systems can be realized in low cost 5G base stations and user devices, if spatial processing algorithms can be reused from the automotive radar industry and RF devices cross-amortized across automotive, fixed point-to-point, and 5G mobile broadband, then several things happen.

Power efficient spectrally efficient user data rates of 1 Gbps become achievable in a wide-area mobile system; the benefits of Wi-Fi get scaled to Big Radio without the cost multipliers implicit in scaling ultradense network topologies. This can only be achieved by combining materials innovation, manufacturing innovation, and mathematical (algorithmic) innovation.

There is a particular need for materials innovation to help realize the high-performance substrates needed for beamforming antenna arrays with high numbers of active elements each with its own dedicated power amplifier or low noise amplifier (LNA). Our personal guess would be that these arrays will support a traditional FDD band plan at millimeter wavelengths to deliver spatial and frequency-domain separation between transmit and receive paths with the uplink and downlink separately beamformed to maximize instantaneous link budgets and minimize intersystem and intrasystem interference. This is 3-D spatial processing and includes horizontal and vertical polarizations.

Theoretically, if system interference could be completely avoided, there would be no need to license or regulate spectrum. This might work in some markets, but the investment scales and return on investment time scales are probably too large and too long for this to be viable. Bid teams will keep their jobs at least for the foreseeable future. Possibly anything that does not move could be lightly licensed, the existing model for point-to-point links in the millimeter band, but it is hard to see how this would work for mobile broadband.

12.11 Technology Investment Time Scales in the Telecoms Industry

This relationship between technology investment and spectrum regulation is, in turn, a function of telecom industry investment and return on investment time scales.

There can be short periods in the telecoms industry where investment returns can be realized in spectacularly short time scales. In the heady days of GSM in Europe in the mid-1990s, an urban base station could be deployed and pay for itself in 3 to 5 months. However, this is an exception in an industry that to all intents and purposes is a utility with the same inherent time scales as other utilities, gas, water, and electricity and equivalent infrastructure investments such as roads and railways. Our ultimate Big Radio example, the Rugby radio coil, had an active service life of 70 years. First generation cellular base station radio masts from the 1980s are still in use. Some of them are Grade 2 listed so mobile broadband operators cannot take them down even if they want to.

The radio industry still manufactures valves 100 years after they were invented, albeit for specialist applications. We still use transistors 50 years after they were invented and will probably still be using them in 50 years' time.

The telecoms industry is 150 years old, the satellite industry is 50 years old, the cellular industry is 30 years old, and the automotive radar industry is about 10 years old. The mobile broadband industry is arguably emerging as a subset of the cellular industry and will have its own investment cycle, although this is unlikely to be shorter than the 30-year to 50-year industry norm. Many of the 3G licenses will never be fully amortized.

The IoT industry is probably another emerging subset industry but once you blow away the market fluff you would probably come to the conclusion that it is a market still in gestation, conceived but yet to undergo parturition.

12.12 5G Technology Economics

Economics obeys the laws of physics. That great prewar invention, the Smith Chart [10], the workhorse of the RF engineer for the past 70 years has all you need to become an instant expert on economics including telecom economics. For inductance substitute quantitative easing, for resistance substitute interest rates, for capacitance substitute international cash and debt ratios, find the point of conditional stability and you will have solved the world's economic problems.

A similar analysis can be applied to social engineering and politics. This might seem trite, but it is important to understand the relative mechanics and evolutionary time scales and coupling processes (interdependency) of developed and developing economies and their associated regulatory regimes.

This also applies to the stock market. Warren Buffet, the sage of Omaha, regards his biggest mistake as buying shares in the U.K. supermarket chain Tesco. If he had bought the shares in 1960, he would have been pleased. He would have achieved the same return buying shares in British Control and Communications, the parent company of Racal Electronics, and, ultimately (under different ownership), the Vodafone Group. We labor this point to remind people that some companies last for a long time. Siemens and BT, for example, trace their origins back well over 100 years.

12.13 Quantum Telecom: Quantum Physics and 5G Scientific Time Scales

Briefly referenced at the end of Chapter 7, we may as well use the telecommunications space and time continuum as an excuse to revisit quantum physics. Going back 100 years takes us back to Einstein in 1915 and his General Theory of Relativity. The relevance of the general theory might seem initially remote to telecommunications, but within 15 years of publishing the paper, Einstein became embroiled in an argument about how the theory could be reconciled with quantum mechanics. The debate was not resolved until the 1960s when the process of entanglement became more fully understood.

Quantum mechanics has been credited as the basis for many of the most significant technology innovations of the twentieth century including semiconductors, microprocessors, lasers, nuclear energy, and thermal imaging. In the scientific community this is described as Quantum 1.0 [11]. The argument is that the next 100 years of scientific innovation (Quantum 2.0) will be defined by an emerging ability to harness quantum physics.

In the telecommunications industry we rely on atomic clocks that are effectively quantum mechanical devices exploiting the properties of atomic rather than nuclear physics. The SI second, the base unit of time in the International System of Units (SI), is defined by a cesium fountain atomic clock in which six lasers are fired at a group of cesium atoms. This creates an optical trap in which the cesium atoms are pushed closer together to the point at which they more or less stop vibrating, at which point they become very cold, a tiny fraction of a degree above absolute zero (–273.15°C). The lasers above and below the optical trap then launch the cesium atoms upwards into a microwave chamber and the atoms fall back down under gravity. Microwave radiation is used to make the electrons in the cesium atom move between energy levels as they move up and down and the energy levels are measured through fluorescence. This fountain process takes approximately a second and is repeated with different microwave frequencies until the frequency that causes the maximum number of cesium electrons to change level is realized. This is the resonant frequency that defines a second as the amount of time it takes for the radiation from the transition to complete 9,192,631,770 full-wave transitions [12].

Quantum computing similarly exploits a combination of quantum mechanical and quantum physical properties. Classical computers, based on Alan Turing's work in the 1930s, work on two logic states, a 0 or a 1. Quantum computing exploits the ability of quantum bits (qubits) to exist in three states, a 0 or a 1 or a superposition of 0 and 1.

Quantum bits are subatomic particles in which a change in energy state can be stimulated and measured by a control device. The control device can be an ion trap using optical or magnetic fields or a combination of both, optical traps using light waves and microwave radiation (atomic clocks), quantum dots using semiconductor material to manipulate electrons, the use of semiconductor impurities producing electrons from unwanted atoms and or superconducting (atomic clocks again).

Quantum computers also exploit the property of entanglement (Einstein's spooky action at a distance) in which the application of an external force on two atoms induces the second atom to take on the property (energy state) of the first atom. If left on its own, a single atom will spin in all directions. If disturbed, it chooses one spin or one value at which apparently instantaneous moment the second atom adopts the opposite spin or value irrespective of how far the two atoms are apart.

Frustratingly, rather like 2-D materials, these semimagical properties are hard to harness in real-world, low-cost compact devices. In terms of performance, the enemy of quantum computers is noise and what are known as quantum decoherence phenomena.

This influences both the hardware used and the maths used. The hardware can be exotic and unfamiliar and chemically or organically based. Quantum computers in the late 1990s, for example, used amino acids to analyze quantum state decay and chlorinated hydrocarbon to spread out the qubits to make them more resistant to corruption.

Over the last 15 years quantum computing research has been motivated by its potential application to cryptography and cryptoanalysis. Theoretically, superposition and entanglement together enable quantum computers to perform massively parallel computing. A 30-qubit quantum computer would be equivalent to a conventional computer working at 10 teraflops per second (trillions of floating point operations per second). This would allow existing factoring algorithms used in cryptography, for example, RSA encryption, to be easily deencrypted. Inevitably much work is now ongoing on new cryptosystems that are secure from quantum computers. This is described as post quantum cryptography.

The applicability of quantum computing to 5G depends substantively on whether the hardware noise problems can be resolved and on cost and size issues; a highly accurate quantum clock would need a small truck and a very large power supply.

The associated maths are seductive. The ability to support three states rather than two allows quantum computers to run polynomial time algorithms (algorithms based on three or more state inputs).

Mathematical operations such as addition, subtraction, multiplication, and division, as well as computing square roots, powers, and logarithms, can be performed in polynomial time. Algorithms such as the Shor algorithm are efficient because of the efficiency of the quantum Fourier transform.

Present academic research suggests that an efficient quantum Fourier transform may not realize any improvement in the classical Fourier transform, but there is clearly scope for further mathematical and algorithmic innovation. Research would appear to be coalescing around a subset of algorithmic research known as adiabatic quantum computation. The definition of adiabatic is a process that occurs without loss or gain of heat. Given that heat can be equated directly to noise, this is potentially useful but depends on a combination of hardware and innovation to which there is presently limited visibility. Progress will also be dependent on different disciplines working together efficiently including physicists, chemists, hardware, software and system engineers, and any physical layer applicability is likely to be in base station processing rather than user or IoT products. Even in base stations, it is worth remembering that mast-mounted superconducting filters were promoted as a mechanism for achieving selectivity in larger macro cells but have to date not succeeded in achieving significant market uptake. Partly this is due to maintenance cost considerations in outdoor environments.

12.14 Positive and Negative Technology and Economic Touch-Points: Science Is Not Going to Be a Short-Term Answer

Being realistic, within the next 15 years, quantum computing may change or influence cryptography in telecommunications but is unlikely to transform 5G delivery economics.

Delivery economics are more likely to be determined by improving existing technologies and the efficiency with which we exploit these technologies.

Let us summarize the positive and negative economic touch-points between the mobile broadband industry and the rest of the radio industry across our three wavelength bands of interest.

In the meter band, the dominant touch-points have been with the TV and radio industry. There are technology lessons we probably should have learned from the TV industry. The harmonization of DVB-T, DVB-S, and DVB-C was a neat trick and has been a significant factor in reducing receiver decoder cost. In practice an adversarial auction process has frustrated potential technology and economic translation opportunities and common interests. It has been a limited mutual interest model constrained to site sharing and backhaul.

In the centimeter band the dominant technical and economic touch-points are with the satellite industry with a common commercial interest in 4G backhaul. The satellite industry has successfully deployed spot beam antenna arrays, beams within beams, to maximize link budgets and manage interference and have deployed innovative relay and repeater technologies that span LEO, MEO, and GSO radio systems.

At the component level the space industry has implemented some remarkable technology innovation. The *Voyager* spacecraft launched in 1977 still manages its onboard systems with a radiation-hardened silicon on sapphire-based microcontroller, a man-made brain in space now on its way to the next solar system and we can still talk to the spacecraft by radio.

The space industry has many technology and system efficiency innovations with relevance to 5G including the mechanics of building hardware and software systems that can withstand a massive once-in-a-century electrical storm. Commercial innovation in the satellite industry includes mixed payloads and integrated L-band, S-band, Ku-band, and Ka-band service offers. The risk is that an adversarial auction process of centimetre-band spectrum will make it harder to realize these potential touch-point benefits.

Pragmatically it might be easier to work with the backhaul industry. Backhaul technologies scale across the meter band, centimeter band, and millimeter band. Bringing 5G mobile wide-area and backhaul technologies together would allow the 5G industry to benefit from already amortized or partly amortized RF hardware and software investment. Conversely, the backhaul industry would benefit from additional scale and harmonized technical standards.

Antenna arrays with high numbers of active elements will be needed to match the spatial focus of dish antennas. As our deep-space friends say, it is all about aperture.

The signal processing algorithms needed and the front-end RF components needed for millimetre-band transceivers coupled to adaptive phased array antenna systems have already been developed by the automotive industry and it is clear that there are many common interest points between the automotive industry and 5G development community.

Existing 4G smart phones understand their physical orientation: vertical, horizontal, stationary, or mobile. A 5G user device will have 3-D RF spatial awareness. A 5G smart phone will not attempt to transmit or receive centimeter-wavelength or millimeter-wavelength RF energy into a human head or through a human hand. In other words, it will be more energy-efficient.

Next generation user devices and next generation IoT devices together define network value. User devices and IoT device physical layer form factor and functionality is defined by RF hardware and digital signal processing and processors in the same way that application value is defined by display technology and microcontroller and application processor bandwidth. User devices and IoT devices together dictate the spectrum that will be needed for 5G and the standards sweet spot, the compromise point between complexity and cost.

There is a need in 5G to avoid the over complexity of meter-band spectrum, particularly the lack of regional harmonization and a need for ultrasparse as well as ultradense networks, a 5G physical layer that can scale from small cells to big cells, from small radio to big radio. This suggests that FDD and contiguous bandwidth will be needed with channels of between 1 and 2 GHz to deliver multiplexing gain and high peak data rates.

It will be possible to build antenna arrays with a range of active element wavelengths, but fundamental bandwidth ratios still apply and there is no reason to challenge the existing rule of thumb of 10% antenna bandwidth ratios for a maximally efficient overall bandpass bandwidth. Why fight physics if you do not need to?

This suggests a natural differentiation between the centimeter band and millimeter band on a simple scaling model (see Table 12.1).

Irrespective of the coexistence issues in the centimeter band, the millimeter band and specifically the fixed link allocations at 71–76 GHz and 81–86 GHz are the only allocations which would tidily divide into, for example, five 1-GHz bandwidth channels or the 2-GHz channel bandwidths referenced in the Nokia/NTT DoCoMo case study.

This is not to say that 5G could not be deployed into longer wavelength bands. The Nokia/DoCoMo case study clearly showed the cell edge throughput benefits of combining millimeter and centimeter physical layers from colocated base stations. The 70-GHz and 80-GHz bands look particularly attractive technically and commercially on the basis of bandwidth availability, availability of

Table 12.1

Bandwidth Scaling in the Centimeter and Millimeter Bands

	Centimeter Band		Millimeter Band						
Frequency	3 GHz	30 GHz	40 GHZ	50 GHz	60 GHz	70 GHz	80 GHz	90 GHz	100 GHz
Wavelength	10 cm	1 cm	7.5 mm	6 mm	5 mm	4.2 mm	3.75 mm	3.33 mm	3 mm
Pass band	300 MHz	3 GHz	4 GHz	5 GHz	6 GHz	7 GHz	8 GHz	9 GHz	10 GHz
5G bands?						71–76	81–86		
Auto radar						77–81			

fixed link hardware that could be repurposed or cross-amortized, adjacent 5G relevant automotive radar technologies (signal processing, spatial awareness, component reliability, and cost benefits) and military system and associated RF component investment.

Similar arguments are made for immediate spectral proximity to the 60-GHz Wi-Fi band, but this ignores the need to differentiate wide-area mobile 5G from local area Wi-Fi. This suggests a need to steer clear of the oxygen resonance band. Similarly, it would seem sensible to keep the regulatory domain clearly differentiated, the unlicensed model for Wi-Fi and licensed or lightly licensed for wide-area mobile where investment profiles require certainty and control.

It is reasonable to put a value on spectrum through an auction process, particularly if the auction structure facilitates rather than frustrates the mutual interest opportunities that exist between the mobile broadband industry, the point-to-point industry, the fixed access industry, the automotive industry satellite industry, and other radio stakeholders, a mechanism for arbitrating common interest, a mutual interest model in which all parties gain equal benefit.

The U.S. Incentive Auction at 600 MHz is a crucial first test of this approach, although it will need refinement to be useful and efficient as a mechanism for facilitating economic and efficient 5G deployment in the centimeter and millimeter bands.

12.15 A Final Word from Mr. Ford

We have spent a significant part of this book making the case for producing a closer coupling between 5G innovation and automotive innovation. In 1970, there were 200 million cars worldwide. By 2020, this will have increased to at least 2 billion. If China had the same ownership density as the United States, there would be over 1 billion cars in China on its own; that is a lot of mobile base stations and repeaters and relays.

Given that we started with the Model T as an exemplar of cross-industry technology translation, we will also finish with a quote from Henry Ford, a man who understood that competition and cooperation were not mutually exclusive. As Mr. Ford put it, "If everyone moves forward together, then success takes care of itself."

References

[1] METIS 2 Horizon 2020, EU Framework Programme for Research and Innovation.

[2] http://spectrum.ieee.org/tech-talk/telecom/wireless/fujitsu-makes-a-terahertz-receiver-small-enough-for-a-smartphone.

[3] http://www.freespaceoptics.org/.

[4] http://www.lifi-centre.com/.

[5] http://www.microwavejournal.com/articles/22950-maximizing-rf-spectrum-utilization-with-simultaneous-transmit-and-receive.

[6] http://www.photonicsinc.com/star_fm_demo.html.

[7] http://www.microsoft.com/en-gb/mobile/phone/lumia1020/.

[8] https://itunes.apple.com/us/app/uvmeter-check-your-uv-index/id662827178?mt=8.

[9] http://www.ccmoss.com/.

[10] http://smithchart.org/phsmith.shtml.

[11] http://www.newelectronics.co.uk/electronics-technology/harnessing-quantum-technology-for-uk-plc/111031/.

[12] http://www.npl.co.uk/reference/measurement-units/si-base-units/the-second.

Acronyms and Abbreviations

3GPP Third Generation Partnership Project (standards group for 3G and 4G)

ACLR Adjacent channel leakage ratio (measure of channel to channel interference)

ADSL Asymmetric digital subscriber line

AGC Automatic gain control

AIS Automatic Identification System

AM Amplitude modulation (used in medium-wave and long-wave broadcasting)

A-MPR Adaptive Maximum Power Relaxation

AMPS Advanced Mobile Phone System (first generation cellular system in the United States)

APT Asia Pacific Telecommunity

ARCEP Autorité de Régulation des Communications Électroniques et des Postes (French regulator)

ASIC Application-specific integrated circuit

ATM Asynchronous transfer mode

ATSC Advanced Television Systems Committee (TV standard in the United States)

AWS Advanced Wireless Service (description for wireless networks in the United States below and above 2 GHz)

AXI Advanced Extensible Interface

BAR Bulk acoustic resonator

BOPS Billion operations per second

BSSID Base station system identity (MAC address for a Wi-Fi access point)

CCIR Consultative Committee for International Radio

CDMA Code division multiple access

CDPD Cellular digital packet data

CEN Centre European Nationale (European technical standards body)

CEPT Committee European Post and Telecommunications

CMOS Complementary metal-oxide semiconductor

CoMP Coordinated multipoint transmission

COST Committee on Science and Technology

CPE Consumer premises equipment

CQI Channel quality indicator

CTE Coefficient of Thermal Expansion

DAB Digital Audio Broadcasting

DARPA Defense Advanced Research Projects Agency

DDR Double data rate memory

DFT Discrete Fourier transform

DOD U.S. Department of Defense

DRM Digital Radio Mondiale

DSP Digital signal processor

DSRC Digital Short-Range Communication (Standard for automotive connectivity in the 5-GHz ISM band)

DTT Digital Terrestrial Television

DVB-C Digital Video Broadcasting-Cable

DVB-S Digital Video Broadcasting-Satellite

DVB-T Digital Video Broadcasting-Terrestrial

EBITDA Earnings Before Interest Tax and Depreciation

ECM Electronic countermeasure

EDGE Enhanced Data Rates for GSM Evolution

EIRP Effective isotropic radiated power

EM Electromagnetic

eMBMS Enhanced Multimedia Broadcast Multicast Services

EMC Electromagnetic compatibility

EMEA Europe, Middle East, and Africa

EMI Electromagnetic interference

eNB Enhanced node B (LTE base station)

ETSI European Technical Standards Institute (European technical standards body)

EVM Error vector magnitude

FBAR Film bulk acoustic resonator

FBMC Filter bank multicarrier

FCC U.S. Federal Communications Commission

FDD Frequency division duplex

FDMA Frequency division multiple access

FFT Fast Fourier transform

FIR Finite impulse response (e.g., digital filters also known as discrete time filters based on a tapped delay line)

FM Frequency modulation

fmax Maximum oscillation frequency

FMCW Frequency-modulated, continuous-wave radar

FPGA Field programmable gate array

FR4 Flame-retardant woven fiberglass

FSK Frequency shift keying

FSS Fixed services satellite

ft Cutoff frequency

GaAS Gallium arsenide

GDP Gross domestic product

GFET Graphene field effect transistor

GMSK Gaussian minimum shift keying (Constant envelope modulation)

GPRS General Packet Radio Services

GPS Global Positioning System

GSM Global System Mobile (Second generation cellular standard

GSO Geostationary orbit

H ARQ Hybrid Automatic Repeat Request

HAP High-altitude platform

HSDPA High-speed downlink packet access

HSPA High-speed packet access

HSUPA High-speed uplink packet access

ICIC Intercell interference coordination

ICT Information and communications technology

IEEE Institute of Electrical and Electronics Engineers

IIR Infinite impulse response (e.g., analog filters composed of resistors, capacitors and inductors where the internal state never completely relaxes following an impulse)

IMT International Mobile Telecommunications

IoT Internet of Things

ISDB-T Sistema Brasileiro de Televisão Digital (digital television standard for Brazil, Peru, Argentina, Chile, Honduras, Venezuela, Ecuador, Costa Rica, Paraguay, Philippines, Bolivia, Nicaragua)

ISM Industrial Scientific Medical (unlicensed bands also used for Wi-Fi)

ITU International Telecommunication Union

LAA Licensed Assisted Access for Unlicensed Spectrum

LAN Local area network

LBT Listen before talk (polite protocol used in Wi-Fi)

LDPC Low-density parity checks

LED Light-emitting diode

LEO Low Earth orbit

LiDAR Light Detection and Ranging (use of a pulsed laser to measure distance)

LNA Low noise amplifier (amplifier for receive path)

LO Local oscillator

LOS Line of sight

LTCC Low temperature cofired ceramic

LTE Long Term Evolution (fourth generation cellular standard)

LTE A LTE Assisted

LTE-U LTE in unlicensed spectrum

MAC Medium Access Control

MBSFN Multicast Broadcast Single Frequency Network

MEO Medium Earth orbit

MIMO Multiple input multiple output (use of multipath to support multiple data streams)

MIPS Millions of instructions per second

MMIC Monolithic microwave integrated circuit

MOPS Millions of operations per second

MPEG Motion Picture Experts Group

MPR Maximum power relaxation

MSS Mobile services satellite (e.g., Iridium and Inmarsat)

MTC Machine-type communication. See also IoT

NAND Non-volatile FLASH memory

NASA National Aeronautics and Space Administration

NATO North Atlantic Treaty Organization

NEO Near-Earth objects

NLOS Nonline of sight

NOMA Nonorthogonal multiple access

OFDM Orthogonal frequency division multiplexing

OFDMA Orthogonal frequency division multiplexing access

OIRT Organisation Internationale de Radiodiffusion et de Télévision

OVSF Orthogonal variable spreading factor

PCB Printed circuit board

PCS Personal Communications Service

PDCP Packet Data Control Protocol

PDSCH Packet data shared channel

Phemt Pseudomorphic high electron mobility transistor

PLL Phase locked loop

PMMA Polymethyl methacrylate

PMSE Programme Making and Special Events

PPDR Public Protection and Disaster Relief

PRACH Packet Random Access Channel

PSR Public Safety Radio

PTFE Polytetrafluoroethylene (material for printed circuit boards)

PUCCH Packet uplink control channel

PUSCH Packet uplink shared channel

QAM Quadrature amplitude modulation

QPSK Quaternary phase shift keying

RAM Random access memory

RAN Radio access network

RF Radio frequency

RF MEMS Radio frequency microelectromechanical system

RFFE Radio frequency front end

RFIC Radio frequency integrated circuit

RLC Radio link control

ROI Return on investment

RRC Radio resource controller

RTOS Real-time operating system

RX Receive path

SAW Surface acoustic wave

SC-FDMA Single-carrier FDMA (used in 4G uplink)

SEU Single event upset

SFN Single-frequency network

SiC Silicon carbide

SiGe Silicon germanium

SISO Single input single output

SOI Silicon on insulator

SOS Silicon on sapphire

SV LTE Simultaneous voice and data over LTE

TACS Total Access Communications System (first generation cellular system in the United Kingdom)

TCM Tightly coupled memory

TDD Time division duplex

TDM Time division multiplexed

TDMA Time division multiple access

TEM Transverse electromagnetic (propagation mode)

TETRA Terrestrial Trunked Radio

TID Total ionization dosage

TIS Total isotropic sensitivity

TPSK Ternary phase shift keying

TRDS Tracking and data relay satellites

TRP Total radiated power

TRS Total radiated sensitivity

TTI Transmission time interval

TX Transmit

U NNII Unlicensed National Information Infrastructure Bands (e.g., the 5-GHz ISM band for Wi-Fi)

UHF Ultrahigh frequency (includes the TV bands)

UMTS Universal Mobile Telephone System

USB Universal Serial Bus

USSR Union of the Soviet Socialist Republics

UWB Ultrawideband

VCO Voltage controlled oscillator

VHF Very high frequency (includes radio FM broadcasting)

VLBI Very long baseline interferometer

VLIW Very long instruction word

VoLTE Voice over LTE

W-CDMA Wideband code division multiple access (descriptor for 3G physical layer)

WiGIg Wireless Gigabit Alliance

WiMAX Worldwide Interoperability for Microwave Access

WRC World Radio Congress (Congress every 4 years setting global spectrum policy)

About the Author

Geoff Varrall joined RTT in 1985 as an executive director and shareholder to develop RTT's international business as a provider of technology and business services to the wireless industry.

He codeveloped RTT's original series of design and facilitation workshops including RF Technology, Data over Radio, Introduction to Mobile Radio, and Private Mobile Radio Systems and developed the Oxford Program, a 5-day strategic technology and market programme presented annually with the Shosteck Group. Over 20 years, several thousand senior-level delegates attended these programs. Mr. Varrall has undertaken spectrum and standards consultancy work for a broad cross-section of operators and vendors and regulatory and standards agencies in the United States, Europe, and Asia.

Mr. Varrall is a coauthor of the *Mobile Radio Servicing Handbook* (Heinemann Butterworth), *Data Over Radio* (Quantum Publishing), and *3G Handset and Network Design* (John Wiley), and the author of *Making Telecoms Work: From Technical Innovation to Commercial Success* (John Wiley). He has written regularly for a number of European trade journals and has been a regular chair at industry conference and trade events.

As a former director of Cambridge Wireless, Mr. Varrall remains actively involved in a number of wireless heritage initiatives that aim to capture and record past technology and engineering experience in the telecommunications industry.

Index

Recent Titles in the Artech House Mobile Communications Series

William Webb, Series Editor

3G CDMA2000 Wireless System Engineering, Samuel C. Yang

3G Multimedia Network Services, Accounting, and User Profiles, Freddy Ghys, Marcel Mampaey, Michel Smouts, and Arto Vaaraniemi

5G Spectrum and Standards, Geoff Varrall

802.11 WLANs and IP Networking: Security, QoS, and Mobility, Anand R. Prasad and Neeli R. Prasad

Achieving Interoperability in Critical IT and Communications Systems, Robert I. Desourdis, Peter J. Rosamilia, Christopher P. Jacobson, James E. Sinclair, and James R. McClure

Advances in 3G Enhanced Technologies for Wireless Communications, Jiangzhou Wang and Tung-Sang Ng, editors

Advances in Mobile Information Systems, John Walker, editor

Advances in Mobile Radio Access Networks, Y. Jay Guo

Applied Satellite Navigation Using GPS, GALILEO, and Augmentation Systems, Ramjee Prasad and Marina Ruggieri

Artificial Intelligence in Wireless Communications, Thomas W. Rondeau and Charles W. Bostian

Broadband Wireless Access and Local Network: Mobile WiMax and WiFi, Byeong Gi Lee and Sunghyun Choi

CDMA for Wireless Personal Communications, Ramjee Prasad

CDMA Mobile Radio Design, John B. Groe and Lawrence E. Larson

CDMA RF System Engineering, Samuel C. Yang

CDMA Systems Capacity Engineering, Kiseon Kim and Insoo Koo

CDMA Systems Engineering Handbook, Jhong S. Lee and Leonard E. Miller

Cell Planning for Wireless Communications, Manuel F. Cátedra and Jesús Pérez-Arriaga

Cellular Communications: Worldwide Market Development, Garry A. Garrard

Cellular Mobile Systems Engineering, Saleh Faruque

Cognitive Radio Interoperability through Waveform Reconfiguration, Leszek Lechowicz and Mieczyslaw M. Kokar

Cognitive Radio Techniques: Spectrum Sensing, Interference Mitigation, and Localization, Kandeepan Sithamparanathan and Andrea Giorgetti

The Complete Wireless Communications Professional: A Guide for Engineers and Managers, William Webb

Digital Communication Systems Engineering with Software-Defined Radio, Di Pu and Alexander M. Wyglinski

EDGE for Mobile Internet, Emmanuel Seurre, Patrick Savelli, and Pierre-Jean Pietri

Emerging Public Safety Wireless Communication Systems, Robert I. Desourdis, Jr., et al.

The Future of Wireless Communications, William Webb

Geographic Information Systems Demystified, Stephen R. Galati

Geospatial Computing in Mobile Devices, Ruizhi Chen and Robert Guinness

GPRS for Mobile Internet, Emmanuel Seurre, Patrick Savelli, and Pierre-Jean Pietri

GPRS: Gateway to Third Generation Mobile Networks, Gunnar Heine and Holger Sagkob

GSM and Personal Communications Handbook, Siegmund M. Redl, Matthias K. Weber, and Malcolm W. Oliphant

GSM Networks: Protocols, Terminology, and Implementation, Gunnar Heine

GSM System Engineering, Asha Mehrotra

Handbook of Land-Mobile Radio System Coverage, Garry C. Hess

Handbook of Mobile Radio Networks, Sami Tabbane

High-Speed Wireless ATM and LANs, Benny Bing

Inside Bluetooth Low Energy, Second Edition, Naresh Gupta

Interference Analysis and Reduction for Wireless Systems, Peter Stavroulakis

Internet Technologies for Fixed and Mobile Networks, Toni Janevski

Introduction to 3G Mobile Communications, Second Edition, Juha Korhonen

Introduction to 4G Mobile Communications, Juha Korhonen

Introduction to Communication Systems Simulation, Maurice Schiff

Introduction to Digital Professional Mobile Radio, Hans-Peter A. Ketterling

Introduction to GPS: The Global Positioning System, Ahmed El-Rabbany

An Introduction to GSM, Siegmund M. Redl, Matthias K. Weber, and Malcolm W. Oliphant

Introduction to Mobile Communications Engineering, José M. Hernando and F. Pérez-Fontán

Introduction to Radio Propagation for Fixed and Mobile Communications, John Doble

Introduction to Wireless Local Loop, Broadband and Narrowband, Systems, Second Edition, William Webb

IS-136 TDMA Technology, Economics, and Services, Lawrence Harte, Adrian Smith, and Charles A. Jacobs

Location Management and Routing in Mobile Wireless Networks, Amitava Mukherjee, Somprakash Bandyopadhyay, and Debashis Saha

LTE Air Interface Protocols, Mohammad T. Kawser

Metro Ethernet Services for LTE Backhaul, Roman Krzanowski

Mobile Data Communications Systems, Peter Wong and David Britland

Mobile IP Technology for M-Business, Mark Norris

Mobile Satellite Communications, Shingo Ohmori, Hiromitsu Wakana, and Seiichiro Kawase

Mobile Telecommunications Standards: GSM, UMTS, TETRA, and ERMES, Rudi Bekkers

Mobile-to-Mobile Wireless Channels, Alenka Zajić

Mobile Telecommunications: Standards, Regulation, and Applications, Rudi Bekkers and Jan Smits

Multiantenna Digital Radio Transmission, Massimiliano Martone

Multiantenna Wireless Communications Systems, Sergio Barbarossa

Multi-Gigabit Microwave and Millimeter-Wave Wireless Communications, Jonathan Wells

Multipath Phenomena in Cellular Networks, Nathan Blaunstein and Jørgen Bach Andersen

Multiuser Detection in CDMA Mobile Terminals, Piero Castoldi

OFDMA for Broadband Wireless Access, Slawomir Pietrzyk

Personal Wireless Communication with DECT and PWT, John Phillips and Gerard MacNamee

Practical Wireless Data Modem Design, Jonathon Y. C. Cheah

Prime Codes with Applications to CDMA Optical and Wireless Networks, Guu-Chang Yang and Wing C. Kwong

Quantitative Analysis of Cognitive Radio and Network Performance, Preston Marshall

QoS in Integrated 3G Networks, Robert Lloyd-Evans

Radio Engineering for Wireless Communication and Sensor Applications, Antti V. Räisänen and Arto Lehto

Radio Propagation in Cellular Networks, Nathan Blaunstein

Radio Resource Management for Wireless Networks, Jens Zander and Seong-Lyun Kim

Radiowave Propagation and Antennas for Personal Communications, Third Edition, Kazimierz Siwiak and Yasaman Bahreini

RDS: The Radio Data System, Dietmar Kopitz and Bev Marks

Resource Allocation in Hierarchical Cellular Systems, Lauro Ortigoza-Guerrero and A. Hamid Aghvami

RF and Baseband Techniques for Software-Defined Radio, Peter B. Kenington

RF and Microwave Circuit Design for Wireless Communications, Lawrence E. Larson, editor

RF Positioning: Fundamentals, Applications, and Tools, Rafael Saraiva Campos and Lisandro Lovisolo

Sample Rate Conversion in Software Configurable Radios, Tim Hentschel

Signal Processing Applications in CDMA Communications, Hui Liu

Signal Processing for RF Circuit Impairment Mitigation, Xinping Huang, Zhiwen Zhu, and Henry Leung

Smart Antenna Engineering, Ahmed El Zooghby

Software Defined Radio for 3G, Paul Burns

Spread Spectrum CDMA Systems for Wireless Communications, Savo G. Glisic and Branka Vucetic

Technologies and Systems for Access and Transport Networks, Jan A. Audestad

Third-Generation and Wideband HF Radio Communications, Eric E. Johnson, Eric Koski, William N. Furman, Mark Jorgenson, and John Nieto

Third Generation Wireless Systems, Volume 1: Post-Shannon Signal Architectures, George M. Calhoun

Traffic Analysis and Design of Wireless IP Networks, Toni Janevski

Transmission Systems Design Handbook for Wireless Networks, Harvey Lehpamer

UMTS and Mobile Computing, Alexander Joseph Huber and Josef Franz Huber

Understanding Cellular Radio, William Webb

Understanding Digital PCS: The TDMA Standard, Cameron Kelly Coursey

Understanding GPS: Principles and Applications, Second Edition, Elliott D. Kaplan and Christopher J. Hegarty, editors

Understanding WAP: Wireless Applications, Devices, and Services, Marcel van der Heijden and Marcus Taylor, editors

Universal Wireless Personal Communications, Ramjee Prasad

WCDMA: Towards IP Mobility and Mobile Internet, Tero Ojanperä and Ramjee Prasad, editors

Wireless Communications in Developing Countries: Cellular and Satellite Systems, Rachael E. Schwartz

Wireless Communications Evolution to 3G and Beyond, Saad Z. Asif

Wireless Intelligent Networking, Gerry Christensen, Paul G. Florack, and Robert Duncan

Wireless LAN Standards and Applications, Asunción Santamaría and Francisco J. López-Hernández, editors

Wireless Sensor and Ad Hoc Networks Under Diversified Network Scenarios, Subir Kumar Sarkar

Wireless Technician's Handbook, Second Edition, Andrew Miceli

For further information on these and other Artech House titles,including previously considered out-of-print books now available through our In-Print-Forever® (IPF®) program, contact:

Artech House
685 Canton Street
Norwood, MA 02062
Phone: 781-769-9750
Fax: 781-769-6334
e-mail: artech@artechhouse.com

Artech House
16 Sussex Street
London SW1V 4RW UK
Phone: +44 (0)20 7596-8750
Fax: +44 (0)20 7630-0166
e-mail: artech-uk@artechhouse.com

Find us on the World Wide Web at: www.artechhouse.com